DIDACTICS OF MATHEMATICS AS A SCIENTIFIC DISCIPLINE

Mathematics Education Library

VOLUME 13

The titles published in this series are listed at the end of this volume.

DIDACTICS OF MATHEMATICS AS A SCIENTIFIC DISCIPLINE

Edited by

ROLF BIEHLER
ROLAND W. SCHOLZ
RUDOLF STRÄSSER
BERNARD WINKELMANN

Institute for Didactics of Mathematics,
University of Bielefeld, Germany

KLUWER ACADEMIC PUBLISHERS
DORDRECHT / BOSTON / LONDON

A C.I.P. Catalogue record for this book is available from the Library of Congress.

ISBN 0-7923-2613-X

Published by Kluwer Academic Publishers,
P.O. Box 17, 3300 AA Dordrecht, The Netherlands.

Kluwer Academic Publishers incorporates
the publishing programmes of
D. Reidel, Martinus Nijhoff, Dr W. Junk and MTP Press.

Sold and distributed in the U.S.A. and Canada
by Kluwer Academic Publishers,
101 Philip Drive, Norwell, MA 02061, U.S.A.

In all other countries, sold and distributed
by Kluwer Academic Publishers Group,
P.O. Box 322, 3300 AH Dordrecht, The Netherlands.

Printed on acid-free paper

Printed in the Netherlands

Dedicated to Hans-Georg Steiner.

R. B., R. W. S., R. S., B. W.

TABLE OF CONTENTS

PREFACE

DIDACTICS OF MATHEMATICS AS A SCIENTIFIC DISCIPLINE

Since the work of the International Commission for Mathematics Instruction (ICMI) at the beginning of this century, nobody can challenge the fact that scientific work has been done in the field of teaching and learning mathematics. This research work has been carried out by mathematicians, psychologists, educational scientists, mathematics teacher trainers, and mathematics teachers themselves. However, scientific communication on these issues long remained in its infancy, particularly on an international level; much work was done in isolation; and it was rare to find people who considered that they belonged to a separate scientific discipline, independent from mathematics or educational science.

In the late 1960s, a societal debate on the values and organization of a large number of industrialized countries (such as Germany, France, and the United States of America) stimulated a new concern for education and for the related educational sciences. In the 1970s and 1980s, these developments led to a certain breakthrough for research in mathematics education. The revival of international organizations such as ICMI and regular global conferences known as ICMEs (since 1969) has led to the formation of an international community of mathematics educators. We call the scientific discipline related to this research and the research-based development work didactics of mathematics – a notion that is common at least in German- and French-speaking countries and has become increasingly popular in the English-speaking world. Didactics of mathematics certainly exists as a discipline, at least in a social sense, as can be seen from journals, research and doctorate programs, scientific organizations, and conferences. However, didactics of mathematics is fairly young compared to other sciences such as mathematics or psychology. As a fairly young discipline, its system of objects, methodologies, and criteria for valid knowledge exhibits more variability and less consensus. Its role among other sciences at the university is still disputed.

This book has been written for the international scientific community of researchers in mathematics education. It provides a state-of-the-art portrait of a new branch of science. The reader will find a structured sample of original contributions from researchers in the field of didactics of mathematics.

The book will be of interest to all researchers in the field. However, mathematics educators who are interested in the theory of their practice and teacher trainers will also appreciate this survey and the diverse stimulations and reflections it provides. Prospective and practicing teachers of mathematics will find a variety of interesting spotlights on their practice that focus on different age groups and ability ranges among their students. In addition to persons directly engaged in mathematics education, the book as a whole and/or individual papers should be of interest to researchers from neighboring disciplines, such as mathematics, general education, educational psychology, and cognitive science.

The basic idea was to start from a general perspective on didactics of mathematics, to identify certain subdisciplines, and to suggest an overall structure of its field of research. This book should provide a *structured* view, or a "topology," of the breadth and variety of current research in didactics of mathematics by presenting authentic and vivid contributions of individual authors on their current research in certain subdisciplines. The subdisciplines are represented by the chapters of this book. The volume provides a sample of 30 contributions from 10 countries. The authors were asked to present an example of their research in a way that would also make the broader research fields represented by the individual contributions accessible for other colleagues in didactics of mathematics.

We use *chapter introductions* to provide a synthesis and an orientation for the research domain represented by the contributions. The individual contributions are related to the overall idea of the chapter, and the readers' attention is focused on relations and differences between the different papers in a chapter as well as their relation to other chapters. This makes it clear that our aim is not to provide a *handbook of didactics of mathematics* with authoratively written subchapters synthesizing research from one author's point of view. The organization of the book places more emphasis on a variety and multiplicity of perspectives. It is through the readers' (re-) construction and rethinking of our discipline – which we hope to stimulate with this book – that we can contribute to further reflection on and interest in our discipline.

The reader will find the following chapters:

1. Preparing Mathematics for Students
2. Teacher Education and Research on Teaching
3. Interaction in the Classroom
4. Technology and Mathematics Education
5. Psychology of Mathematical Thinking
6. Differential Didactics
7. History and Epistemology of Mathematics and Mathematics Education
8. Cultural Framing of Teaching and Learning Mathematics

The first five chapters are widely accepted as subdisciplines in the sense of the existence of many cross-references, intensive communication, and a common object of study. The other three "subdisciplines" seem to be less well-structured up to now. We include them because we regard them as important. This may be a certain bias due to our involvement with the IDM and its research tradition. We invented the concept of "Differential Didactics" in analogy to "Differential Psychology" in order to create a focus for research on gender, cultural minorities, and different groups of learners in contrast to what may be considered as "mathematics for all."

Didactics of mathematics is an applied area of activity: As in engineering, (applied) psychology, and medicine, the boundary between scientific work and (constructive) practice is – to say the least – "fuzzy." Didactics of mathematics shares a certain type of (social) problem with the above-mentioned disciplines, namely mathematics education; and it uses a multiplicity of methods. The topics of the first four chapters are often conceived of as practical concerns requiring constructive work, namely, the preparation of curricula and textbooks, the development of programs in teacher education, the formulation of guidelines for classroom interaction and learning, and the development of software. A major recent development has been the attempt to establish a rationalization, theorization, and reflection of these practical activities. Rationalization is understood in the twin sense of reflecting on the rationality of goals as well as improving instrumental efficiency. Sometimes this has led to work that is more comparable to basic science than applied science, because researchers felt that it was necessary to deepen theory and methodological reflection in order to improve our understanding of practical problems. Research on teachers' cognition and on classroom interaction presents an example of this trend.

We can also group the chapters into those that are closer to classroom teaching and learning (chapters 1 to 4) and those that reflect and analyze problems of learning, thinking, knowledge, and culture from a more general perspective, though still related to problems in mathematics education (chapters 5 to 8). In the first four chapters, the reader will find papers ranging from a mere analytical stance to papers with research-based constructive implications. Chapters 5 to 8 place more emphasis on analytical aspects. Didactics of mathematics has to be structured from a systemic point of view. Even work on subsystems such as the learner or the teacher have to bear in mind the relation to other components. The chapters concentrate mostly on subsystems in this sense. Starting from the knowledge to be taught, namely mathematics, we first try to assemble research on the didactical system in a strict sense: the "didactical triangle" of mathematics – teacher – learner.

Chapter 1 discusses principles of *preparing mathematics for students.* Concepts like "didactical transposition," "elementarization" of mathematics, and "didactical engineering" are analyzed. Consequently, the focus of the chapter is on the content of teaching, on knowledge to be taught. Nonetheless, the influence of other factors and institutions is revealed. Chapter 2 concentrates on *teacher education and research on teaching.* Its link to the preceding chapter obviously is the knowledge to be taught. Its main topic is the knowledge a teacher has or should have, the structure of this knowledge, and ways to influence and develop the teachers' knowledge. Chapter 3 on *interaction in the classroom* focuses on research that analyzes the complex "social interaction" of teachers and learners in the classroom and in small groups. The analysis of language and discourse in the classroom is an important issue. Chapter 4 on *technology and mathematics education* can be viewed from a systematic point of view as "educational technology" including textbooks and assessment schemes. These form an important product of the didactics of mathematics that is handed on to the practice of teaching. The design and use of such "products" is an important research topic. The focus on problems and potentials of the use of computers and software was chosen because this technology represents a critical issue in the current development of the teaching and learning of mathematics as well as an important research field in didactics. Chapter 5 on *psychology of mathematical thinking* concentrates on the organization of knowledge and mathematical thinking processes in individual learners and presents a variety of methodological approaches to mathematical thinking and cognitive processes. Chapter 6 on *differential didactics* presents an analysis of the accessibility of mathematics for specific subgroups of the population. It studies the impact of teaching and learning mathematics on these subpopulations. Chapter 7 on *history and epistemology of mathematics and mathematics education* comprises research and reflection about mathematics from different perspectives: philosophical, epistemological, historical, and cultural, and their relevance and impact on mathematics education. Chapter 8 on *cultural framing of teaching and learning mathematics* analyzes constraints and cultural influences, the actual and possible scientific, political, and cultural powers that have a deep influence on the teaching/learning process. This provides more depth on a topic relevant to preparing mathematics for students, because it is not taught in a vacuum, but in a social context that cannot be overlooked in a scientific analysis of this process. Although mathematics educators cannot control these factors to any large extent, they have to be aware of them. The mathematics to be taught is not viewed as a free-floating knowledge that is easy to digest for the learner, but as something that is socially shaped. An analysis of political and social boundaries of mathematics education is offered.

The classification into chapters is not intended as a disjunctive partition of the field. Inevitably, the reader will find mutual overlaps, some subdisciplines will lie nearer or further away from each other, and they will be linked in different ways. Obviously, the topics presented in these chapters touch upon a variety of different neighboring sciences. Primary links to specific sciences can be identified by relating chapter 1 on *preparing mathematics for students* to mathematics; chapter 2 on *teacher education and research on teaching* and chapter 3 on *interaction in the classroom* to social science and pedagogy. Chapter 5 on *psychology of mathematical thinking* draws heavily upon cognitive psychology, and chapters 7 on *history and epistemology,* and 8 on *cultural framing of teaching and learning mathematics* are tied in with sociology, history, and philosophy. From the reasoning as a whole, it should be clear that these disciplinary links are in no way exclusive; all these fields of research are closely linked to mathematics.

Aspects of mathematics education are also being analyzed in a multitude of other disciplines, such as educational science, psychology, epistemology, and the history of mathematics. Didactics of mathematics can draw upon these various disciplines, and, consequently, a variety of methodological approaches can be considered to be adequate methods. Taken as the scientific endeavor to describe and analyze the teaching and learning of mathematics, didactics of mathematics has to organize its own approach to the problem and exploit the knowledge available in neighboring disciplines. The systematic self-reflection of didactics of mathematics is a necessary element of its further development. Hans-Georg Steiner founded the international working group of "Theories of Mathematics Education (TME)" in Adelaide in 1984 in order to promote such research, and he continues to be a major supporter of such a systematic view on didactics of mathematics as a scientific discipline. This intellectual context contributed to the genesis of this book.

GENESIS OF THIS BOOK

The birth of every book has its occasion, its reasons, and its history. The occasion for this book is two anniversaries: 20 years of work at the Institut für Didaktik der Mathematik (IDM), Bielefeld University, and Professor Hans-

Georg Steiner's 65th birthday on November 21, 1993. The rise of didactics of mathematics as a scientific discipline has been fostered through exemplary scientific work, through reflections on the status of the discipline, and through organizational, institutional, and promotional work. This development has been closely connected both with the work and the activities of Hans-Georg Steiner and the work of the IDM. It was the editors' desire to commemorate these two events by presenting the object of Hans-Georg Steiner's work and the IDM's field of research by showing the process of doing scientific work *in actu*. We wanted not only to demonstrate the level reached and the maturity gained but also to indicate questions that are still open and tasks that need be solved in the future. Both Professor Steiner and the IDM may be honored by showing that the object of their promotion is alive and well in both its international connections and its disciplinary diversions.

Let us take a brief look at the history of the IDM. The idea of setting up an IDM as a national center was born in the mid-1960s. As in many other countries, research on mathematics education and thus knowledge about this object was seen as underdeveloped and ill-reputed at universities. This was why the Volkswagen foundation decided to promote the development of didactics of mathematics as a scientific discipline by funding a central institute. The main tasks of this institute were (a) to promote the contruction of curricula through research and development; (b) to develop a theoretical framework for research in didactics of mathematics in interdisciplinary collaboration with mathematics and other related disciplines; (c) to educate scientific successors; and (d) to build up an international center for documentation and communication. The IDM was founded in 1973. Together with Hans-Georg Steiner, Heinrich Bauersfeld and Michael Otte were appointed as professors and directors of the IDM. The status of the IDM as a scientific institute at the university was not undisputed during its existence. The biggest crisis came in 1991, when it was questioned whether a single university still has the resources to support a central institute like the IDM. However, the institute received so much national and international support that the university decided to confirm the institutionalization of the IDM and continue to support it for at least another 8 years, that is, until the year 2000.

Clearly, the differentiation of the theoretical framework of didactics of mathematics, the diversification of methods used and of the objects of interest in the international discussion, and the research work done at the IDM is reflected in the structure of this book. In some respects, the increasing differentiation of the body of knowledge available in didactics of mathematics has opened up more general and fundamental perspectives for future research on mathematics education at the IDM. Perhaps this perspective is reflected by the central questions in the IDM guidelines for research during the current period: How do people acquire mathematics? How does it affect their thinking, their work, and their view of the world?

Professor Steiner accompanied and guided the IDM from its very beginning. All four editors have been cooperating with Hans-Georg Steiner in a continuous working group that stretches back for more than 15 years. We have all benefited very much from his personal friendship and his generous support. His interests and influence have not been confined to work in this group. He did not join the other members in their trend toward definite specialization and always looked at the whole of didactics of mathematics, which he promoted continuously, for instance, by organizing and structuring international meetings such as the Third International Conference on Mathematics Education (ICME3) in Karlsruhe, 1976, as well as many bilateral symposia, and founding and leading TME, the international working group on Theories of Mathematics Education.

Hans-Georg Steiner is one of the rare persons who possesses an overview of a whole discipline that has developed parallel to his own research and partly under his influence. Presumably, this makes him one of the few scientists who can constructively criticize nearly all the chapters in this book. Without doubt, one criticism will be the almost total omission of explicit discussions on theories of mathematics education. However, the very concept of this book is to show just how these theories may be applied.

ACKNOWLEDGEMENTS

The concept of this book was born in early summer 1992. The chapters were divided among the editors and contacts were initiated with possible authors of specific articles in summer and autumn 1992. In contrast to experiences with other edited books, the vast majority of answers to our call for papers were positive. Many authors named their friendship to Hans-Georg Steiner and their appreciation of his and the IDM's work as decisive motives for their decision to collaborate, even if there were serious difficulties in joining the book project due to other commitments. We are very grateful to all our authors and hereby thank them for their excellent work.

All the authors provided abstracts of their papers, which were reviewed by the chapter editors and exchanged between authors of the same chapter. The full papers reached the editors in spring and early summer 1993 and were reviewed by the editors. The articles were revised or partly rewritten till the end of June, 1993.

We want to thank Herta Ritsche, secretary at the IDM, who was responsible for producing the camera-ready copies. She was at the center of the production of the book. She carefully managed the many successive versions of the papers and coordinated the editorial work.

We want to thank Jonathan Harrow and Günter Seib for translating some of the chapters. We are indebted to Jonathan Harrow not just for his perfect language checks and stylistic polishing of most of the papers, including some papers by English native speakers. His professional approach to the final formal editing and his remarks and suggestions on many formulations helped to clarify many texts and has made them more easily accessible for the reader. Without such generous and dedicated help, this book would not have been possible. However, the editors are fully responsible for any remaining printing errors and mistakes due to the editing process.

We wish the IDM and Hans-Georg Steiner a good and productive future in their continued efforts to promote the didactics of mathematics as a scientific discipline!

Rolf Biehler
Roland W. Scholz
Rudolf Sträßer
Bernard Winkelmann

CHAPTER 1

PREPARING MATHEMATICS FOR STUDENTS

edited and introduced
by
Bernard Winkelmann
Bielefeld

For many didacticians of mathematics, reflections on and improvements in the process of the curriculum development and implementation of mathematics teaching are both the starting point and motivating goal of their research. They serve as a main goal of research in mathematics teaching and learning and as a bridge between various social groups engaged in mathematics education such as teachers, parents, employers, and educationalists. The process of preparing mathematics for students can be described from different viewpoints and with different theoretical frameworks in mind. *Mogens Niss* (this volume) uses a concise formulation when he names the solving of the following problems as necessary actions in this process:

1. The problem of *justification*. Why should some specific part of mathematics (considered in a broad sense) be taught to a specific group of students?

2. The problem of *possibility*. Given the mental abilities of the group of students in question, can the mathematical subject be taught, and, if so, how?

R. Biehler, R. W. Scholz, R. Sträßer, B. Winkelmann (Eds.),
Didactics of Mathematics as a Scientific Discipline, 9-13.

3. The problem of *implementation:* preparing material and immaterial means to make possible the teaching of the mathematical subject given the constraints of society, the school system, the qualifications of teachers, and so forth.

These three problems could be handled in sequence only in a very idealized theoretical setting; in more involved theories and in practice, they have to be dealt with simultaneously or in a quasi-spiraling process.

So, let me turn to the notorious statement attributed to J. Bruner, "the fundamental ideas of each subject can be taught to any individual at any age in some honest manner" (cf. the critical remarks in J. Fey's article and also the discussion of conceptualizations of "fundamental" in U. Tietze's paper, both in this chapter). Even if it could be understood as belonging to the problem of possibility, the term "fundamental" certainly has to do with justification, and the "honest manner" combines justification with implementation. Perhaps, also, the statement is just a kind of axiom, implicitly defining the meaning of "fundamental," "honest," and "subject" within the realm of the problems mentioned.

Another way of describing or rather conceptualizing the process of curriculum formation, which is much referred to especially but by no means exclusively in the French didactics of mathematics, is the theory of didactical transposition (cf. the hints in Artigue's article, this chapter; and, for an English source, Chevallard, 1992). It describes the inevitable processes of change by which (mathematical) knowledge is transformed on its way from the academic realm through various negotiation processes over knowledge that is socially considered as important for school education, over different elaborations according to specific circumstances (knowledge to be taught), to the knowledge induced in the minds (and hearts) of the students (taught knowledge). The theory of didactical transposition concentrates on the constraints the diverse agents are subject to, and claims to unmask the transparency illusion of curriculum developers who tend to think of their decisions as scientific and deliberately chosen, whereas, in this theory, they are kinds of unconscious elements in a system obeying its own rules.

In a more self-confident setting, preparing mathematics for teaching can be conceived of as elementarization, that is, "the translation of mathematical concepts, principles, techniques, and reasoning methods from the forms in which they are discovered and then verified by formal reasoning to forms that can be learned readily by a broad audience of students," as Jim Fey describes it at the beginning of his paper. His concept includes the steps of implementation – such as development of materials, training teachers, convincing decision makers – and assessment. It is applied science that relies not only on research in basic sciences but also on its own methodologies and principles.

In another conceptualization, which is rooted in German didactical traditions, elementarization is conceived as the constructive version of the first step in the process of didactical transposition: It means the *active* transformation of mathematical substance to more elementary forms. Here "elementary" has the double meaning of being fundamental and accessible for the intended groups of students; it includes elements of all three problems mentioned above: justification, possibility, and implementation. In such a conception, the negotiation process described by the theory of didactical transposition is left to the necessary second step, namely, that of proper implementation.

Elementarization in this narrow sense has a long tradition in mathematics teaching, since every teacher and every textbook author teaching a new topic, a new aspect of a topic, or the same topic to a different group of students naturally tries to present his or her ideas in an elementary way. The topic has to be presented as something *accessible* to the intended learners, that is, not too complicated technically, understandable through links to previous knowledge, and as a path leading to some *general goals* like mathematical thinking, understanding the role of mathematics, or solving important problems. The successful teacher or textbook author has to develop the art of elementarization, and mathematics education benefits from such art, even if it is not reflected scientifically. As an art, it includes also elements of simplicity, elegance, and salience. In didactics of mathematic as a scientific discipline, this art and, furthermore, the whole process of reorganizing mathematical knowledge for the purposes of schools and teaching are described and methodologically reflected. The art is refined by methodically elaborating didactical principles or specific operations and procedures (cf. *Uwe-Peter Tietze,* this chapter), and the process is guided by systematically including insights yielded by other, related disciplines, thereby exposing the unavoidable shortcomings and lurking pitfalls of the whole process.

As may be deduced from this introduction, there are different traditions in different cultures and different didactical schools of handling this process of choosing, preparing, and evaluating mathematical topics for teaching purposes. These traditions differ in their emphasis on specific elementarization strategies, students' needs, fundamental ideas of mathematics, topic levels (examples, concepts, methods, or general ideas such as model building), description levels and the like, and degrees of elaboratedness. This is reflected only partly in the set of three articles in this chapter, which to a certain extent represent part of the French, the North-American, and the German tradition. They intentionally show not only the strong interconnections within such a tradition, which naturally can be traced to own education and language barriers, but also tendencies to absorb or critically discuss influences of other national schools as well.

In his paper on *eclectic approaches to elementarization, James T. Fey* asks about the prospects for making elementarization a rational activity in the science of didactics of mathematics. In the form of a fictitious naive approach to curriculum reform, he describes facts, insights, and methods to be learned for careful curriculum design in mathematics when different communities contributing to the necessary knowledge required by those design processes are taken seriously: mathematicians, psychologists, and classroom teachers. Elementarization is seen as a complex interdisciplinary enterprise that cannot be described as a deductive science but contains strong elements of scientific and creative work. He describes the real influences on the reform and organization of mathematics teaching exerted by different groups of society such as those mentioned above and by mathematics education researchers, general educators, politicians, supervisors, and the lay public. In an analysis of recent reform movements in mathematics teaching in the USA, he shows the mutual argumentations, rhetoric strategies, and means of exerting influence that occur, but also the strengths and weaknesses that are the result of such negotiating processes. In this report, essential factors of elementarization are dealt with in a seemingly spontaneous but indeed well-organized manner, such as choice of representation, use of technology, role of applications, role of assessments, formal mathematics versus intuitive understanding, but also dangers and possible pitfalls of elementarization resulting from the overemphasis of specific viewpoints.

Michèle Artigue illustrates the *concept of didactical engineering* and its theoretical background. This systemic approach is connected to theoretical ideas prevalent in the French didactics of mathematics but also introduces many "engineering" elements. These are decisionist and practical elements that are based on scientific research and theories but necessarily have to extend to more complex, concrete objects than the simplified objects of the theories. The author describes the concrete studies and developments a curriculum reformer has to undertake in order to cope constructively with a specific perceived teaching problem; her concrete case is the inadequateness of a traditional part of university mathematics teaching (differential equations) due to modern developments in mathematics, sciences, technology, and society. She clearly and explicitly elaborates the tension between the theoretical ideals of the researcher, whose teaching aims at researchable results in strictly controlling as many variables as possible, and the practical needs of the constructive developer, whose measure of success is a sound, accepted, and adaptable teaching sequence. The systemic approach consists in a careful analysis of the teaching situation to be acted upon, of the epistemological, cognitive, and didactical obstacles against change, and of the possibilities for global (macrodidactic) and local (microdidactic) choices. The complexity of the object requires repeated application of the design - experimental teaching - redesign cycle on increasingly higher levels, and also consideration of the obstacles when the product of the engineering is to be distributed – obstacles not only in the students but also in the teachers who tend to adapt new ideas to their old teaching styles and thereby to destroy them.

In the course of reforming mathematics teaching in connection with the new-math movement, the question of justification became very virulent; it had to be dealt with in a scientific debate that, to a certain extent, was independent from the question of realization in practical mathematical teaching. This is the theme of *Uwe-Peter Tietze's* paper. He describes the historic development in the efforts of the community of mathematical educators in Western Germany and Austria to cope with the problem of defining and justifying *mathematical curricula and the underlying goals*. How can we decide which part of mathematics, which insights, applications, and methods of mathematics are worth being taught and learned? The author explains the logical difficulties of argumentations about normative aspects. In a *tour de force* on the German didactical discussion about the problems of elementarization and justification, he describes and criticizes many constructive concepts dealing with the problem, such as the formulation of didactic principles, the development of general objectives, the efforts to identify fundamental ideas in mathematics as a whole or in specific domains, the idea of exactifying as teaching goal and teaching process, and the role of applications in justifying goals of mathematics teaching. (The historical introduction to his section on applications should be compared to the more detailed account in *Jahnke*'s article, this volume.) The survey is very condensed and rich in content, arguments, criticisms, and even constructive examples, mostly taken from the debate on calculus teaching in German upper secondary schools (Gymnasium).

All three authors mark in different ways the tension exerted on curriculum designers between the practical question "what can be taught and what can be done to make it happen?" and the connected but somehow independent theoretical question "what should be taught, and why, how, to whom?" It is the tension between the ideal of knowing and taking into account the real possibilities and constraints as described in other chapters of this book, and the necessity to develop argumentations and theories of an applied scientific or engineering character in order to prepare for the necessary decisions in domains that are only partly known.

REFERENCES

Chevallard, Y. (1992). A theoretical approach to curricula. *Journal für Mathematikdidaktik*, *13*(2/3), 215-230.

ECLECTIC APPROACHES TO ELEMENTARIZATION: CASES OF CURRICULUM CONSTRUCTION IN THE UNITED STATES

James T. Fey
Maryland

1. INTRODUCTION

Translation of mathematical concepts, principles, techniques, and reasoning methods from the forms in which they are discovered and verified to forms that can be learned readily by a broad audience of students involves at least two fundamental tasks: (a) choosing the mathematical ideas that are most important for young people to learn, and (b) finding ways to embed those ideas in learning experiences that are engaging and effective.

At first glance, it would seem that, for a highly structured discipline like mathematics, design of curricula and instructional strategies would be straightforward tasks that are dealt with routinely by experts in mathematics and its teaching. But American school mathematics programs are developed in a complex and loosely structured process involving a wide variety of people with different values, expertise, interests, and experiences. While there are mathematics educators and educational policymakers who attempt to guide curriculum development and implementation through application of thoughtful content analyses and coherent research-based theories of learning and teaching, it seems fair to say that American school mathematics is actually the result of compromises that emerge from informal competition among many opinions. Furthermore, the competing opinions are usually formed by intuitive reflection on personal experiences with mathematics and teaching, not by systematic didactical analysis.

Over the past decade, curriculum advisory reports for American mathematics education have been offered from groups representing classroom teachers (NCTM, 1989, 1991), research mathematicians (Pollak, 1982; Steen, 1990), scientists and science educators (AAAS, 1989), educational psychologists (Linn, 1986), and political groups without any special expertise in education (Bush, 1991). Those recommendations, and the changes in school mathematics programs to which they have led, have been widely debated in a variety of professional and public political forums. Analysis of this lively but eclectic process shows something of the effects of curriculum

R. Biehler, R. W. Scholz, R. Sträßer, B. Winkelmann (Eds.),
Didactics of Mathematics as a Scientific Discipline, 15-26.

building in an educational system without central control of such activity. Listening to the voices in those forums also raises questions about the feasibility of developing elementarization as a scientific activity in the didactics of mathematics. The ferment of American debate about goals and methods of school mathematics has led to production of imaginative curriculum materials and teaching ideas, but very modest and uneven implementation of the possible innovations.

In this paper I will analyze, with examples from recent American experiences, the influences of various factors in formation of school curricula. The underlying goal is improving translation of mathematics as a discipline of human knowledge and reasoning to a subject for school learning. But the immediate question is how a broad range of interests and expertise can be organized to perform that task effectively. What are the prospects for making elementarization a rational activity in the science of mathematical didactics?

2. INSIGHTS FROM MATHEMATICS

One of the most obvious places to look for guidance in construction of school mathematics curricula is in the structure and methods of the root discipline itself. It seems reasonable that the mathematical education of young people should provide them, in some appropriate way, with the basic understandings and skills that enable mathematicians to reason so effectively about quantitative and spatial problems. Who could be better qualified to identify the core concepts, principles, and techniques of mathematics and the paths by which mastery of those ideas can be most naturally reached than professional mathematicians?

As Kilpatrick (1992) notes, "mathematicians have a long, if sporadic, history of interest in studying the teaching and learning of their subject." This concern for the content and organization of school mathematics curricula was especially acute during the reform era of the 1950s and 1960s when hundreds of research mathematicians engaged in curriculum development and teacher education projects designed to update school programs. The influence of many of those mathematicians led to emphasis in the new programs on underlying abstract structures of mathematical domains, increased attention to precision of language for expressing mathematical ideas, and introduction to school mathematics of topics previously viewed as part of collegiate study (NACOME, 1975).

In summarizing a conference of prominent research mathematicians and scientists who gathered to think about directions for improvement of school curricula and teaching, the psychologist Jerome Bruner (1960) recorded the brave assertion that, "any subject can be taught to anybody at any age in some form that is honest." He, and many others, went on to suggest that school mathematics should give students an understanding of the discipline and its methods that parallels (albeit in a weaker form) that of mathematicians who are active at the frontiers of pure and applied research.

Unfortunately, proposals to use the structure and methods of advanced mathematics as a guide to school curricula have proven problematic at best. The concepts and principles of the major branches of mathematics can, in some sense, be derived logically from a small set of primitive assumptions and structures. However, the formal logical coherence of the subject masks quite varied aspects of the way the subject is actually developed and used by mathematicians. Almost as soon as the first *new math* reform projects got underway in the United States, there were debates about the proper mathematical direction of that reform. Differences of opinion on the balance of pure and applied mathematics, the role of deduction and intuition in mathematical work, and the importance of various mathematical topics reflected the diversity of the discipline itself. There was little unanimity in the advice about school mathematics coming from the professional mathematics community. Consequently, if school curricula are to convey images of mathematics that faithfully represent the content and methods of the subject as practiced in mathematical research and applications, it seems likely that they will include a combination of topics chosen from many options, as a result of competition among opinions that reflect the mathematical taste and experience of concerned individuals, not scientific analysis.

In retrospect, promises that the content and organization of school mathematics curricula could be guided by following the deductive structure of formal mathematics seem incredibly naive. While there is a certain plausibility to the idea that all students can profit by acquiring something of the mathematical power possessed by experts in the field, a little thought on the subject reminds us that many people use mathematical ideas and techniques in ways quite different than those taught in school and in settings quite different from formal scientific and technical work. Thus it seems quite reasonable to ask whether school mathematics should be designed with an eye on formal academic mathematics alone, or in consideration of the varied ways that people actually use mathematics in daily life and work. This tension between images of formal and practical mathematics has always been a factor in curricular decision-making. Research over the past 20 years has added intriguing insights into the mathematical practices of people in various situations (e.g., Rogoff & Lave, 1984), adding a new dimension to the debate over what sort of mathematics is most worth learning and what should be in school curricula.

In the past decade, the task of selecting content goals for school curricula has been further complicated by a dramatic revolution in the structure and methods of mathematics itself. Electronic calculators and computers have become standard working tools for mathematicians. In the process, they have fundamentally altered the discipline. For centuries, if not millennia, one of the driving forces in development of new mathematics has been the search for algorithmic procedures to process quantitative and geometric information. But execution of those procedures was always a human activity, so school mathematics had to devote a substantial portion of its program to training students in rapid and accurate execution of algorithms. With calculators now universally available at low cost, few people do any substantial amount of arithmetic computation by traditional methods; with powerful personal computers also widely available to anyone engaged in scientific or technical work, few people do algebraic symbolic computation by traditional methods. Furthermore, the visual representations provided by modern computers provide powerful new kinds of tools for mathematical experimentation and problem-solving. The effect of these changes in the technological environment for mathematics is to change, in fundamental ways, the structure of the subject and its methods. For those who look to the structure and methods of mathematics as guides to school curricula, it is time for reconsideration of every assumption that underlies traditional curriculum structures (Fey, 1989; NRC, 1990). Of course, this fundamental change in mathematics wrought by emergence of electronic information-processing technology underscores another factor in the curriculum design process – we plan curricula to prepare students for lives in a future world that will undoubtedly evolve through continual and rapid change. Our experience of the recent past suggests that we can hardly imagine what that future will hold, and this uncertainty itself must be a factor in the curriculum decision-making process.

What then are the insights from mathematics that play a role in the task of elementarization for school curriculum design? The structure of mathematics obviously provides some guidance to selection and organization of topics in school curricula. However, it now seems clear that, in making content choices, we must consider a very complex web of insights into the ways that the subject can and will be used by our students. Those judgments can be informed by analyses of alternative conceptual approaches to the content, by assessments of how the subject is used, and by implications of new technologies. However, such analyses will ultimately be blended into personal judgments by people who must make choices based on incomplete evidence, not by following an algorithm for curriculum design.

3. INSIGHTS FROM PSYCHOLOGY

When mathematicians become concerned about school curricula, their first instinct is usually to focus on the content of textbooks and instruction at various grade levels. Quite reasonably, they feel most expert at judging the relative importance and correctness of the topics and their presentation. However, anyone who remains engaged with the reform process long enough to work on the production and testing of alternative curricula for schools will soon realize that selection of content goals is only the easy part of the task. The naive faith expressed in Bruner's assertion that any child can learn any mathematics in some honest form led many curriculum innovators to try some daring experiments. However, those who watched the classroom experiments carefully and listened to voices of teachers and students soon found that the search for accessible honest representations of mathematical ideas is a deep problem that gets entangled quickly in questions of how young people learn.

It is natural to turn to psychology for insight into the mechanisms by which humans learn facts, concepts, principles, skills, reasoning processes, and problem-solving strategies. There is a long tradition of research by American and European psychologists on questions related to mathematics learning and teaching (Kilpatrick, 1992; Schoenfeld, 1992). Sometimes that research has focused on mathematics, because the subject appears to offer a domain of well-defined content in which knowledge can be objectively measured, but psychological investigations have also addressed questions that are fundamental in mathematics education.

In the heyday of connectionist and behaviorist psychology, studies of arithmetic learning examined questions in the procedural aspects of arithmetic and algebra. Psychologists in the Gestalt tradition were more interested in problem-solving and concept formation, with mathematical subject matter useful in both types of investigation. Developmental psychologists have used mathematical tasks in their studies aimed at understanding stages and rates of cognitive development. The work of Piaget and his descendants in the constructivist school of learning and teaching has been enormously influential in thinking about school mathematics teaching and learning. Psychologists exploring the contemporary information-processing models of learning have found it convenient to use mathematical procedural knowledge in their studies.

There is now a very strong and active collaboration of research psychologists and mathematics educators that has resulted in focusing investigations of human learning on issues that are central to mathematics education in school. Several examples illustrate that collaboration and its potential for productive influence on design of mathematics curricula and teaching.

For instance, in modern cognitive theories, one of the central issues is the representation of knowledge in memory. Representation of facts and relationships is a very important aspect of mathematical thinking and learning, so mathematics educators have become vitally interested in psychological research that contributes to understanding of representations. At the same time, many mathematics educators, stimulated by the notion of representation, have launched independent work in curriculum development and research on teaching that tests hypotheses about representation in practical settings. The capability of computers for simultaneously displaying graphic, numeric, symbolic, and verbal representations of mathematical information and relationships has led to important work aimed at helping students acquire better mathematical understanding and problem-solving power. Furthermore, the computer representations have made deep ideas and difficult problems accessible to students in new ways – altering traditional curriculum assumptions about scope and sequence. For example, with the use of inexpensive graphing calculators, students in elementary algebra can solve difficult equations, inequalities, and optimization problems with visual and numerical successive approximation methods, long before they acquire the symbol manipulation skills that have been the traditional prerequisites for such work.

In contemporary psychological research, there is also considerable interest in processes of metacognition and self-regulatory monitoring of mental activity. Since mathematics education is especially interested in developing student ability to work effectively in complex problem-solving situations, there has been considerable interaction between psychological research and mathematical education on that issue.

By any reasonable measure, the power of mathematics as a tool for describing and analyzing patterns and solving problems comes from the fact that common structural concepts and procedures can be recognized and exploited in so many different specific contexts. The central problem of mathematical education is to help students acquire a repertoire of significant conceptual and procedural knowledge and the ability to transfer that knowledge from the specific contexts in which it is presented to new and apparently different settings. The problem of transfer is a central issue in psychological research, and, in a 1989 review, Perkins and Salomon noted that much research suggests, "To the extent that transfer does take place, it is highly specific and must be cued, primed, and guided; it seldom occurs spontaneously." However, they go on to report recent work, much focused in mathematics, which shows that, "When general principles of reasoning are taught together with self-monitoring practices and potential applications in varied contexts, transfer often *is* obtained." On the other hand, recent research on situated cognition (Brown, Collins, & Duguid, 1989) has countered this optimistic conclusion by suggesting that it is impossible to separate what is learned from the activity and context in which learning takes place, that "learning and cognition . . . are fundamentally situated."

What then is the actual and potential contribution of psychological research to the problem of curriculum design in school mathematics? The topics that have been investigated by cognitive and developmental psychologists are relevant to central issues in teaching and learning of mathematics. However, far from providing clear guidance to construction of optimal teaching strategies and learning environments, the results are more suggestive than prescriptive – incomplete and often contradictory. A curriculum developer or teacher who turns to psychology for insight into the teaching of key mathematical ideas and reasoning methods will find provocative theories, but also a substantial challenge to translate those theories into practical classroom practices.

4. INSIGHTS FROM CLASSROOM TEACHERS

Effective mathematics teaching certainly depends on knowledge of mathematics and knowledge of ways that students learn mathematics. But there remains an artistry about superb teaching that weaves mathematical and psychological insights into workable curricula and engaging and effective teaching activities. The findings of scientific research must still be informed and enhanced by wisdom of practice. It is precisely this blending of theoretical and practical knowledge that occurred in the recent National Council of Teachers of Mathematics' efforts to establish and promote *Standards for Curriculum and Evaluation* and *Professional Standards for Teaching Mathematics* (NCTM, 1989, 1991).

Responsibility for public education in the United States is a state and local function, with day-to-day decisions about curriculum and teaching under the control of over 16,000 local school districts. Some of those districts are quite large, with substantial supervisory staffs attending to the quality of instruction in each discipline at each level of schooling. But most are quite small, with limited resources to support curricular innovation or teacher professional development. Therefore, the complex array of advice from the mathematical, psychological, and educational research communities tends to have only modest impact on local decisions. There is no national curriculum. In fact, in most school systems, curriculum development involves only selection of text materials from the offerings of, generally cautious, commercial publishers. That selection is made with strong influence by classroom teachers whose decision criteria are shaped primarily by personal experience in the classroom.

The difficulty of stimulating major reform in the curriculum or teaching of school mathematics has always been a frustration to national professional leaders. The history of American mathematics education in this century is marked by sporadic advisory reports from concerned professional organizations. The recommendations in those reports tend to spur activity at the surface of the profession, but seldom have the innovations been broad and permanent (NACOME, 1975). However, in the last decade, concern about the quality of mathematics and science education has been an issue in state and national political debates. The need for national leadership in reform has gradually overcome the natural American antipathy toward ideas like a national curriculum or national assessments of educational achievement. In this context, the National Council of Teachers of Mathematics undertook two projects to develop professional standards for curriculum, evaluation, and teaching that could guide schools and teachers across the country.

The NCTM *Standards*, published in two volumes (NCTM, 1989, 1991), provide recommendations on three fundamental questions: What mathematics is most important for students to learn? What is the most effective way to teach that mathematics? How should the effects of mathematics teaching be assessed? The processes and products of those standard-setting efforts give interesting insights into the ways that mathematical ideas are transformed into school curricula in a loosely structured system with many different interested parties.

First, membership on the committees to draft standards did not include a single academic mathematician or psychologist active at the research frontiers of mathematics or its teaching and learning. The various subcommittees were made up of outstanding classroom teachers, local and state school system supervisors, and university teacher educators. While each working group included members with broad understanding of mathematics and contemporary research on student learning, that knowledge was applied to design of school mathematics programs with additional insight gained from years of classroom experience.

The *Standards'* emphasis on a practitioner's perspective explains a second noteworthy feature of the proposals – the recognition that it is virtually impossible to separate the mathematical content of a curriculum from the learning experiences by which students acquire understanding and skill in that content. At each level (K-4, 5-8, 9-12) of schooling, the *Standards* recommend important broad mathematical goals (though not so much detail as a syllabus for a national examination might require). But each recommended content topic is elaborated by discussion that includes illustrations of appropriate instructional approaches.

While the *Standards'* documents are clearly influenced by contemporary ideas in mathematics (e.g., attention to stochastics and discrete mathematics) and research on learning and teaching (e.g., emphasis on connections and active student construction of knowledge), that influence is transformed into recommendations clearly related to the classroom. In the *Curriculum and Evaluation Standards*, recommendations about discrete mathematics topics are accompanied by examples of practical situations modeled well by matrices, graphs, and difference equations. In the *Professional Teaching Standards,* each recommendation is accompanied by several vignettes of typical classrooms in action embodying the recommended practices.

The NCTM *Standards'* projects represent a fundamentally new approach to the task of reforming American mathematics education and, in the process, the transformation of new knowledge about mathematics and its learning into school curricula and teaching. While previous reform proposals have often been drafted by groups dominated by research mathematicians, frequently with the imprimatur of a policy-making group like the College Entrance Examination Board, the *Standards'* projects were a grassroots operation led by mathematics educators with strong connections to the mathematical, psychological, and educational research communities, but also with credible knowledge and connections in school practice. Their work was not strongly theory-driven, and their recommendations are not particularly well-supported by hard research evidence, but they have managed a blend of wisdom from many contributors that has gained high praise for the products. Their eclectic approach to elementarization has effectively stimulated and shaped recent debate and innovative activity in mathematics education.

5. IMPACT OF CONTEXT FOR EDUCATION

Despite the broad endorsement of and enthusiasm for the NCTM *Standards,* it is quite reasonable to withhold judgment on their long-term influence in American mathematics education. It is now barely 4 years since release of the curriculum and evaluation *Standards*. It is not uncommon to find schools and teachers who have yet to hear about, much less consider, the proposals in the *Standards'* reports. It is also common to hear schools and teachers who claim that they "did the Standards last year," revealing remarkable naivité about the implications of the proposals. The National Science Foundation has funded at least 10 major curriculum development projects seeking to provide prototypes for school mathematics programs that embody the spirit of the *Standards* in various alternative ways. Those projects have really barely begun work on curriculum development, much less the broad implementation that would be required to realize the *Standards'* goals. Conclusion: It's really too early to tell whether the *Standards* will be a different kind of stimulus for reform.

On the other hand, the *Standards* have emerged from the mathematics education professional community into a national political atmosphere that is unique in the history of American education. Never before have national political figures spoken so boldly about establishing national educational standards and implementing an extensive program of national assessment to measure progress toward achieving those standards. In the debate over this political movement, the NCTM *Standards'* work has been held up as a model of guidelines that would be helpful, and there are now standards-setting projects at work in other disciplines – most notably science.

The attention to NCTM recommendations by governmental agencies and partisan political parties is just one manifestation of an important family of nonprofessional influences on school mathematics in the United States. Our long-standing tradition of broad access to free public schooling and control of school policy by local, often elected, school boards means that many people outside the school and university communities are interested in and express opinions about school matters. Changes in school curricula must generally be approved by lay governing boards. Those same boards are usually interested in quantitative evidence that schools are effective, so they mandate extensive testing programs. The test data commonly makes its way into public media reports on schools, and there are frequent debates about the causes and cures of poor performance. Thus decisions about what mathematics is most important for students to learn and what instructional methods are likely to be most effective are commonly made in an environment that must take account of nonprofessional public opinion.

The classic example of extraschool influences on educational practice is the *new math* movement of the 1960s and early 1970s. While there are certainly significant professional differences of opinion about the wisdom of various innovations from that period, the influence of those differences on the course of the attempted reform is probably modest when compared to the influence of public attitudes and perceptions. The goals of the reform were not clearly understood by or explained to the public constituents of education, and when implementation of the reform agenda coincided with declines in some closely watched national assessment indicators, the public outcry was dramatic. Whether or not *new math* curricula and teaching methods were successful or not, the importance of winning public confidence in educational reform ideas was made very clear.

As important as it is to consider political and public opinion factors, the most important way that the context of mathematics education affects transformation of content goals into effective teaching materials and activities is through our students. The United States is a very large and populous country, but it is also incredibly diverse. Students in typical public schools come with a variety of natural aptitudes and interests in school, from a broad range of family backgrounds, cultural traditions, and conditions of economic advantage or disadvantage. We are a nation of immigrants with dozens of different languages spoken as native tongue by millions of students. We are a transient people, with some large city schools experiencing 50% to75% student turnover in the course of a single school year.

Each of these factors influences the formation of school curricula. For example, with compulsory schooling through at least age 16, our curricula must meet the needs of students whose achievement and interests commonly spread over great ranges; but our commitment to democratic social institutions implies common experiences for most students. One of the most striking statements in the NCTM *Standards,* and a number of other contemporary goals for mathematical education, is the assertion that *all students* are entitled to and capable of a rich and demanding mathematical curriculum. In a subject like mathematics, which has traditionally differentiated curricula for students of different aptitude and prior achievement, this challenge to provide mathematical power to all students has striking implications for curriculum design and teaching.

American schools have also been challenged to provide curricula that respect the diversity of cultural backgrounds of our students (overcoming the common Eurocentric bias of mathematics curricula, for instance) and to make sure that we present mathematics in a way that encourages girls, as well as boys, to high achievement. At the same time, we must organize curricula in a way that is robust enough to develop coherent understanding among students who too often come from unstable and unsupportive home situations. While some of these demands on school mathematics may seem to have little to do with the task of elementarization of subject matter, they are, in fact, very important considerations in the transformation of mathematics for instruction. If, as Brown, Collins, and Duguid (1989) suggest, all learning is essentially situated, it is critical that we embed important mathematical ideas in situations that are meaningful to the full range of students with whom we are working. If it is a fact of school life that many of our students will attend only sporadically, we must be wary of curriculum organizations that present mathematics in tightly structured hierarchies of interdependent skills. In fact, one of the most promising effects of technology on school mathematics is the promise that the traditional litany of detailed computational skills can be superseded by a small number of widely applicable macroprocedures. For instance, in place of the myriad of transformation rules for solving algebraic equations, we can emphasize the macroprocedure of graphing each side and searching by successive approximation for intersection points.

Finally, while we consider the effects of political and student contexts for our mathematics programs, we must also attend to the knowledge, interests, aptitudes, and values of the teachers who will be principal agents of instruction. One of the most obvious features of recent curriculum development in the United States is the fact that some truly imaginative and powerful curriculum materials have been produced, but that the teaching skills required to use those materials effectively are not widely available in schools. The task of transforming mathematical ideas into curriculum materials and plans for teaching activities is challenging. But the task of transforming those materials and activity plans into effective classroom experiences for students is equally demanding. Thus any plan for new curricula must take seriously the teacher capabilities (and school resources) in the settings in which those materials will be used.

6. CONCLUSIONS

What then are the prospects for developing a theory of elementarization – principles of preparing mathematics for students? It seems safe to say that, in the United States, curriculum development is practiced as an art, not a science. Moreover, in the survey of issues and experiences recounted in this paper, we have suggested that the enterprise is so complex that the likelihood of discovering any more than weak principles for a theory of elementarization seems remote.

Does this conclusion imply that curriculum formation is inevitably a hopelessly haphazard and intuitive activity? I think not. American educators tend not, on the whole, to take particularly theoretical approaches to their work. A predominantly practical orientation seems part of our national character.

Nonetheless, while the creative process of forming an engaging mathematics curriculum cannot be reduced to algorithmic application of scientific principles, it seems clear that the creative process is immeasurably enhanced by consideration of insights from analysis of alternative ways to develop mathematical ideas, from studies of conditions that facilitate human learning, and from studies of alternative classroom instructional strategies. Even the implementation of new curricula can be eased by thoughtful consideration of the contextual factors that have been shown to influence acceptance of other innovations.

REFERENCES

American Association for the Advancement of Science (AAAS). (1989). *Science for all Americans.* Washington, DC: The Association.

Brown, J. S., Collins, A., & Duguid, P. (1989). Situated cognition and the culture of learning. *Educational Researcher, 18*(1), 32-42.

Bruner, J. S. (1960). *The process of education.* Cambridge, MA: Harvard University Press.

Bush, G. H. W. (1991). *America 2000: An education strategy.* Washington, DC: U. S. Department of Education.

Fey, J. T. (1989). Technology and mathematics education: A survey of recent developments and important problems. *Educational Studies in Mathematics, 20,* 237-272.

Kilpatrick, J. (1992). A history of research in mathematics education. In D. A. Grouws (Ed.), *Handbook of research on mathematics teaching and learning* (pp. 3-38). New York: Macmillan.

Linn, M. C. (1986). *Establishing a research base for science education: Challenges, trends, and recommendations.* Berkeley, CA: Lawrence Hall of Science.

National Advisory Committee on Mathematics Education (NACOME). (1975). *Overview and analysis of school mathematics K-12.* Washington, DC: Conference Board of the Mathematical Sciences.

National Council of Teachers of Mathematics (NCTM). (1989). *Curriculum and evaluation standards for school mathematics.* Reston, VA: The Council.

National Council of Teachers of Mathematics (NCTM). (1991). *Professional standards for teaching mathematics.* Reston, VA: The Council.

National Research Council (NRC). (1990). *Reshaping school mathematics: A framework for curriculum.* Washington, DC: National Academy Press.

Perkins, D. N., & Salomon, G. (1989). Are cognitive skills context-bound? *Educational Researcher, 18*(1), 16-25.

Pollak, H. O. (1982). The mathematical sciences curriculum K-12: What is still fundamental and what is not. Report from the Conference Board of the Mathematical Sciences. National Science Board Commission on Precollege Education in Mathematics, Science, and Technology. *Educating Americans for the 21st Century* (Source Materials), 1-17.

Rogoff, B., & Lave, J. (Eds.). (1984). *Everyday cognition: Its development in social context.* Cambridge, MA: Harvard University Press.

Schoenfeld, A. (1992). Learning to think mathematically: Problem-solving, metacognition, and sense-making in mathematics. In D. A. Grouws (Ed.), *Handbook of research on mathematics teaching and learning* (pp. 334-370). New York: Macmillan.

Steen, L. A. (Ed.). (1990). *On the shoulders of giants: New approaches to numeracy.* Washington, DC: National Academy Press.

DIDACTICAL ENGINEERING AS A FRAMEWORK FOR THE CONCEPTION OF TEACHING PRODUCTS

Michèle Artigue
Paris / Reims

1. INTRODUCTION

In French research on didactics of mathematics, the issue of preparing mathematics for students, which is the topic of this chapter, is located at a crossroads between two not independent but nonetheless distinct theoretical fields: the *theory of didactical transposition*, developed since the beginning of the 1980s by Y. Chevallard (Chevallard, 1991, 1992), and the *theory of didactical situations*, initiated by G. Brousseau (1986) at the beginning of the 1970s and developed by several different researchers since that time.

My text is located within this perspective. The first part attempts to clarify how the theoretical frameworks mentioned above shape the approach to the preparation of mathematics for students by leading it, in particular, to be placed in a more global systemic perspective than that frequently associated with approaches in terms of the elementarization of knowledge. Then I shall use an example to show how these theoretical frameworks become operational in the development of teaching products through the concept of didactical engineering. In the conclusion, I shall return to more general questions that are still largely unanswered.

2. A SYSTEMIC APPROACH TO THE DEVELOPMENT AND ANALYSIS OF THE CONTENTS OF TEACHING

As pointed out above, this text uses the methodological concepts and tools provided by two distinct theoretical frameworks, the theory of didactical transposition and the theory of didactical situations, to study the issue of the preparation of mathematics for students. Of course, such a short text is unable to launch into an explanation of these theories (the reader is referred to the texts cited in the references); nevertheless, it is clear that these theoretical frameworks shape and determine, to a certain extent, the current approach to this issue. This is precisely the point I shall try to clarify first.

The two theoretical approaches mentioned above concern fundamental but different levels of didactical analysis:

1. The theory of didactical transposition concentrates on the analysis of

R. Biehler, R. W. Scholz, R. Sträßer, B. Winkelmann (Eds.),
Didactics of Mathematics as a Scientific Discipline, 27-39.

those processes that are based on reference knowledge, particularly on the reference knowledge produced by the legitimizing mathematical institution (scholarly knowledge), that lead to objects of teaching (knowledge to be taught) that are found in the daily life of the class (taught knowledge). It naturally tries to go beyond particular studies and highlight certain laws and regularities in these complex transposition processes.

2. To a certain extent, the theory of didactical situations is situated at a more local level. It aims to model teaching situations so that they can be developed and managed in a controlled way.

However, despite their different focuses of interest, these two theories link up on one essential point related to our topic: They emphasize the need to envisage the study of didactical phenomena within a systemic approach. Therefore, in both cases, the preparation of mathematics for students cannot be perceived as a simple process of the elementarization of knowledge established elsewhere, as the simple search for a presentation of some mathematical content adapted to the previous knowledge and cognitive abilities of students. It is perceived as a didactical task requiring a more global systemic analysis.

2.1 The Systemic Approach Via the Theory of Didactical Transposition

If one adopts a "didactical transposition" approach, one introduces an open system to the analysis that includes, in particular, the institutions at the source of the knowledge one aims to teach and the institutions targeted by this teaching. This is done by questioning the constitution and life of this knowledge, while remaining particularly attentive to the economy and ecology of the knowledge to be taught. One questions the possible viability of the content one wishes to promote while considering the laws that govern the functioning of the teaching system. One tries to foresee the deformations it is likely to undergo; one tries to ensure that the object can live and therefore develop within the teaching system without too drastically changing its nature or becoming corrupted.

The reform of modern mathematics has provided excellent ground for the study of these phenomena of didactical transposition, and it is, mainly, the ground chosen by Y. Chevallard in the first reference cited above. The reader is also referred to Arsac's (1992) review analyzing the evolution of the theory through studies undertaken both within and beyond the field of the didactics of mathematics, as well as the following recent doctoral theses:

1. M. Artaud (1993), who studied the progressive mathematization of the economic sphere, the obstacles encountered, the debates and negotiations that arose around this mathematization, and their implications for the contents of teaching itself.

2. P. Tavignot (1991), who used a study of the implementation of a new way of teaching orthogonal symmetry to 11- to 12-year-old students within

the French junior secondary school reforms (commenced in 1986) to develop a schema for the investigation of this type of process of didactical transposition.

I have also used this theoretical framework to study the evolution of the teaching of analysis in "lycées" (senior secondary school) over the last 15 years, through the evolution of a didactical object, "reference functions," which acted as a sort of emblem for the rupture caused by the rejection of the formalized teaching of modern mathematics (Artigue, 1993).

However, it must also be recognized that, up to the present, the theory of didactical transposition has mainly been used to analyze transposition mechanisms *a posteriori*. It has hardly ever been involved in an explicit way in the design of teaching contents or products. For this reason, the rest of this text will concentrate to a greater extent on the more local approach linked to the theory of didactical situations and the operationalization of the latter through didactical engineering.

2.2 The Systemic Approach Via the Theory of Didactical Situations

The present approach will be just as systemic but will concentrate on narrower systems: didactical systems, built up around a teacher and his or her students, systems with a limited life span, plunged in the global teaching system, and open, via the latter, to the "noosphere" of the teaching system and, beyond that, to the society in which the teaching system is located.

The theory of didactical situations, which is based on a constructivist approach, operates on the principle that knowledge is constructed through adaptation to an environment that, at least in part, appears problematic to the subject. It aims to become a theory for the control of teaching situations in their relationship with the production of mathematical knowledge. The didactical systems considered are therefore made up of three mutually interacting components, namely, the teacher, the student, and the knowledge. The aim is to develop the conceptual and methodological means to control the interacting phenomena and their relation to the construction and functioning of mathematical knowledge in the student.

The work involved in the preparation of teaching contents labeled by the expression *didactical engineering*, which is the focus of this text, will be placed in this perspective. Alongside the elaboration of the text of the knowledge under consideration, this needs to encompass the setting of this knowledge in situations that allow their learning to be managed in a controlled manner.

2.3 The Concept of Didactical Engineering

The expression "didactical engineering," as explained in Artigue (1991), actually emerged within the didactics of mathematics in France in the early 1980s in order to label a form of didactical work that is comparable to the work of an engineer. While engineers base their work on the scientific knowledge of their field and accept the control of theory, they are obliged to work with more complex objects than the refined objects of science and therefore to manage problems that science is unwilling or not yet able to tackle.

This labeling was viewed as a means to approach two questions that were crucial at the time:

1. the question of the relationship between research and action on the teaching system,

2. the question of the place assigned within research methodologies to "didactical performances" in class.

This twin function will determine the route that didactical engineering will take through the didactical establishment. In fact, the expression has become polysemous, designating both productions for teaching derived from or based on research and a specific research methodology based on classroom experimentations.

This text focuses particularly on the first aspect. The reader who is interested in the second is directed to Artigue (1989a). Nonetheless, it should be emphasized that didactical engineering for research and didactical engineering for production are closely interrelated for a variety of reasons. In particular, there unfortunately does not exist what, at present and at least in France, could be considered as a body of didactical engineers, and didactical engineering for production is still essentially carried out by researchers. It has developed without becoming independent from research: In production, one simply weakens the methodological constraints of research by integrating them in the form of questioning that guides the conception, but the handling of those problems that are not dealt with by the theory is not mentioned explicitly.

The following section presents an example of how the preparation of teaching contents can be organized from the perspective of didactical engineering. The example is a reform of the teaching of differential equations for first-year university students (in mathematics and physics) undertaken in 1986 (Artigue, 1989b; Artigue & Rogalski, 1990). This presentation will try to bring out the conception of transposition work inferred from the approach chosen and the role played by its theoretical foundations.

3. PRESENTATION AND ANALYSIS OF A PIECE OF DIDACTICAL ENGINEERING

The question to be dealt with here concerns the reform of an element of teaching. The didactician, either a researcher or an engineer, is therefore faced with a teaching object that has already been implemented. Why should it be changed? What aims should be included in this reform? What difficulties can be expected, and how can they be overcome? How can the field of validity for the solutions proposed be determined? This set of questions must be answered. The work will be made up of various phases. These phases will be described briefly.

The first, unavoidable phase consists in analyzing the teaching object as it already exists, in determining its inadequacy, and in outlining the epistemology of the reform project.

3.1 The Characteristics of Traditional Teaching: The Epistemological Ambitions of the Reform Project

In the present case, it had to be noted that, when the study began, the teaching of differential equations for beginners had remained unchanged since at least the beginning of the century, but that it was also at risk of becoming obsolete. In order to describe it, I shall refer to the notion of *setting* introduced in Douady (1984) to diferentiate three essential frameworks for solving differential equations:

1. *the algebraic setting* in which the solving targets the exact expression of the solutions through implicit or explicit algebraic formulae, developments in series, and integral expressions;
2. *the numerical setting* in which the solving targets the controlled numerical approximation of the solutions;
3. *the geometrical setting* in which the solving targets the topological characterization of the set of solution curves, that is to say, the phase portrait of the equation, a solving that is often qualified as being qualitative.

French undergraduate teaching was (and still mainly is) centered on algebraic solving, with an empirical approach that is characteristic of the initial development of the theory. This is a stable object that is alive and well in the teaching system, but it leads students toward a narrow and sometimes erroneous view of this field. For example, most students are convinced that there must be a recipe that permits the exact algebraic integration of any type of differential equation (as they never encounter any others), and that the only aim of research is to complete the existing recipe book.

If one considers the current evolution of the field, of the growing importance of numerical and qualitative aspects, such teaching is, despite its long stability, inevitably threatened with becoming obsolete.

The aim of the work undertaken was to construct a teaching object that was epistemologically more satisfying, mainly by:

1. opening up the teaching to geometrical and numerical solving and by managing the connections between the different solution settings in an explicit way;
2. reintroducing a functionality to this teaching by modeling problems (internal or external to mathematics) and by tackling explicitly the rupture necessitated by the transition from functional algebraic models to differential models (Alibert et al., 1989; Artigue, Ménigaux, & Viennot, 1989).

Beyond a simple elaboration, the conditions for the viability of such an object were studied with an experiment carried out in a reformed DEUG (first two years of university) at the university of Lille I.

3.2 Phase 2 of Engineering: An Analysis of Constraints

In order to better understand and manage the available possibilities, the didactician uses the systemic perspective to view the teaching to be updated as the equilibrium point of a dynamic system. It is this equilibrium that has to be studied in order to obtain an idea of its stability and to analyze the reasons for such stability in terms of constraints. By modifying at least some of these constraints, one may hope to see the system stabilize at another point of equilibrium that is judged to be more satisfying. An inadequate analysis of constraints may lead to failure or more certainly (as experiments have a strong tendency to succeed!) to a more satisfying point of functioning, but one that only appears viable because it corresponds to a maintained equilibrium.

Such an analysis must distinguish between different types of constraint. Classically speaking, three types of constraint can be distinguished:

1. *constraints of an epistemological nature* linked to the mathematical knowledge at stake, to the characteristics of its development, and its current way of functioning;
2. *constraints of a cognitive nature* linked to the population targeted by teaching;
3. *constraints of a didactical nature* linked to the institutional functioning of the teaching, especially in the field concerned and in connected fields.

The identification and analysis of constraints gives rise to the further distinction of constraints that can be qualified as *external,* which are to a great extent unavoidable except in the case of exceptional actions, and of constraints that appear to be constraints because they have been internalized by the actors in the didactical relationship, but are no longer such for the current system. These may be qualified as *internal.*

If one considers the constraints in the present example that are opposed to the extension of the teaching contents to a qualitative approach to the solving of differential equations, the following main constraints can be identified:

1. *On the epistemological level*: (a) the long domination of the algebraic setting in the historical development of the theory; (b) the late emergence at the end of the 19th century of geometrical theory with the work of H. Poincaré; (c) the relative independence of the different approaches, which permits, even nowadays at university level, a certain ignorance regarding the qualitative approach; and, finally, (d) the difficulty of the problems that motivated the birth and subsequently the development of the geometrical theory (the three-body problem, the problems of the stability of dynamic

systems, etc.) and the resulting difficulty on the level of elementary transposition processes.

2. *On the cognitive level:* (a) the permanent existence of mobility between registers of symbolic expression required by the qualitative approach: mobility between the algebraic register of the equations, of the formal expression of the solutions, and the graphic register of curves linked to the solution (isoclinal lines, curves of points of inflexion, solution curves) – increased cognitive difficulty being due to having to work on at least two levels simultaneously: that of functions and that of derivatives; (b) the fact that teaching is aimed at students for whom the concept of function, the links between registers of symbolic expression, are, in fact, in the construction stage; and, finally, (c) the mastering of the elementary tools of analysis required by qualitative proofs.

3. *On the didactical level:* (a) the impossibility of creating algorithms in the qualitative approach, which presents a serious obstacle if one considers the extent of the recourse to algorithms in teaching; (b) the relative ease of traditional algebraic teaching, which can give rise to algorithms, and the status this ease gives it in the DEUG curricula (a time when the pressure caused by new formal and theoretical demands is relaxed, and when even momentary success allows didactical negotiation to be taken up again); (c) the inframathematical status in the teaching of the graphic setting, a framework that is, however, essential here; (c) the need for the teacher to manage situations in which, as is generally the case in qualitative solving, he or she cannot answer all the questions that arise naturally; and (d) the marginal nature of elementary courses that develop a truly qualitative approach and the difficulty, consequently, in finding texts that can be used for reference (currently a text such as Hubbard & West, 1992, could fulfill this role).

The first two phases constitute an essential component of any serious engineering work, even if this component does not often appear in the finished products. In fact, this work, which is fundamental for engineering, is only at its initial stage. It remains constantly present in the background of the conceptual work and will generally be revised after the first experimentation with the engineering, when the hypotheses and choices that guided the conception have been confronted with "reality." As a counterbalance to the analysis of constraints, it allows didacticians to define how much freedom they have, to estimate how much room they have to maneuver: It guides, therefore, in an essential manner, the subsequent choices that can be made.

3.3 The Actual Conception of the Engineering

In line with the preceding section, the conception of the piece of engineering is subject to a certain number of choices. In particular, the constraints, either internal or external, that seem to oppose the viability of the project have to be displaced, at a reasonable cost.

These choices can be distinguished as:
1. macrodidactic or global choices that guide the whole of the engineering;
2. microdidactic or local choices that guide the local organization of the engineering, that is, the organization of a session or a phase.

In the present example, the main choices made on a global level are the following:

1. *Making explicit the contractual change in the status of the graphic setting* through the introduction in the teaching of a work module on functions and their representations that breaks away from normal practice in secondary teaching. Here the didactical and cognitive constraints linked to the status of the graphic framework in teaching have to be tackled, and, at the same time, the students have to be prepared for the mobility between the registers of symbols required by qualitative solving.

2. *Use of computers.* In these situations, computers initially seem to provide a way of breaking up the complexity of qualitative solving. Indeed, they are used in order to embed qualitative solving into a structured set of tasks of varying complexity (tasks of association between equations and phase portraits, tasks of interpretation of phase portraits, tasks of more or less assisted drawing of phase portraits or solutions with given conditions) – a simplification that is more difficult to set up in a traditional environment. Of course, they also appear to be a means of engaging in an approach to numerical solving. Furthermore, as they allow nonelementary situations to be controlled, they help to counter simplistic representations of the field.

3. *The explicit teaching of methods for qualitative solving.* Following the ideas developed in Schoenfeld (1985) or Robert, Rogalski, and Samurcay (1989), this means facilitating the construction of knowledge recognized as being complex by introducing an explicitly metacognitive dimension into the teaching.

4. *The limitation of complexity on the level of the algebraic solution and the transfer of the algorithmic part of this approach to independent aided work.* This last choice is imposed by obvious institutional constraints: The time that can be given over legitimately to this part of the curriculum is limited; new objects cannot be brought in without some losses. Here, the global status of the algebraic approach has been rethought: The cases studied (linear equations, those with separable variables, homogeneous equations) have been conceived as simple, typical examples that will act as a reference in the future and will be used as instruments for comparison or approximation in the study of more complex situations.

Local choices are, of course, subordinate to these global choices and must be compatible with them. It is at their level that the theory of didactical situations is really applied.

At this point, it would seem necessary to distinguish between the functioning of the two types of didactical engineering I have identified above:

didactical engineering of research and *didactical engineering of production*.

The first type constitutes a research methodology. It must therefore allow for validation following explicit rules. Here, the validation is an internal validation based on the confrontation between the *a priori* analysis of the situations constructed and the *a posteriori* analysis of the same situations. Keeping in mind that the theory of didactical situations is based on the principle that the meaning, in terms of knowledge, of a student's behavior can only be understood if this behavior is closely related to the situation in which it is observed, this situation and its cognitive potential have to be characterized before comparing this *a priori* analysis with observed reality. It is clear that such a position on validation is only tenable if the situations involved in the engineering are strictly controlled regarding the contents treated, their staging, the role of the teacher, the management of time, and so forth.

The second type of engineering is more concerned with satisfying the classical conditions imposed on engineering work: effectiveness, power, adaptability to different contexts, and so forth.

Obviously, these demands are not equal. Hence, even if it remains marked by the characteristics of research engineering, production engineering will, in this phase, take on a certain independence.

In both cases, one starts by searching for a reduced set of classes of situations that bring into play, in a way that is both suited to the epistemology of the project and operational, the essential characteristics of the knowledge targeted in the learning. Even if the concept is still under debate, one cannot fail to mention the concept of *fundamental situation* introduced by G. Brousseau (1986).

These classes of situations make up the structure of the engineering by defining its key stages. In effect, the criteria that characterize each class allow an infinite number of situations to be produced. The researcher will therefore choose from each class, concentrating on the variables that have been left free, the specific situation(s) that he or she will integrate into the engineering, and he or she will have to justify the choices made very precisely by linking them to the hypotheses underlying the engineering. The time sequence planned for the situations must also be stated.

Didactical engineers are not expected to provide the same type of construction. They are expected to highlight the core of the engineering and to encourage the construction of products that respect this core in a relatively concise presentation.

This is the type of presentation I attempted, no doubt imperfectly, in Artigue (1989b). After specifying the global choices made and the reasons for them, the engineering is presented in a seven-step structure, each step organized around a few key situations. The seven steps are as follows:

1. What needs does the differential equations tool respond to?

2. Introduction to the qualitative approach.
3. Algebraic solving.
4. The complementarity of the algebraic and qualitative approaches.
5. Introduction to numerical solving.
6. The basic tools of qualitative solving.
7. Integration of the different tools in the solving of more complex problems.

Moreover, each key situation is not described as an isolated object but as one possible representative of a class of situations specified by certain characteristics. In particular, within each class, one can, depending on the population and the time available, adjust the number of situations proposed and their relative complexity.

As an example, I present the text introducing the key situation of Step 4 (translated):

> The key situation retained as a basis for this step is that of forecasting the phase portrait of an equation that can be integrated explicitly and that presents a certain number of characteristics chosen in order to avoid putting one setting at a disadvantage in relation to another and to allow the dialectic between settings to be undertaken at the desired level. In particular: (a) Starting a qualitative study must be easy, as what is at stake in the situation is not located in difficulties at this level. For example, one could arrange things so that the horizontal isoclinal line is made up of straight lines, and so that certain particular solutions, which are relatively easy (e.g., isoclinal lines), allow the research to be organized by providing a regioning of the plane for the solution curves. (b) The algebraic solving, while it does not give rise to any particular difficulties, must not be too easy; in particular, the expressions obtained for the solutions should not be self-evident. (c) The qualitative solving, although easy at the start, allows broad categories of solutions to be determined, to foresee in what way they will vary, but must not allow all the problems set to be solved: for example, the existence of such and such a type of solution, or the nature of such and such an infinite branch. (d) At least some of these properties should, however, be accessible to algebraic solving.

This is followed by the presentation of the example used in the research, showing that it respects the conditions required.

The above description concerns only the mathematical basis of the situation. It is indispensable but notoriously inadequate if, as in the systemic perspective adopted here, one takes up one's position not only on the level of the contents but also on the level of the didactical situations through which they are staged.

As far as the researcher is concerned, he or she now has to clarify how the interaction between the students and this forecasting problem will be organized in the didactical situation, the consequences that can be inferred from the anticipated behavior, and how this can be interpreted. In particular, the researcher must show that the behavior anticipated has a high probability of appearing and prove that it cannot be induced by interference, for example, as phenomena related to the didactical contract.

This analysis, with the hypotheses on which it is based, is tested through experimentation. This is organized around a questioning of the *a priori* analysis of didactical situations. I cannot describe it in detail here. I shall simply point out that it brings into play an interrogation on the knowledge at stake in the situation; on the student and his or her possible relationship to the problem set; on the role of the teacher; how he or she will intervene; and on the possible implications of these interventions.

In production engineering, these demands weaken. The questioning remains present in order to guide the conception, to ensure the necessary didactical vigilance, but it is not directly involved in a process of internal validation. Moreover, once more, a product that is too rigid is not desirable, and, while attempting to avoid changing the nature of the situation, one must take care to leave enough liberty in the management of the situation to allow for necessary adaptability.

3.4 The Regulation of Didactical Engineering

At this point in the process, a teaching project is proposed. Its viability is supposed but not guaranteed. In fact, experience has shown that an engineering product is too complex an object to be able to be perfected at the first attempt. Adjustments will therefore be made during successive experimentations until, in the good cases, one reaches a product that is sufficiently stable and satisfying to be distributed more widely. My work on differential equations did not escape this rule. Three years were necessary to develop the product that is now distributed by the University of Lille 1. In Artigue (1992), I have analyzed the difficulties encountered and emphasized the interweaving between cognitive difficulties and didactical difficulties. These difficulties were finally solved, in particular, through the evolution of the actual teaching contents. In order to face up to the cognitive difficulties encountered in the qualitative justification, it was necessary to develop a set of justifications that operated directly in the graphic setting through relay theorems formulated entirely within this setting. This elaboration allowed wholly satisfactory results to be obtained, but, nevertheless, posed some serious didactical problems due to the institutional status of the graphic setting, highlighting the fact that the distribution of such a product, having nevertheless proved its effectiveness, can only succeed if it takes into account explicitly the in-depth renegotiation of this status, both with the teachers and the students.

4. BEYOND THIS EXAMPLE: SOME PROBLEMS TO CONSIDER

After having used an example to try to illustrate how teaching contents are prepared from a systemic perspective, I would like to return to more general questions in the last part of this text.

The approach developed aims to take into account the reality of the system in which the teaching contents have to exist, and subsequently presents the need for an elaboration that is not reduced to the text of the knowledge. This expresses the wholly reasonable desire to avoid denying the complexity of the didactical aspect. However, it must also be recognized that, at present, the application of this approach at the level of production engineering is not easy, and, moreover, stimulates, through the questions it raises, the theoretical development of research. Artigue and Perrin (1991) have attempted to analyze these difficulties in the construction of engineerings for classes mainly containing learning-disabled students. Working with such classes functioned like a magnifying glass through which the drastical changes of nature accompanying the transmission become particularly visible.

Many of these changes are the result of the gaps between the teachers' beliefs about learning and their role as teacher and the representations underlying the engineering: the teacher's desire to construct a smooth progression without any breaks, made up of little steps, in which nothing is proposed to the student that has not already been prepared, to anticipate any possible errors, which is opposed to the theoretical approaches in terms of obstacles and cognitive conflicts but allows a comfortable management of the didactical contract – everything is done so that the student who cooperates can show the exterior signs of success; if the student fails, the teacher is not in question. In all good faith, the teachers will therefore twist the proposed engineering in order to adapt it to their representations and, while believing that they have altered only a few details, will in fact have changed its nature.

In fact, these difficulties are indirectly related to failings in the theoretical framework on which the engineering is based. For too long, the theoretical framework has not considered the teacher wholly as an actor in the situation in the same way as the student, and modeling has remained centered on the relations of the student to the knowledge. This level of modeling is inadequate to take into account the problems of engineering outside the strictly experimental framework, and it is not by chance that, at present, research concerning the teacher is expanding at a rapid rate.

Finally, besides these questions, designers of an engineering are faced with delicate problems in writing up their work: What level of description should they use? How can the underlying epistemology be maintained? How can conciseness and accuracy be reconciled? How can conciseness and the presentation of the product be reconciled? These problems, which can already be seen appearing in any manual that attempts to stray from the beaten track, are multiplied here, and it must be recognized that, for the moment, we do not have the means to provide satisfactory answers.

The work accomplished up to now is certainly helpful for a better understanding of the problems linked to the preparation of teaching contents, for the identification of the points on which efforts should be concentrated, and it has also allowed the creation of a set of functional products that are compatible with the theoretical frameworks. However, no more than any other approach, it does not provide a miraculous solution to these highly complex problems.

REFERENCES

Alibert A., Artigue M., Hallez M., Legrand M., Menigaux J., & Viennot L., (1989). *Différentielles et procédures différentielles au niveau du premier cycle universitaire.* Research Report. Ed. IREM Paris 7.

Artaud, M. (1993). *La mathématisation en économie comme problème didactique: Une étude exploratoire.* Doctoral dissertation, Université d'Aix-Marseille II.

Artigue, M. (1989a). Ingénierie didactique. *Recherches en Didactique des Mathématiques,* 9(3), 281-308.

Artigue, M. (1989b). Une recherche d'ingénierie didactique sur l'enseignement des equations différentielles. *Cahiers du Séminaire de Didactique des Mathématiques et de l'Informatique de Grenoble.* Ed. IMAG.

Artigue, M., Menigaux, J., & Viennot, L. (1990). Some aspects of student's conceptions and difficulties about differentials. *European Journal of Physics, 11,* 262-272.

Artigue, M., & Rogalski, M. (1990). Enseigner autrement les équations différentielles en DEUG première année. In *Enseigner autrement les mathématiques en DEUG A première année* (pp. 113-128). ed. IREM de Lyon.

Artigue, M., & Perrin Glorian, M. J. (1991) Didactical engineering, research and development tool, some theoretical problems linked to this duality. *For the Learning of Mathematics, 11,* 13-18.

Artigue, M. (1992). Functions from an algebraic and graphic point of view: Cognitive difficulties and teaching practices. In *The concept of function: Aspects of epistemology and pedagogy.* (pp. 109-132). MAA Notes No. 28.

Artigue, M. (1993). *Enseignement de l'analyse et fonctions de référence.* Repères IREM 11, 115-139.

Arsac, G. (1992). L'évolution d'une théorie en didactique: L'exemple de la transposition didactique. *Recherches en Didactique des Mathématiques, 12*(1), 33-58.

Brousseau, G. (1986). *Les fondements de la didactique des mathématiques.* Doctoral dissertation, Université de Bordeaux I.

Chevallard, Y. (1991). *La transposition didactique* (2nd ed.). Grenoble: La Pensée Sauvage

Chevallard, Y. (1992). Concepts fondamentaux de la didactique: Perspectives apportées par une perspective anthropologique. *Recherches en Didactique des Mathematiques, 12*(1), 73-112.

Douady, R. (1984). *Dialectique outil / objet et jeux de cadres, une réalisation dans tout le cursus primaire.* Doctoral dissertation, Université Paris 7.

Hubbard, J, & West, B. (1992). *Ordinary differential equations.* Heidelberg: Springer.

Robert, A. (1992). Projet longs et ingénieries pour l'enseignement universitaire: Questions de problématique et de méthodologie. Un exemple: Un Enseignement annuel de licence en formation continue. *Recherches en Didactique des Mathématiques, 12*(2.3), 181-220.

Robert, A., Rogalski, J., & Samurcay, R. (1987). *Enseigner des méthodes.* Cahier de didactique No. 38. Ed. IREM Paris 7.

Schoenfeld, A. (1985). *Mathematical problem solving.* Orlando, FL: Academic Press.

Tavignot, P. (1991). *L'analyse du processus de transposition didactique: L'exemple de la symétrie orthogonale au collège.* Doctoral dissertation, Université Paris V.

MATHEMATICAL CURRICULA AND THE UNDERLYING GOALS

Uwe-Peter Tietze
Göttingen

1. CURRICULUM DEVELOPMENT: A SURVEY

In the early 1960s, the so-called Sputnik shock led to a radical reform of the American curriculum. This reform had, after a delay of several years, a strong impact on education in Germany. Discussions by the OECD (Organization for Economic Cooperation and Development) were also influential. Education was no longer seen merely as a way of cultivating the personality, but – like capital and labor – was then regarded as a crucial production factor, one that determines whether there will be economic growth in a country or not. While the OECD stressed training to improve the qualifications of future users of mathematics, the leading mathematics educators in the Federal Republic of Germany deemed it crucial to bridge the wide gap between the school and the university. As a result, mathematics education was decisively influenced by a structural mathematics initiated by Bourbaki, which had become generally accepted at the universities. The reformers attempted a fundamental revision of the curriculum by emphasizing a set-theoretical approach to primary school mathematics and by stressing algebraic and logical structures in the lower secondary school. The reconstruction of calculus in terms of an extensive formalization and the transformation of analytic geometry into linear algebra was a later step. Although the OECD furnished convincing arguments for the necessity to emphasize teaching of stochastics in school as early as 1959, they were ignored almost until the middle of the 1970s. One explanation could be that the predominant way of thinking in formal mathematical structures had blocked the insight into other possibilities.

When developing new curricula, mathematics educators for a long time took little notice of the general educational discussion on the main goals guiding German school reform, far less so than educators of other school subjects. In this comprehensive discussion, questions concerning "science propaedeutics" and "exemplary teaching" were of great importance (see Klafki, 1984). The new mathematical curricula were mainly oriented toward a modern, highly formalized, pure mathematics. In addition to the conception of new math, curriculum development concerning the German high school ("Gymnasium") was influenced by a teaching technology based on

R. Biehler, R. W. Scholz, R. Sträßer, B. Winkelmann (Eds.),
Didactics of Mathematics as a Scientific Discipline, 41-53.

behavioristic teaching theories. The subject matter was to be broken down into operationalistic goals. These goals were then to be organized into so-called taxonomies.

In elementary school teaching, the "structural conception" was of great importance in developing curricula in addition to new math. Based on cognitive psychology (e.g., the works of Piaget), the structural conception stresses the analogy between scientific structures and learning structures (cf. Keitel, 1986). It asserts that basic mathematical structures are best fitted to further mathematical learning. "Spiral curriculum" and "explorative learning with structured material" were basic methodical principles.

The structures of the German educational system, which allows basic changes only within an administrative framework, have hindered any independent curriculum development on a rather major scale. There were no equivalents to the extensive British or American curriculum projects such as SMSG, SMP, and SSMCIS (cf. Howson, Keitel, & Kilpatrick 1981). Curriculum development in Germany meant, and still means, that the general curricular plan of the KMK (Conference of the Federal Secretaries of Education) is concretized and adapted to the special conditions of the federal states ("Bundesländer"). This (scarcely inquired) process is influenced by existing teaching practice and an extensive published didactic discussion treating the analysis of subject-matter problems ("Stoffdidaktik"). Stoffdidaktik mainly deals with the subject matter under the aspects of mathematical analysis and of transforming mathematical theories into school mathematics. *Elementarizing*, *simplifying*, and *visualizing* are central issues in this process. The question of choice concerning subject matter is often traced back to the question of what is characteristic and/or fundamental in mathematics. When discussing curricula and the underlying goals, it seems appropriate to view the question on *elementarizing* and *fundamental ideas* as one focal point (cf. section 3). Another field of growing interest in curriculum development concerns the application of mathematics (cf. section 4). Due to limitations of space, I shall focus on high school curricula especially those of senior high school (and the specific sociocultural background); I shall not discuss textbooks and syllabi (cf. Tietze, 1992, and the references there).

1.1 Curriculum Development: Innovative Forces; Goals, Content, Methods, and their Justification

This short survey shows that societal and political forces prompt and direct innovation. There is also pressure that is exerted by the scientific mathematical community (mostly unconsciously and in a sociologically complex way). Howson, Keitel, and Kilpatrick (1981, p. 4) stress that there are also forces rooted in the educational system as a result of research, new educational theories, or the pioneering work of individuals (e.g., Piaget, Bloom). The existence of new technologies that can be applied to education must likewise be subsumed under these innovating factors. The expected rewards of innovation may also be a powerful impetus. Innovation is exciting, attracts the attention of others to one's work, foments approval, and, not seldom, contributes to the professional advancement of the educator.

Curriculum means more than a syllabus or textbook – it must encompass aims, content, methods, and assessment procedures. In developing curricula, one must justify aims, content, and methods with rational and intersubjective argument. In the German pedagogical discourse, one can primarily distinguish two methods: (a) deriving aims from highly general normative statements, which serve as axioms, by using the rules of a deontic logic or – and this method is predominant and more convincing – (b) by goals-means arguments (cf. König, 1975). The goals-means arguments consist of systems of prescriptive and descriptive statements. Such goals-means arguments allow us to transfer the justification of a certain objective to objectives of greater generality – step by step. The question remains of how to justify the highest aims in such a hierarchy. This question was not a problematic one in mathematics education, as there is strong consensus on several general objectives (see below).

The validation of a goals-means argument requires: (a) a clarification of semantics and syntax, and (b) an empirical validation of the descriptive part. From a pragmatic point of view, the clarification of the involved concepts is of great importance, but is often neglected. Statements such as "students shall learn to perform mathematical proofs" or "the student shall acquire qualifications in applying mathematics" can mean a great variety of objectives. The argument often used to justify mathematics in school, "mathematics trains logical thinking," is not only nebulous in its semantics but also based on a transfer hypothesis that does not withstand closer examination. The idea that starting off with very general concepts (e.g., a general concept of variable) will facilitate the learning process reveals an implicit learning theory that lacks scientific sanction. This implicit learning theory influenced curriculum development especially in algebra and has increased learning difficulties in this subject, which is quite difficult as is.

1.2 Principles in Mathematics Education

Normal curriculum development, the writing of schoolbooks and syllabi, is not guided by sophisticated goals-means arguments – if explicit arguments exist at all – but is rather based on so-called "didactic principles." Such principles, which are prescriptive statements based on descriptive assumptions (factual knowledge from psychology, pedagogics, mathematics, experience, etc.) and normative postulates (educational goals and objectives, societal goals, etc.) – for the most part implicit – say what should be done in mathematics teaching (Winter, 1984).

The importance and acceptance of such principles changes over the course of time. The central (underlying) principle in traditional mathematics education, for example, was that of isolating difficulties. The subject matter was divided into poorly integrated sections, each of which was characterized by a special type of exercise. Integrative ideas and strategies were neglected. Mathematics appeared to the students as a collection of isolated types of exercise. This, in its essence, originally correct idea has turned into something false by exaggeration and oversimplification – a critical tendency inherent in most didactic principles.

Although several authors feel that principles in mathematics education are of fundamental significance (e.g., Wittmann, 1975), there are empirical and other considerations that advise us to be careful in dealing with them. Several didactic principles, for example, recommend the intensive use and variation of visual representations. Empirical studies show, however, that iconic language can cause considerable additional difficulties in comprehension (Lorenz & Radatz, 1980). Further principles that are problematic in a related respectively similar way are the operative principle and the principle of variation that demands the use of a variety of models for learning mathematical concepts. The main problem with didactic principles is the lack of a sound analysis of their descriptive and prescriptive components, which are often compounded.

2. NEW MATH AND COUNTERTENDENCIES

The reform of the mid-1960s – often called the *new math* – adopted many characteristics of modern pure mathematics. The textbooks on calculus or linear algebra resembled, to a certain extent, university lectures in content, sequence, and diction. Subjective aspects such as the students' experiences, knowledge specific to their age group, and inner representation of concepts were scarcely taken into account. One consequence of the similarity of this approach to the systematic structure of formal scientific mathematics was that important subject matter had to be elementarized. This fact stimulated several interesting analyses and works in mathematical fields adjacent to school mathematics, such as the construction and characterisation of real numbers and the development of the function concept (cf. Steiner, 1966, 1969). At that time, a formalistic-logistic mathematical science had established itself at the universities, a mathematics that was not interested in a theory concerned with the meaning of mathematical concepts and that almost completely ignored any reflection on mathematics and its application. In the beginning, this narrow scientific program was adopted by mathematics educators. It soon provoked opposition. The main reason for this opposition was the fact that highly abstract and formalized mathematical concepts proved impracticable in school. In high school, this effect became more and more pronounced the more the German Gymnasium lost its status as an elite school and became an educational institution for a significant part of the population. The higher vocational and technical schools, which had teachers who differed in their academic backgrounds, were not as strongly affected at that time by the wave of mathematical rigor as the general high schools were. The critique of new math resulted in fruitful research and discussion from two perspectives that do not exclude each other, but represent different focal points.

1. The first position focuses on the idea that mathematics education should further an undistorted and balanced conception of mathematics, including the aspects of theory, application, and mathematical modeling. It should also emphasize the learning of meaningful concepts (in the semantic sense) and the teaching of the fundamental ideas of mathematics. (a) Interesting papers have been published dealing with the question of how mathematical theories and concepts can be simplified and elementarized without falsifying the central mathematical content. Others focus on fundamental ideas either for mathematics in general or for a specific field. (b) Some mathematics educators made it their objective to analyze epistemologically the process of mathematical concept and theory formation. They then tried to derive didactic consequences from this.

2. The other position considers the students and the benefits that mathematics can render to them. In the mid-1970s, (high school) mathematics educators were asking how *curricula* could be *justified* – mainly as a consequence of the lack of justification in the new math. Some authors referred to Wagenschein and Wittenberg, well-known educators in mathematics and natural sciences. They pleaded for the Socratic teaching method to encourage students to discover mathematical ideas and theories by themselves. This also means teaching by examples without being pressured by a voluminous canon of subject matter. Winter greatly influenced this discussion with his catalog of *general objectives*. This catalog is based on the question of "basic mathematical activities, which are rooted in normal everyday thinking and therefore can influence general cognitive abilities." (1975, p. 107, translated). Winter stresses: (a) the ability to argue objectively and to the point; (b) the ability to cognitively structure situations of everyday experience, to detect relationships, and describe them in mathematical terms, or to develop mathematical tools and concepts with this in mind; and (c) creativity; that is, to acquire and use heuristic strategies to cope with unknown problems, especially strategies for developing and examining hypotheses. This research and the implied curricular suggestions cited above can be regarded as a late but substantial attempt to explicate the central pedagogical objective of school reform, that is, science propaedeutics in a way specific to the subject.

Theories and results obtained from the *psychology of learning* were gradually introduced into mathematics education in high school. In elementary mathematics education, such questions and issues have had a long tradition. Didactic principles derived from the psychology of motivation and learning became important in developing curricula. Along with recognizing that didactic principles often proved to be problematic in their descriptive parts (cf. section 1), attempts were undertaken to inquire into the processes of learning mathematics in general and those specific to certain topics.

3. ELEMENTARIZATION, FUNDAMENTAL IDEAS

3.1. Formation of Concepts and Theories Taking Calculus as an Example

The question of how to facilitate the learning of mathematical theories by *elementarizing* them is of central importance, especially in the upper classes of secondary school. One can roughly discriminate three ways of doing this:

1. by suitably choosing basic definitions and axioms; for example, the foundation of differential calculus on the concept of continuity instead of on the concept of limit or taking the intermediate value property as a completeness axiom;
2. by using stronger postulates; for example, one does not base calculus on the classical concepts of Cauchy continuity and limit, but on the concepts of Lipschitz continuity and differentiability;
3. by pursuing a so-called gradual development of exactness; the objectives are exact but not fully formalized concepts.

The first two points of view have been the subject of controversy in educational circles for many years. They are nevertheless considered outmoded today. The main critique of the second form of elementarization points out that it furthers the tendency to simplify merely in a technical way (such as for proofs); on the other hand, intuitive aspects of the concept could be neglected and the entire mathematical situation falsified. As regards the third way, Blum and Kirsch (1979) have suggested a curriculum (for basic courses) that stresses at the beginning the *calculation* of derivatives and not the question of their *existence*. One starts out with an "intuitive" idea of limit. This is then challenged, when the occurrence of a problem makes this desirable, for example, in the context of the product formula or of $\lim_{x \to 0} \frac{\sin x}{x}$. Kirsch (1976) has pleaded for an introduction to the integral concept that uses the naive idea of measure of area as its basis. Sequential *steps of exactitude* could be achieved by (a) formulating the properties of the area function, (b) making the students aware of the problem of existence, and (c) proving it. This conception can also be applied to proofs. As regards the derivative of $\exp_b$, one can start by calculating $\lim_{h \to 0} \frac{b^h - 1}{h}$ and by leaving the well-defined question of existence to a later step.

This curricular idea shows that mathematical precision is not necessarily sacrificed when the axiomatic-deductive method is renounced. Exactitude is not needed here, however, at the beginning, but occurs as the result of a long process of questioning and clarifying. This process, which Fischer (1978) called *exactifying*, is also characteristic of many historical developments in calculus. Exactifying means in calculus – also historically – the process of grappling with the original naive ideas of function, number, and limit. In arriving at the modern concepts, the question of existence plays an important role.

The historical starting point of many mathematical concepts – this is especially true for school mathematics – is a more-or-less practical problem. It has always been an objective of mathematics to find exact definitions of such concepts in order to avoid contradictions, and also to make possible communication between mathematicians. On the way to a precise (and formal) concept, many of the originally involved aspects are lost. For a mathematician, this is not a problem, because he or she is mainly interested in working with the precise, up-to-date form of the concept and is not concerned with its historical and epistemological origin. For the nonmathematician, especially the high school student, it is the other way around; in particular, when the naive concept is to serve as an introduction to the mathematical concept. For the nonmathematician, for example, it does not make sense that a square cannot be divided into two (disjoint) congruent parts.

The development of the function concept is of central interest in school. The common formal definition that uses sets of pairs is the result of a long historical process and has lost much of the original naive idea of drawing an uninterrupted curve by hand. Some of the original aspects emerge in additional concepts like continuity, differentiability, integrability, and rectifiability, and constitute, as such, essential parts of differential and integral calculus. The function concept is fundamental in modern school mathematics and is taught at all levels. In Grades 1 to 6, students work propaedeutically with tables, arrow diagrams, and simple geometric mappings. In Grades 7 and 8, they become acquainted with important examples such as linear functions. In Grades 9 and 10, they learn a formal definition and a great variety of empirical and nonelementary functions (e.g., the square and its inverse, exponential, logarithmic, and trigonometric functions). The objective is to enable the students to develop a well-integrated scheme including graphs, tables, curves, arrow diagrams, and set-theoretical and algebraic aspects and to discriminate between function, function value, term, equation, and graph. There has been research on concept formation, especially concerning the function concept (cf. Vollrath, 1989, and the references there).

Exactifying is significant in the development of calculus curricula for two reasons: On the one hand, it is a central epistemological and methodological aspect and is therefore an important aim of teaching; on the other hand, it can and should be a leading idea in sequencing.

New curricula in calculus usually accept the didactic principle of acknowledging the student's previous knowledge and preconceptions. From a didactic point of view, it does not make sense to expect the student to forget all about angle measure, for example, and then accept a definition by a bilinear form. Such "antididactical inversions" are: defining convexity by first and second derivative or introducing the integral by the antiderivative, thereby reducing the Fundamental Theorem of Calculus to a mere definition and hindering applications. The student's formation of concepts can further be facilitated by the appropriate representation and by a suitable change in the representation mode (cf. Kirsch, 1977). Thus, some modern textbooks begin with graphical differentiation and integration.

3.2. Fundamental Ideas

The conception "fundamental idea" can be seen as a response to the present-day flooding by extremely isolated and detailed knowledge. Since Bruner stressed the importance of fundamental ideas in his widely distributed book *The Process of Education* (published in German in 1970), this conception has raised concern under German mathematics educators. In Wittmann's widely read book *Grundfragen des Mathematikunterrichts,* the request that mathematics teaching should center on fundamental ideas is one of the central didactic principles. Wittmann follows Bruner also in the didactic principle that the fundamental ideas of mathematics, adequately adapted, can be conveyed to students of any age. Bruner wanted the best scientists of each discipline to work out the fundamental ideas. This conception suffered from the fact that no consensus on the central ideas could be achieved. The general educator Blankertz objected that the choice of fundamental ideas in a subject cannot be abstracted from the educational objectives and should not therefore merely be assigned to the specialists.

Along with Bourbakism, there has been a strong trend in *mathematical science* to structure mathematics with the help of basic conceptions such as composition/order/topology or set/structure/mapping. These conceptions have had a tremendous impact on modern mathematics, but – from the perspective of school mathematics – are related mainly to mathematics as a product. Their explicatory and ordering power exceeds the realm of the school, especially after the retreat from new math. Halmos (1981) tried to evolve basic ideas that refer also to the process of doing mathematics as a researcher. He ended up with the following catalogue: (a) universal algebra: structure, categories, isomorphism, quotients; (b) size: primes, duality, pigeonhole, infinity; (c) composition: iteration, cross-section, exponential; (d) analogy: commutativity, symmetry, continuity. He is aware of his speculative attempt: "Is what I have been saying mathematical mysticism, or is it possible that there really are some underlying guiding principles in mathematics that we should try to learn more about?" (Halmos, 1981, p. 152).

There have been several attempts to cope with the question of fundamental ideas in *math education* (for a historical survey, cf. Schweiger, 1992). Some inquiries try a general approach, others focus on special subject matter such as calculus. Although there are differences in the methodological approach and the philosophical background, one nevertheless can find much correspondence when comparing inquiries on *fundamental ideas for mathe-*

matics as a whole. Algorithm (mechanical procedure for calculation or decision-making, the idea of calculus, computability, programming), *approximation, function* (assignment, mapping, transformation, operator), and *modeling* are well accepted as central aspects of mathematics in school.

Linear functions are of importance in many fields of secondary mathematics. In junior high school, proportionality prevails, but also geometrical topics such as area and similarity can be treated fruitfully under the aspect of linearity. In senior high, differentiation, integration, and the mapping of convergent series to their limits can be seen as linear operators. Linearity is of course central to linear algebra (linear mappings, linear and multilinear forms such as scalar products and determinants). Linearity can also mean *linearization*. Thus differentiation can be looked at under the aspect of local linear approximation (instead of local rate of change). Special linear approximations of certain functions (e.g. $\sin x \approx x$ for small x) are of importance. Linearization is also relevant to Newton approximation and to the theory of errors. In stochastics, linear regression is a powerful tool. But linearity has not become an organizing idea for the students. This seems also to be true for quite a few teachers.

Invariance is a central and fruitful idea in mathematical research (e.g., structural isomorphisms, characterization by invariants, Klein's Erlanger program, Galois theory, etc.). It has temporarily gained some attention in school mathematics during the wave of mapping-oriented geometry ("Abbildungsgeometrie"), but seems to be too abstract an idea to be helpful for learning mathematics in school.

Schreiber (1983) proposes very general ideas such as exhaustion (e.g., successive approximation, mathematical modeling, also real approximation), idealization, abstraction, representation as basic and universal. It is unquestionable that these ideas are universal, but I doubt – and here I rely on modern research on learning – that these ideas are powerful tools and/or have a special explanatory power in the realm of learning mathematics.

Other mathematics educators have proposed extracting fundamental ideas more in an inductive and pragmatic way *for specific subject matter*. Fundamental ideas are seen as central points in a relational net and/or as powerful tools for mathematical problem-solving or mathematical modeling in a certain field. One distinguishes between: (a) *central concepts* that refer to mathematics as product, (b) *subject specific strategies*, and (c) *patterns of mathematization*, the last two stressing the processual aspect (cf. Tietze, 1979). An idea can be fundamental in more than one sense. As modern transfer research shows, it is not the general heuristic strategies that are powerful in problem-solving, but strategies that are specific to a certain matter.

The *central concepts* of a subject matter depend on the perspective from which one looks at it. If one takes Bourbaki's perspective on linear algebra, then vector space, linear mapping, scalar product, and Steinitz exchange theorem are central. If one looks at it from the angle of "linear algebra and its applications" (e.g., Strang, 1976), then linear equation and Gaussian algorithm are fundamental. We shall discuss some *subject-specific strategies* and *patterns of mathematization*. The "analogy between algebra and geometry" (geometrization of algebraic contexts and vice versa) is a powerful tool in coping with mathematical questions. The analogy between geometric theorems such as Pappos, Desargues, cosine law, ray law, and so forth, and the corresponding theorems/axioms in the language of vector spaces are powerful in solving problems and/or gaining an adequate understanding. By interpreting the determinant as oriented volume, many complicated proofs "can be seen." In the latter example, another fundamental idea is involved, the idea of "generalized visual perception," which means translating geometric concepts and "carrying names" of the perceptual 3-dimensional space to the abstract n-dimensional space. This idea allows, for example, a normal applicant of complicated statistical procedures, such as factor analysis or linear progression, to get an adequate idea of the tool, its power, and its limits.

Fischer analyses fundamental ideas of calculus in an influential work (1976). He particularly stresses the idea of *exactifying*, which was described in section 3.1. He further accentuates the following ideas in addition to others: approximation, rate of change, and the potential of a calculus (in a general sense).

4. APPLICATION-ORIENTED TEACHING TAKING CALCULUS AS AN EXAMPLE

By the turn of the century, the question was already in dispute as to what emphasis should be given to application-oriented problems in calculus teaching. This discussion took place against the backdrop of the magnificent technical and industrial development occurring at that time. The opinions ranged from "application means providing an inferior service" to "mathematics should only be taught on behalf of its applications." The central idea of the formal education of the traditional and dominant German "Humanistisches Gymnasium," with its major interest in ancient languages, was an important issue in this discussion. Klein attempted to reconcile the conflicting positions in this dispute by pleading for "practical calculus, which limits itself to the simplest relationships and demonstrates these to the students by modeling familiar processes in nature" (1904, p. 43, translated).

There is an intensive discussion on teaching applied mathematics and mathematical modeling in Germany today. This must be seen, in part, as a reaction to the extreme structure orientation of the late 1960s and 1970s. One can distinguish three main trends in the argument (cf. Kaiser-Messmer, 1986): (a) an *emancipatory* trend, (b) a *science-oriented* trend, and (c) an *integrative* trend.

These trends differ from each other mainly with respect to the aims associated with applied mathematics and mathematical modeling. Representatives of the first trend plead for an *emancipatory* education. They demand the use of mathematical methods in realistic situations, where this use serves to elucidate situations that are really important to the student. This conception can be illustrated by teaching units such as analyzing unemployment and the effect of a reduction of weekly working hours, comparing special train fares for young people, and discussing the effects of speed limits in cities and on highways. In calculus courses, one can treat problems dealing with the planning of freeways (e.g., the alignment of crossings) and the ecological implications. This is not only to develop problem-solving qualifications, but primarily to enhance the students' general political abilities (cf. Böer & Volk, 1982).

The second trend in the argument aims at developing the central ideas of mathematics and its epistemology. Students should gain basic *epistemological and methodological experiences* and insights, so that they acquire a broad and flexible understanding of mathematics (cf. e.g., Steiner, 1976). Calculus seems to be too complex to meet the requirements for these objectives in school.

The integrative trend demands a balanced relation between utilitarian, methodological, epistemological, and internal mathematical objectives. This trend is strongly influenced by the pedagogical aims of mathematics teaching formulated by Winter (see section 2). Blum (1988) illustrates how such objectives can be reached in applied calculus by analyzing the problem of constructing functions for income tax as a teaching example.

The natural sciences provide numerous opportunities for teaching applied calculus. Physics yields a great variety of examples appropriate for teaching purposes in senior high school. In the 1970s, several applied problems from biology were developed as teaching units, especially those problems concerning processes of growth. Other important fields for the teaching of applied calculus are the social sciences and economics (e.g., relations between cost, profits, prices, supply, and demand; the modeling of markets).

While the textbooks of so-called traditional mathematics contained a great variety of applied problems and exercises from physics that could be solved by calculus, and that were actually covered in class, applied problems were avoided in the textbooks of the new math period. But during the last 5 years, many examples of mathematical modeling in the fields of economics, the social sciences, and biology have been incorporated into calculus textbooks. Economic problems are especially stressed in special senior high schools for economics ("Wirtschaftsgymnasium"). The importance of physics in applied mathematics teaching has faded, since today's students, especially in basic courses, lack knowledge and interest. Before the school reform, physics was a compulsory subject in senior high school; now it is optional and very few students take it, an exception being students in technical senior high schools. Another reason lies in the diminished number of teachers who teach both subjects.

Kaiser-Messmer (1986) investigated the question of whether and to what extent the general objectives of an application-oriented mathematics teaching can be realized. She carried out extensive case studies on classes exposed to application-oriented calculus teaching. Most students in her sample improved considerably their ability to understand and cope with everyday situations; they acquired simple abilities of applying mathematics. But there were only a few students who gained or improved their general abilities to cope with mathematical modeling problems. The development of component skills was more easily achieved. The students' motivation and attitude with regard to mathematics improved in nearly all cases.

5. CONCLUSION

New empirical research shows the limits of curriculum development in principle. The teacher alone determines the effectiveness of curriculum by his or her decisions, behavior, attitudes, and cognitive processes, no matter how carefully the curriculum has been developed. The high expectations educators once had about the benefits of scientifically developed curricula have been supplanted by a more modest assessment. Recent research has placed more emphasis on everyday curriculum in the classroom, on teachers' ideas and subjective theories concerning their quotidian preparation of classes, their subjective learning theories, implicit and explicit objectives, philosophy of mathematics, and the influence of these cognitions on their teaching.

6. REFERENCES

Blum, W. (1988). Analysis in der Fachoberschule. In P. Bardy, F. Kath, & H.-J. Zebisch (Eds.), *Umsetzen von Aussagen und Inhalten. Mathematik in der beruflichen Bildung.* Alsbach: Leuchtturm (Technic didact Bd. 3).

Blum, W., & Kirsch, A. (1979). Zur Konzeption des Analysisunterrichts in Grundkursen. *Der Mathematikunterricht, 25*(3), 6-24.

Böer, H., & Volk, D. (1982). *Trassierung von Autobahnkreuzen - autogerecht oder ... .* Göttingen: Gegenwind.

Fischer, R. (1976). Fundamentale Ideen bei den reellen Funktionen. *Zentralblatt für Didaktik der Mathematik, 8*(4), 185-192.

Fischer, R. (1978). Die Rolle des Exaktifizierens im Analysisunterricht. *Didaktik der Mathematik, 6*(3), 212-226.

Halmos, P. (1981). Does mathematics have elements? *The Mathematical Intelligencer, 3,* 147-153.

Howson, G., Keitel, Ch., & Kilpatrick, J. (1981). *Curriculum development in mathematics.* Cambridge: Cambridge University Press.

Kaiser-Messmer, G. (1986). *Anwendungen im Mathematikunterricht* (Vols. 1 & 2). Bad Salzdetfurth: Franzbecker.

Keitel, CH. (1986). Lernbereich: Mathematik und formale Systeme. In H. D. Haller & H. Meyer (Eds.), *Ziele und Inhalte der Erziehung und des Unterrichts* (pp. 258-269). Stuttgart: Klett-Cotta.

Kirsch, A. (1976). Eine "intellektuell ehrliche" Einführung des Integralbegriffs in Grundkursen. *Didaktik der Mathematik, 4*(2), 87-105.

Kirsch, A. (1977). Aspekte des Vereinfachens im Mathematikunterricht. *Didaktik der Mathematik,* 5(2), 87-101.

Klafki, W. (1984). Thesen zur " Wissenschaftsorientierung" des Unterrichts. *Pädagogische Rundschau, 38*(1), 79-87.

Klein, F. (1904). Bemerkungen im Anschluß an die Schulkonferenz von 1900. In F. Klein & E. Riecke (Eds.), *Neue Beiträge zur Frage des mathematischen und physikalischen Unterrichts an den höheren Schulen* (pp. 33-47). Leipzig: Teubner.

König, E. (1975/1978). Theorie der Erziehungswissenschaft. München: Fink.

Lorenz, J.-H., & Radatz, H. (1980). Psychologische Aspekte des Mathematikunterrichts. In D. H. Rost (Ed.), *Unterrichtspsychologie für die Grundschule* (pp. 134-149). Bad Heilbronn: Klinkhardt.

Schreiber, A. (1983). Bemerkungen zur Rolle universeller Ideen im mathematischen Denken. *mathematica didactica, 6*(2), 65-76.

Schweiger, F. (1992). Fundamentale Ideen. Eine geisteswissenschaftliche Studie zu Mathematikdidaktik. *Journal für Mathematik-Didaktik, 13*(2/3), 199-214.

Steiner, H. G. (1966). Äquivalente Fassungen des Vollständigkeitsaxioms für die Theorie der reellen Zahlen. *Mathematisch-Physikalische Semesterberichte, 13,* 180-201.

Steiner, H. G. (1969). Aus der Geschichte des Funktionsbegriffs. *Der Mathematikunterricht, 15*(3), 13-39.

Steiner, H. G. (1976). Zur Methodik des mathematisierenden Unterrichts. In W. Dörfler & R. Fischer (Eds.), *Anwendungsorientierte Mathematik in der Sekundarstufe II* (pp. 211-245). Klagenfurt: Heyn.

Strang, G. (1976). *Linear algebra and its applications.* New York: Academic Press.

Tietze, U.-P. (1979). Fundamentale Ideen der linearen Algebra und analytischen Geometrie. *mathematica didactica, 2*(3), 137-165.

Tietze, U.-P. (1992). Der Mathematikunterricht in der Sekundarstufe II. Curriculumentwicklung und didaktische Forschung. *mathematica didactica, 15*(2), 3-37.

Vollrath, H.-J. (1989). Funktionales Denken. *Journal für Mathematik-Didaktik, 10,* 3-37.

Winter, H. (1975). Allgemeine Lernziele für den Mathematikunterricht? *Zentralblatt für Didaktik der Mathematik, 7*(3), 106-116.

Winter, H. (1984). Didaktische und methodische Prinzipien. In H. W. Heymann (Ed.), *Mathematikunterricht zwischen Tradition und neuen Impulsen* (pp. 116-147). Köln: Aulis.

Wittmann, E. (1975). *Grundfragen des Mathematikunterrichts.* Braunschweig: Vieweg.

CHAPTER 2

TEACHER EDUCATION AND RESEARCH ON TEACHING

edited and introduced
by
Rolf Biehler
Bielefeld

Teacher education and teacher training aim at developing teachers' knowledge and practical competence, ideally not only to reproduce existing practice but also to prepare for an improved practice on the basis of recognized deficiencies in current mathematics education.

The knowledge of teachers, their attitudes, beliefs, and personalities are essential factors for the success of mathematics teaching, although this success also depends on the social conditions of schooling and the available tools. Teachers' professional work is situated in a social context that constrains their activities. The contraints such as syllabi, textbooks, media, software, 45-minute lessons, structures of classroom interaction, assessment as a necessity, students' intellectual capabilities and motivation, and so forth are supportive and limiting at the same time. An awareness of not only these constraints but also the real freedom for teachers' actions and decisions should be an important part of teachers' knowledge. In this sense, the dimensions of mathematics education and all the scholarly knowledge pre-

R. Biehler, R. W. Scholz, R. Sträßer, B. Winkelmann (Eds.),
Didactics of Mathematics as a Scientific Discipline, 55-60.

sented in the other chapters of this book are relevant to teacher education and to teachers' knowledge.

However, teacher education has its own constraints, and the variation between and within countries seems to be much larger than in mathematics education itself. Different systems are in action: The relative function of university studies in mathematics and in mathematics education, institutionalized training on the job, and in-service education of experienced teachers varies. The process of giving life to research results and innovative curricula in everyday classroom practice through communication with teachers is itself a complex process whose success has often proved to be fairly limited. That is why the following three topics have become domains of research and reflection within the didactics of mathematics:

1. *teachers' cognitions and behavior;*
2. *the relation between theory and practice*;
3. *models and programs of teacher education.*

In other words, these three problem domains have shifted from being merely practical problems to problems at a theoretical level. The four papers in this chapter discuss all three problem domains from different perspectives and with different emphases. However, the major concern of all papers is teachers' knowledge: its structure and its function in teaching practice, descriptive models of teachers' knowledge, normative requirements based on theoretical analyses, and possibilities and failures to influence and develop teachers' knowledge.

Teachers' beliefs and teachers' knowledge are increasingly considered as research topics in didactics of mathematics. Two chapters of the *Handbook of Research on Mathematics Teaching and Learning* (Grouws, 1992) are devoted to this topic and provide a review of research mainly from a North American perspective. Hoyles (1992) analyzes how research on teachers has developed from isolated papers to a new major direction at the international conferences of the group of Psychology in Mathematics Education (PME). One of the recent conferences on Theory of Mathematics Education (TME) organized by Hans-Georg Steiner was devoted to the topic of *Bridging the gap between research on learning and research on teaching* (Steiner & Vermandel, 1988).

> Compared with other professions, the special structural problem of the teaching profession is that it does not have one "basic science" such as law for the lawyer, medicine for the physician . . . scientific theory is related in two utterly different ways to the practical work of mathematics teachers: first, scientific knowledge and methods are the subject matter of teaching; second, the conditions and forms of its transmission must be scientifically founded. (Otte & Reiss, 1979, p. 114-115)

These two kinds of scientific knowledge have always played different roles with regard to teacher education for different school levels. Whereas, in primary teacher education, the mathematical content knowledge was often regarded as trivial compared to the emphasis on educational knowledge, the situation for high school teacher education was the reversed. Although this sharp distinction has become blurred, the different emphases still exist and can be explained partly by the complexity of the knowledge on the respective level.

Didactics of mathematics in its relation to teachers can be viewed in two ways: First, as an endeavor to bridge the gap between theoretical knowledge (mathematics, educational theories, psychology, etc.) and the practice of mathematics teaching. However, second, didactics of mathematics as a discipline sometimes regards itself as the "basic science" for the mathematics teaching profession. In this sense, didactics of mathematics itself creates a theory-practice problem insofar as it has developed scholarly knowledge of its own.

Teachers' knowledge related to mathematics is crucial. The question what kind of knowledge, experience, and understanding of mathematics a mathematics teacher should have has turned into a research question for the didactics of mathematics. A symposium of ICMI at the ICM in Helsinki, 1978 (Steiner, 1979), offered a perspective on this topic based on the assumption that mathematics has to be interpreted within its larger cultural role and in relation to other subjects, and not only as an academic subject. For primary teacher education, Wittmann (1989) argued for a type of course on elementary mathematics that should have a quite different character than usual academic mathematics courses, for instance, it should be rich in relationships to history, culture, and the real world; it should be organized in a problem and process-oriented way; it should involve a variety of representations (concrete materials, diagrams, symbolic language, etc.); and it should allow for a variety of teaching/learning formats. Dörfler and McLone (1986) provide a differentiated analysis on relations between academic mathematics, school mathematics, and applied mathematics with regard to the knowledge teachers should have about the different characteristics and natures of mathematics. (see, also, Niss, this volume).

Hans-Joachim Vollrath describes a course in pre-service teacher education for high school teachers in mathematics that should enable teacher students to reanalyze, restructure, and evaluate the academic mathematical knowledge they have already learned from a didactical point of view. *Reflections on mathematical concepts as starting points for didactical thinking* are taken as a focus, because problems of mathematical concept definition and meaning can be related to psychological aspects of concept learning, principles of teaching concepts, and the historical development of concepts. The examples are taken from calculus, a field of advanced mathematical thinking that recently has received more attention from researchers in mathematics education (Tall, 1991). Vollrath discusses the possibilities and needs for integrating historical and epistemological aspects of mathematics (see chapter 7, this volume) in teacher education. His contribution relates to a tradition in German didactical thinking of trying to get teachers to reflect on the relation between school mathematics and university mathematics in order to enable them to make conscious choices instead of simply reproducing either of them. In other words, teachers should be enabled to reflect on, understand, and actively shape the process "of preparing mathematics for students" as it is also analyzed in chapter 1 of this volume.

Vollrath's paper can be read an as example of how teachers' knowledge related to mathematics should be extended, enriched, and transformed from a didactical point of view, even if teachers have had a high-quality academic mathematics education. Complementary to this normative conception is a descriptive-empirical orientation toward the question how the knowledge of practicing teachers can be modeled and whether and how their knowledge does affect their classroom behavior. Empirical research on this question should, at least in the long run, inform teacher educators with reliable knowledge on how to overcome mere intuitive priorities and content selection in their courses.

Rainer Bromme develops a *psychological topology of teachers' professional knowledge* that distinguishes between several kinds of knowledge related to mathematics, namely, mathematical content knowledge, school mathematical knowledge, philosophy of school mathematics, and subject-matter-specific pedagogical knowledge. On the basis of this model, he reviews and reinterprets empirical research concerned with identifying and analyzing the function of teachers' knowledge and beliefs for teaching practice. The paper is situated in an increasingly important research tradition concerned with modeling teachers' knowledge and beliefs (Fennema & Franke, 1992; Thompson, 1992). Instead of the notion of teachers' *beliefs,* the author prefers the notion of *philosophy of school mathematics,* similar to Ernest (this volume) who theoretically extends this topic. By this, the interindividual aspects of this knowledge and its interwovenness with subject-matter aspects as compared to mere subjective belief systems should be stressed. Bromme reinterprets research results that have found deficiencies concerning teachers' knowledge about individual students' understanding and concerning the subject-matter-specific pedagogical knowledge of teachers, showing that, nonetheless, teachers' have shown practical competence to cope with the demands of the classroom that indicates the richness in intuitive knowledge that teachers have developed during their professional life.

Bromme's approach of considering *teachers as experts* from the perspective of an educational psychologist establishes a certain tension to those reflections in didactics of mathematics that criticize teachers too easily but do not take sufficient account of their concrete working conditions, the limits to rationality in everyday acting.

Heinz Steinbring's conception of a *dialogue between theory and practice in mathematics education* takes this perspective of "teachers as experts" into account. He provides an introduction to the discussion on reconceptualizing the relation between theory and practice in the didactics of mathematics and summarizes insights from projects under the heading of *Systematic cooperation between theory and practice,* in which teachers and researchers have been trying to establish new kinds of relations: Overcoming the widespread "teaching as telling" (the "broadcast metaphor") in the classroom is related to overcoming the broadcast metaphor in teacher education as well. With respect to teachers' knowledge, the paper is based on the assumption that a deeper understanding of the epistemological nature of mathematical knowledge as theoretical knowledge with its specific relation between objects, symbols, and concepts is necessary if teachers are to cope adequately with problems in the classroom. The author gives examples from the teaching and learning of fractions. The role of diagrams for communicating and working with theoretical knowledge is one focus. In this respect, the paper relates to the analysis of representations for mathematical teaching, learning, and thinking by Kaput (this volume). With regard to in-service teacher education, the important function of shared situations (in the shape of lesson transcripts), besides theoretical knowledge, is elaborated for stimulating reflection and communication between researchers and teachers. Steinbring respects *teachers as experts* with a lot of intuitive knowledge but tries to transform and elaborate this knowledge by means of a dialogue.

Tom Cooney's analyses *on the application of science to teaching and teacher education* are concerned more explicitly with overcoming the unsatisfactory practice of mathematics teaching. Complementary to Steinbring's contribution, he discusses what kind of didactical research and didactical theory is necessary in order to not just mirror existing practice but open up ways for innovations. Research is necessary to broaden our understanding of how teachers come to believe and behave as they do, where and how their attitudes toward mathematics and its teaching are created, and how this may be changed toward a more adaptive and reflective teacher with a "scientific attitude" to his or her own teaching practice. From this point of view, research on teachers' cognitions as well as on the efficiency of in-service programs is reviewed. Research points to the limited view of mathematics that teachers communicate in the classroom and the lack of that mathematical sophistication (especially in elementary teachers) that would be needed to implement innovative mathematics teaching such as described in the NCTM standards. However, a simple extension and broadening of the knowledge related to mathematics in teacher education can hardly be sufficient, because of the complex social situation of the teachers' work place and longstanding habits. For Cooney, it is necessary to "create contexts in which teachers . . . can envision teaching methods that reflect reasoning, problem-solving, communicating mathematics, and connecting mathematics to the real world . . . and yet feel comfortable with their role as classroom managers." Discussing with teachers new forms of problems for assessment that reflect the above innovative ideas are seen as an important possibility of a *shared situation* in the sense of Steinbring that may foster the dialogue between theory and practice and develop the teacher in the direction of an intellectual leader rather than the determiner of mathematical truth.

The papers in this chapter elaborate the complex demands on teachers spanning from the teacher's role of being a representative of the mathematical culture outside school to being a confident manager of classroom interaction. In doing this, the papers have analyzed the teacher's role as a subsystem of the complex system of mathematics education, which is elaborated in the other chapters of this book.

REFERENCES

Dörfler, W., & McLone, R. R. (1986). Mathematics as a school subject. In B. Christiansen, A. G. Howson, & M. Otte (Eds.), *Perspectives on mathematics education* (pp. 49-97). Dordrecht, Netherlands: Reidel.

Fennema, E., & Franke, M. L. (1992). Teachers' knowledge and its impact. In D. Grouws (Ed.), *Handbook of research on mathematics teaching and learning* (pp. 147-164). New York: Macmillan.

Grouws, D. (Ed.). (1992). *Handbook of research on mathematics teaching and learning.* New York: Macmillan.

Hoyles, C. (1992). Mathematics teaching and mathematics teacher: A meta-case study. *For the Learning of Mathematics, 12*(3), 32-45.

Otte, M., & Reiss, V. (1979). The education and professional life of mathematics teachers. In International Commission on Mathematical Instruction (ICMI) (Ed.), *New trends in mathematics teaching* (Vol. IV, pp. 107-133). Paris: UNESCO.

Steiner, H.-G. (Ed.). (1979). *The education of mathematics teachers. IDM Materialien und Studien 15.* Bielefeld: Universität Bielefeld.

Steiner, H.-G. & Vermandel, A. (Eds.). (1988). *Investigating and bridging the teaching-learning gap. Proceedings of the 3rd International TME Conference.* Antwerp: University of Antwerp.

Tall, D. (Ed.). (1991). *Advanced mathematical thinking.* Dordrecht, Netherlands: Kluwer.

Thompson, A. G. (1992). Teachers' beliefs and conceptions: A synthesis of research. In D. Grouws (Ed.), *Handbook of research on mathematics teaching and learning* (pp. 127-146). New York: Macmillan.

Wittmann, E. C. (1989). The mathematical training of teachers from the point of view of education. *Journal für Mathematikdidaktik, 10*(4), 291-308.

REFLECTIONS ON MATHEMATICAL CONCEPTS AS STARTING POINTS FOR DIDACTICAL THINKING

Hans-Joachim Vollrath
Würzburg

1. INTRODUCTION

1.1 Mathematics Didactics in Teacher Education for Gymnasium

In Germany, the Gymnasium comprises Grades 5 to 13 and is oriented toward preparing students for university studies. Nowadays, about 20% to 40% of an age group attend the Gymnasium. Students preparing to teach mathematics at the Gymnasium (see Weidig, 1992) traditionally have to master a complete university education in mathematics. This means that they are introduced to calculus, linear algebra, analytical geometry, theory of functions, algebra, number theory, differential geometry, differential equations, probability and statistics, numerical mathematics, and so forth. This mathematics is far beyond the elementary mathematics they will have to teach as future teachers. But the idea of this type of education is that teachers can only present elementary mathematics at the Gymnasium in a valid manner if they are familiar with the higher mathematics behind it. *Elementary Mathematics from an Advanced Standpoint* by F. Klein (1968) made this notion explicit: A mathematics education of this type should make the future teachers think *mathematically*.

But F. Klein also saw the need for lectures about the didactics of mathematics in teacher education to help student teachers to think *didactically*. This was supported by other university mathematicians such as A. Pringsheim. As a result, lectures in didactics of mathematics were offered at some universities (Griesel & Steiner, 1992). This development was continued in the 1960s by mathematicians such as H. Behnke, H. Kunle, D. Laugwitz, and G. Pickert, who invited experienced teachers to offer lectures in didactics of mathematics. It turned out that these lectures stimulated research in didactics of mathematics, and that the growing didactical research helped to improve these lectures. Very typical were H.-G. Steiner's lectures at Münster. His lecture on the foundations of geometry from a didactical point of view was published in 1966 (Steiner, 1966a). During the following decades, didactical theories for most of the mathematical subject areas of the Gymnasium in Germany were developed, for example, algebra

R. Biehler, R. W. Scholz, R. Sträßer, B. Winkelmann (Eds.),
Didactics of Mathematics as a Scientific Discipline, 61-72.

(Vollrath, 1974); calculus, linear algebra, and stochastics (Tietze, Klika, & Wolpers, 1982); calculus (Blum & Törner, 1983); numerical mathematics (Blankenagel, 1985); geometry (Holland, 1988); and stochastics (Borovcnik, 1992).

1.2 Reflecting on Concepts in Lectures on Didactics of Mathematics

In their mathematical education, student teachers are expected to acquire hundreds of mathematical concepts, to become acquainted with properties of these concepts through hundreds of theorems, and to solve problems involving these concepts. Relatively few of these concepts are relevant for their future teaching. It turns out that their knowledge of these concepts is often as vague as their knowledge of concepts in general. But for teaching, their metaknowledge about concepts is absolutely insufficient. Lectures on didactics of mathematics therefore have to reflect on concepts, because they affect teaching. And this can be a starting point for didactical thinking.

Questions should be discussed with student teachers that can help them to arrive at central problems of didactics of mathematics. This paper reports about questions on *concept teaching and learning*. It will show how students' reflections about their experience with mathematics lead to basic problems of concept learning and teaching, and how elements of a theory of concept teaching can give the student teachers a perspective for their future work.

Elements of a theory of concept teaching, as I understand it, were offered in my book *Methodik des Begriffslehrens im Mathematikunterricht* (Vollrath, 1984), which was the result of empirical and analytical research on concept teaching. This research has been continued in recent years. In this paper, I want to show how it was stimulated by discussions with student teachers, and, vice versa, how this research has stimulated the discussions.

Many student teachers contributed to this research by investigations connected with a thesis for their examination. As a side effect, most of my student teachers felt that the lectures in didactics of mathematics also helped them to understand their "higher" mathematics better.

2. STARTING POINTS FOR DIDACTICAL THINKING

2.1 Evaluation of Mathematical Concepts

At the beginning of my lectures on didactics of calculus, I usually ask my student teachers: "What are the central concepts of calculus?" They suggest concepts like real number, function, derivative, integral, limit, sequence, series, and so forth. At some point, a discussion starts on whether a certain concept is "central." This can happen with concepts such as boundary, monotony, accumulation point, and so forth. Ultimately, the students feel a need for a discussion about the meaning of the term "central concept." Obviously there is no definition for this term. But one can argue for a certain concept to be central or not. For example, calculus is about functions. But calculus deals with functions in a specific manner: One is interested in the derivative and in the integral of functions. Forming these concepts was the beginning of calculus in history. But for a certain class of functions, the derivative and the integral can be found algebraically. Calculus really starts at functions that need limits to find the derivatives and the integral. Therefore one could say that the central concept is the concept of limit (although calculus without limits is possible to some extent, e.g., Laugwitz, 1973). On the other hand, the concept of limit needs the concepts of real number and function, which can therefore also be called "central concepts."

One might think that this is a rather academic discussion. But questions like this are essential when one plans a calculus course for the Gymnasium. A key problem then is the *choice of concepts* that have to be taught in this course. This calls for an evaluation of concepts in the context of teaching (this might lead to different results!).

There seems to be a tendency to put too much emphasis on the use of a concept. But Otte has pointed out that concepts have to be seen both as objects and tools. Therefore concepts offer both *knowledge and use*. An adequate evaluation of concepts from the standpoint of teaching therefore has to take into account both these properties and how they complement each other. Otte and Steinbring (1977) worked this out for the concept of continuity; Fischer (1976) compared the concepts of continuity and derivative from this point of view. One important approach to the evaluation process is through historical analysis of the development of the concept, which incorporates intentions, definitions, properties, applications, and so forth.

For example, concept formation is very often embedded in problem-solving. A historical analysis of the relationship between concept formation and problem-solving can reveal different roles that concepts can play (Vollrath, 1986). Infinite series were introduced as *instruments for solving problems* of calculating areas of surfaces. But infinite series also became *solutions of problems* when they were used to develop functions into series, for example, sine, logarithms. When the concept of infinite series was established in calculus, it turned out to be a *source of new problems*. The critical conceptual work in infinite series became an *aid for precisely specifying the problem* of "infinite addition." The concept of absolutely convergent series served as *means for guaranteeing a certain method*, namely, the possibility of rearranging the terms.

This analysis shows different possibilities for embedding concept teaching into problem-solving processes. Obviously this gives rise to specific conceptual images through the process of teaching. Through these considerations, student teachers can get an idea of a genetic problem-oriented approach to the teaching of concepts. The perspective of *different roles of concepts* can help them to build up a repertoire of different modes of concept teaching in mathematics education.

When a mathematical concept is taught in school, the students are expected not only to understand it but also to know its importance (Winter, 1983). Investigations show (Vollrath, 1988) that there are different ways for teachers to express their own appreciation of a concept. Explicit expressions based on reasons seem to be most effective. But future teachers must also learn to accept students' evaluations as expressions of their personality when they differ from their own appreciation of a concept.

2.2 Relationships Between Mathematical Concepts

During our discussion on the central concepts of calculus, we refer to relationships between concepts. This can be the starting point for further investigations (Vollrath, 1973). For example, I ask my student teachers for the different types of sequence. A possible collection is: rational sequence, real sequence, constant sequence, arithmetical sequence, geometrical sequence, convergent sequence, zero-sequence, bounded sequence, increasing sequence, decreasing sequence, finally constant sequence, Cauchy-sequence, convergent sequence with rational limit, and so forth. We then try to get an overview. Theorems such as:

Every convergent sequence is bounded

or:

Every increasing and bounded sequence is convergent

lead to a hierarchy of concepts (Vollrath, 1973). Student teachers discover that knowledge of calculus means not only knowledge of concepts but also of relationships between concepts. They become aware of the importance of *networked learning*.

The study of the hierarchy of concepts leads to the didactical problem of *arranging the concepts* for teaching in school. In a first approach, different teaching sequences are formed and discussed from the point of view of teaching and learning. But it is also necessary to provide opportunities for the students to discover relationships between concepts.

From a systematic point of view, it seems convenient to start with the most general concept and to arrive at special concepts. But there can also be reasons for taking the opposite path. There has been a long discussion in pedagogics on whether one should proceed from the general to the specific or vice versa. Didacticians know that this question is too general. Didactics of mathematics is looking for more precise answers. More particularly, didacticians agree that there are many different ways of learning a network of concepts so that the concepts are understood and mastered, and so that the relationship between them is known and can be used.

2.3 Structural Analysis of Mathematical Concepts

Our discussions about the essentials of calculus lead to the real numbers as the basis of calculus. One can then continue the investigation by asking which property of the real numbers is needed to satisfy the specific requirements of calculus. Analyzing the central concepts, theorems, and proofs of calculus leads to the discovery of the well-known fact that the real number system is "complete." For most students, this means that nested intervals always contain one real number. Student teachers will perhaps learn that completeness can also be expressed in terms of Dedekind-sections or Cauchy-sequences. But Steiner (1966b) has shown that completeness has to do not only with the method by which the real numbers are constructed in terms of rational numbers. His paper revealed that completeness is equivalent to the propositions of the fundamental theorems of calculus, for example, the intermediate value property, the Heine-Borel property, or the Bolzano-Weierstrass property. This study helps student teachers to understand the fundamentals of calculus better.

But the great variety of the 12 different properties expressing completeness in Steiner's paper raises questions relevant to teaching. A first question could be: Which property should be used in mathematics instruction (Grade 9) to introduce the completeness of real numbers? And, again, it is not just the answer that matters, but, more importantly, the reasoning. Moreover, reasons can refer to both knowledge and use. One can discuss which property offers most knowledge and best use in the easiest way. But although didactics tries to optimize teaching and learning (Griesel, 1971, p. 73), it must not be neglected that each property reveals a certain aspect of real numbers that emerged during a certain period in the history of the development of the concept.

Although there are different possible approaches, which are equivalent from a systematical point of view, "easy" ways can be misleading. For example, defining convexity of a function by its derivatives, or defining logarithm as an integral of $1/x$, is "putting the cart before the horse" (Kirsch, 1977).

We took this discussion about completeness as an example of a structural analysis that was an interesting didactical problem in the 1960s. Things change; nowadays, problems of applications of calculus seem to be more interesting. Certainly this change of interest can also be a point of reflection.

2.4 Logical Analysis of Definitions

When we talk about the definitions of the central concepts of calculus, most of my student teachers confess that they have had difficulties in understanding these definitions. We then want to find out the reasons for these difficulties.

Certainly one problem is the complex logical structure of the definitions. Take for example continuity:

A function f is said to be continous at x_0 iff
for all $\varepsilon > 0$ there exists a $\delta > 0$, such that for all x,
if $|x - x_0| < \delta$ then $|f(x) - f(x_0)| < \varepsilon$.

It is especially the "tower of quantifiers" "for all" . . . , "there exists" . . . "for all," and the implication "if . . . then" that causes the difficulties.

Therefore one would look for equivalent but less complex definitions. Different calculus books help my students to find a lot of definitions and to compare them from the perspective of logical structure. Obviously the difficulties are only shifted by the "simpler" definition:

A function f is said to be continuous in x_0 iff $\lim_{x \to x_0} f(x) = f(x_0)$

Now the problems are contained in the definition of the limit.

Discussions like these have a long tradition in the didactics of calculus. There are some psychological findings (e.g., disjunctive definitions are more difficult to learn than conjunctive definitions; see Clark, 1971) that can support judgments. But they are not very surprising.

Another possibility is to restrict the concepts of calculus. A very interesting approach is the Lipschitz-calculus (Karcher, 1973), in which, for example, the definition of *L*-continuity is logically simpler then the definition of continuity in general.

But finally, the whole problem of generalization and formalization in calculus teaching has become problematic. Historical considerations make clear that the epsilon-delta form of the definition is the result of a long process of rigorization that was completed by the end of the last century (Fischer, 1978). Teaching should give students a chance to experience a similar process in concept learning. For this reason, there is a renewed interest in more intuitive approaches to calculus in the Gymnasium (e.g., Blum & Kirsch, 1979). A historical discussion about the development of rigor in calculus can help students to understand better the use of all the "epsilon-delta stuff" of calculus.

As an excellent example of a stepwise, increasingly precise approach to the concepts of calculus, I present to my student teachers the introduction to continuity by Ostrowski (1952) in which a sequence of trial, critique, further trial, . . . finally leads to the epsilon-delta definition.

2.5 Understanding of Concepts

Didactical discussions about concepts soon arrive at the problem of understanding. What does it mean to understand a concept? The first answer of student teachers is usually "to know a definition." But this answer can easily provoke a discussion. A definition can be learnt by heart without being understood. They soon find out that one has to describe understanding of a concept by means of abilities; for example, to be able to give examples - to give counterexamples - to test examples - to know properties - to know relationships between concepts - to apply knowledge about the concept. Abilities like these can be tested. But it is more difficult to describe what we mean by "having images of a concept," "to appreciate a concept," or "knowing the importance of a concept."

Discussions soon lead to the insight that there are *stages of understanding*. This view has a long tradition. And there are also "masterpieces" on presenting concepts in stages. A good example is Mangoldt and Knopp's (1965) introduction to integration. It starts with an intuitive approach on the basis of area functions. After this, integrals are calculated. And in a third stage, a lot of conceptual work on defining integrals is done.

Considerations like these help the students to understand stage models of understanding (see Dyrszlag, 1972a, b; Herscovics & Bergeron, 1983; Vollrath, 1974).

The need for better understanding leads to the discovery that there is no final understanding. This is a sort of paradox: Understanding is both a goal and a process. And there are further paradoxes of understanding (Vollrath, 1993). They have their origin in the nature of mathematical knowledge (see Jahnke, 1978; Keitel, Otte, & Seeger, 1980; Steinbring, 1988).

2.6 Forming Mathematical Concepts

The strangest question for my student teachers is: "Have you ever formed a new mathematical concept on your own?" They are generally very puzzled by this question. I always get the answer: "No!" And sometimes they ask me: "Should we have done so?"

For most student teachers, university education in mathematics means receptive learning. They can be creative to some extent in problem-solving when they find a solution, perhaps on the basis of an original idea. But they will never be asked to form a new concept. Some students have perhaps written poems on their own, they have painted pictures, composed melodies, and made biological, chemical, or physical experiments. But why do they not develop mathematics on their own? We all feel that they will have no real chance of inventing an important piece of mathematics. But is this not also true for their poetry, their painting, their music, their biology, chemistry, or physics? Perhaps it is "the power of the mathematical giants" that discourages students from making mathematics.

As an example, I try to encourage my student teachers to invent a new type of real sequence just by thinking out a certain property. Maybe one chooses as the property of a sequence (a_n):

$$a_n = 0 \text{ for infinitely many } n.$$

At first, one will think of a suitable name for this type of sequence. Let us call it a "stutter sequence." Does a stutter sequence exist? Is every sequence a stutter sequence? These questions ask for examples and counterexamples.

What about the sum or the product of stutter sequences? Are they stutter sequences too? What is the relationship to other sequences? Answers can be formulated as theorems that form a small piece of theory. These steps are routines. But most of my students are not familiar with these routines. How then will they adequately teach their future students about concept formation? Students in general do not think of mathematics as a subject in which they can be creative. Concept formation offers the possibility of *creative thinking* in mathematics (Vollrath, 1987).

2.7 Thinking in Concepts

From a formalistic point of view, the names of mathematical concepts are arbitrary. But to some extent the name often expresses an image. "Continuous" is a term that bears intuitions. This is also true for terms like "increasing," "decreasing," "bounded," and so forth. On the other hand, "derivative" and "integral" give no hints to possible meanings. Most of my student teachers are familiar with the fact that a name does not give sufficient information about a concept. But there is some research suggesting that most students in school refer to the meaning of the concept name and not to a definition. There is also research indicating that images evoked by the everyday meaning of the name are responsible for misunderstanding the concept (Viet, 1978; Vollrath, 1978).

On one hand, students have to learn that the *meaning of a mathematical concept* has to be defined. On the other hand, it is true that certain images, ideas, and intentions lead to definitions that stress certain aspects but disregard others. The concept of sequence can be defined as a function defined on the set of natural numbers. This stresses the image of mapping, whereas the idea of succession is left in the background. The same is true for many of the central concepts of calculus. This was pointed out very clearly by Steiner (1969) in his historical analysis of the function concept, and it was investigated for many of these concepts by Freudenthal in his *Didactical Phenomenology* (1983).

2.8 Personal Shaping of Mathematical Concepts

When a mathematician wants to define a concept, then there is not much freedom for him or her to formulate the defining property. Some authors prefer to use formal language, others try to avoid it as much as possible. A comparison of textbooks from the same time shows rather little variety of styles. A comparison between textbooks with similar objectives published at different times reveals more differences. But again, this is more a congruence of developing standards than the expression of different personalities.

However, during the development of an area of mathematics, concept formation is strongly influenced by the leading mathematician at the time. This has been true for calculus. There are fundamental differences in the ways Leibniz and Newton developed calculus. A historical analysis can still identify their different fundamental ideas in modern calculus. The same is true for the theory of functions of a complex variable. One can still see today the different approaches of Riemann and Weierstrass in a modern presentation of the theory. It is possible to speculate with Klein that their different "characters" are responsible for the different ways of building up the theory (1926, p. 246). But it is more helpful to concentrate on the differences in experience, intention, and image as the decisive influences on concept formation.

A lecture on the didactics of calculus should give the student teachers an opportunity to recognize different sources of central parts of the theory, to get acquainted with the mathematicians who pushed forward the development, and to become aware of their motives and images.

Although mathematics has a universal quality when presented in highly developed theories, one should not forget the fact that there are women and men behind it who have influenced the development.

When mathematicians want to learn a new theory, they read or hear definitions and at once use certain routines to understand the new concepts. They are at ease when they find that the new concept fits into their existing network of concepts, when it corresponds with their own images, knowledge, and experience. They feel resistant to the new concept when they encounter discrepancies. In any case, learning a new concept involves an active process of concept formation. Very often this is accompanied by feelings of interest or resistance. And this is something that the student teachers will often have experienced in their own mathematical education at the university.

However, many of them have the idea that teaching concepts means to present as much knowledge about the concept as they can in as interesting a manner as possible. This is a point at which student teachers can encounter results of communication analysis (Andelfinger, 1984; Voigt, 1991), which show that students often resist when they are expected to learn new concepts. As a consequence, they often form "personal concepts" that differ from their teacher's concepts. And it is surprising that this may occur even though they can solve a lot of problems about the concept correctly. This should sensitize the student teachers to comments made by the students that they will hear when they observe mathematics instruction in their school practice.

3. STRATEGIES OF CONCEPT TEACHING

Finally, we arrive at a rather delicate problem. When the student teachers look at their own experience as learners of mathematics, they all know that there are teachers, professors, and authors who are very effective in teaching concepts, whereas others raise many difficulties for the learners. What is the mystery of successful teaching? Is there an optimal way of teaching concepts?

The preceding discussions will protect the student teachers from giving simple answers. They are aware that learning concepts is rather complex. It is not difficult for them to criticize empirical studies testing the effectiveness of "Method A" versus "Method B." They can also easily identify the weaknesses of investigations about the effectiveness of artificial methods such as those used in psychological testing (e.g., Clark, 1971). They soon find out that one needs a theory of teaching in the background as a basis for making decisions. A good example of such a theory is *genetic teaching* (e.g., Wittmann, 1981), which can be used to give a sense of direction.

To master the complexity of concept teaching, students find that they need to look at the relevant *variables.*

Teaching mathematical concepts has to take into consideration:

1. the students: their cognitive structures, their intellectual abilities, their attitudes, and their needs;
2. the concepts: different types of concept, logical structure of definitions, context, development of concepts;
3. the teachers: their personality, their intentions, their background.

Behind each of these variables there is a wide variety of theories (see Vollrath, 1984). It is impossible to present these theories to the students. However, they can be sensitized to the problems and can get references to literature for further study. Some of these problems can also be touched on in exercises and at seminars.

These considerations help student teachers to get a differentiated view of teaching: Concept teaching has to be planned with respect to these variables.

A reasonable plan for teaching a concept in a certain teaching situation is called a *strategy.* My practice is to look at strategies for teaching concepts by considering different *ranges* of strategies (Vollrath, 1984). *Local strategies* refer to the plan of a *teaching unit,* which is applicable for *standard concepts* like rational function, bounded function, step-function, and so forth.

Regional strategies serve for planning the teaching of *key concepts* in *teaching sequences* such as the concept of limit, derivative, or integral of a function.

Global strategies are needed for *leading concepts* that permeate the whole *curriculum,* for example, the concept of function is a candidate for such a leading concept.

Student teachers get the opportunity to study models of these types of strategy from "didactical masterpieces" (see, also, Wittmann, 1984). And they are invited to develop strategies on their own for some examples of different ranges.

Finally, student teachers should get some hints on how to evaluate certain strategies. The most important goal is that they can reason without being dogmatic. It would be a disaster if didactics of mathematics as a science were to prop up educational dogma.

REFERENCES

Andelfinger, B. (1984). Arithmetische und algebraische Lerner-Konzepte in der S I. In *Beiträge zum Mathematikunterricht 1984* (pp. 71-74). Bad Salzdetfurth: Franzbecker.

Blankenagel, J. (1985). *Numerische Mathematik im Rahmen der Schulmathematik.* Mannheim: Bibliographisches Institut.

Blum, W., & Kirsch, A. (1979). Zur Konzeption des Analysisunterrichts in Grundkursen. *Der Mathematikunterricht, 25*(3), 6-24.

Blum, W., & Törner, G. (1983). *Didaktik der Analysis.* Göttingen: Vandenhoeck & Ruprecht.

Borovcnik, M. (1992). *Stochastik im Wechselspiel von Intuitionen und Mathematik.* Mannheim: Wissenschaftsverlag.

Clark, D. C. (1971). Teaching concepts in the classroom: A set of teaching prescriptions derived from experimental research. *Journal of Educational Psychology, 62*(3), 253-278.

Dyrszlag, Z. (1972a). Zum Verständnis mathematischer Begriffe 1. *Mathematik in der Schule, 10*(1), 36-44.

Dyrszlag, Z. (1972b). Zum Verständnis mathematischer Begriffe 2. *Mathematik in der Schule, 10*(2), 105-114.

Fischer, R. (1976). Fundamentale Ideen bei den reellen Funktionen. *Zentralblatt für Didaktik der Mathematik, 8*(4), 185-192.

Fischer, R. (1978). Die Rolle des Exaktifizierens im Analysisunterricht. *Didaktik der Mathematik, 6*(3), 212-226.

Freudenthal, H. (1983). *Didactical phenomenology of mathematical structures.* Dordrecht, Netherlands: Reidel.

Griesel, H. (1971). Die mathematische Analyse als Forschungsmittel in der Didaktik der Mathematik. In *Beiträge zum Mathematikunterricht 1971* (pp. 72-81). Hannover: Schroedel.

Griesel, H., & Steiner, H.-G., (1992), The organization of didactics of mathematics as a professional field. *Zentralblatt für Didaktik der Mathematik, 24*(7), 287-295.

Herscovics, N., & Bergeron J. (1983). Models of understanding. *Zentralblatt für Didaktik der Mathematik, 15*(2), 75-83.

Holland, G. (1988). *Geometrie in der Sekundarstufe.* Mannheim: Wissenschaftsverlag.

Jahnke, H. N. (1978). Zum Verhältnis von Wissensentwicklung und Begründung in der Mathematik-Beweisen als didaktisches Problem. *IDM Materialien und Studien 10.* Bielefeld: Universität Bielefeld.

Karcher, H. (1973). Analysis auf der Schule. *Didaktik der Mathematik, 1*(1), 46-69.

Keitel, Ch., Otte, M., & Seeger, F. (1980). *Text, Wissen, Tätigkeit.* Königstein: Scriptor.

Kirsch, A. (1977). Aspects of simplification in mathematics teaching. In H. Athen & H. Kunle (Eds.), *Proceedings of the Third International Congress of Mathematical Education* (pp. 98-120). Karlsruhe: Organizing Committee of the 3rd ICME.

Klein, F. (1926). *Vorlesungen über die Entwicklung der Mathematik im 19. Jahrhundert* (Vol. 1). Berlin: Springer.

Klein, F. (1968). *Elementarmathematik vom höheren Standpunkte aus* (Vols. 1-3, Reprint). Berlin: Springer.

Laugwitz, D. (1973). Ist Differentialrechnung ohne Grenzwertbegriff möglich? *Mathematisch-Physikalische Semesterberichte, 20*(2), 189-201.

Mangoldt, H., & Knopp, K. von (1965). *Einführung in die höhere Mathematik.* (Vol. 3, 12th ed.). Leipzig: Hirzel.

Ostrowski, A. (1952). *Vorlesungen über Differential- und Integralrechnung* (Vol. 1). Basel: Birkhäuser.

Otte, M., & Steinbring, H. (1977). Probleme der Begriffsentwicklung - zum Stetigkeitsbegriff. *Didaktik der Mathematik, 5*(1), 16-25.

Steinbring, H. (1988). "Eigentlich ist das nichts Neues für Euch!" - Oder: Läßt sich mathematisches Wissen auf bekannte Fakten zurückführen? *Der Mathematikunterricht, 34*(2), 30-43.

Steiner, H.-G. (1966a). *Vorlesungen über Grundlagen und Aufbau der Geometrie in didaktischer Sicht.* Münster: Aschendorff.
Steiner, H.-G. (1966b). Äquivalente Fassungen des Vollständigkeitsaxioms für die Theorie der reellen Zahlen. *Mathematisch-Physikalische Semesterberichte, 13*(2), 180-201.
Steiner, H.-G. (1969). Aus der Geschichte des Funktionsbegriffs. *Der Mathematikunterricht, 15*(3), 13-39.
Tietze, U.-P., Klika, M., & Wolpers, H. (1982). *Didaktik des Mathematikunterrichts in der Sekundarstufe II.* Braunschweig: Vieweg.
Viet, U. (1978). Umgangssprache und Fachsprache im Geometrieunterricht des 5. und 6. Schuljahres. In H. Bauersfeld (Ed.), *Fallstudien und Analysen zum Mathematikunterricht* (pp. 13-23). Hannover: Schroedel.
Voigt, J. (1991). Das Thema im Unterrichtsprozeß. In *Beiträge zum Mathematikunterricht 1991* (pp. 469-472). Bad Salzdetfurth: Franzbecker.
Vollrath, H.-J. (1973). Folgenringe. *Der Mathematikunterricht, 19*(4), 22-34.
Vollrath, H.-J. (1974). *Didaktik der Algebra.* Stuttgart: Klett.
Vollrath, H.-J. (1978). Lernschwierigkeiten, die sich aus dem umgangssprachlichen Verständnis geometrischer Begriffe ergeben. *Schriftenreihe des IDM.* Bielefeld: Universität Bielefeld, *18*, 57-73.
Vollrath, H.-J. (1984). *Methodik des Begriffslehrens im Mathematikunterricht.* Stuttgart: Klett.
Vollrath, H.-J. (1986). Zur Beziehung zwischen Begriff und Problem in der Mathematik. *Journal für Mathematikdidaktik, 7*(4), 243-268.
Vollrath, H.-J. (1987). Begriffsbildung als schöpferisches Tun im Mathematikunterricht. *Zentralblatt für Didaktik der Mathematik, 19*(3), 123-127.
Vollrath, H.-J. (1988). Mathematik bewerten lernen. In P. Bender (Ed.), *Mathematikdidaktik: Theorie und Praxis, Festschrift für Heinrich Winter* (pp. 202-209). Berlin: Cornelsen.
Vollrath, H.-J. (1993). Paradoxien des Verstehens von Mathematik. *Journal für Mathematikdidaktik, 14*(1), 35-58.
Weidig, I. (1992). On the school system in Germany and the regulation of mathematics teaching. *Zentralblatt für Didaktik der Mathematik, 24*(7), 214-219.
Winter, H. (1983). Über die Entfaltung begrifflichen Denkens im Mathematikunterricht. *Journal für Mathematikdidaktik, 4*(3), 175-204.
Wittmann, E. (1981). *Grundfragen des Mathematikunterrichts.* (6th ed.). Braunschweig: Vieweg.
Wittmann, E. (1984). Teaching units as the integrating core of mathematics education. *Educational Studies in Mathematics, 15*(1), 25-36.

Acknowledgements

The considerations in this paper are strongly influenced by the experience of teaching and research in didactics of mathematics for 25 years that I was able to gain through the promotion of D. Laugwitz and through stimulating discussions with H.-G. Steiner. With this paper, I want to acknowledge Steiner's influence on my work. I have to thank D. Quadling for shaping my English.

BEYOND SUBJECT MATTER: A PSYCHOLOGICAL TOPOLOGY OF TEACHERS' PROFESSIONAL KNOWLEDGE

Rainer Bromme
Frankfurt

1. INTRODUCTION

In both educational psychology and mathematical education, the professional knowledge of teachers is increasingly becoming an object of research. In recent years, it has become clear that innovations in the curriculum and in teaching methods are successful only when what the teacher does with these innovations is taken into account (Steiner, 1987). However, this depends on which conceptual tools teachers possess in order to deal with their work situation. The professional knowledge of teachers is, in part, the content they discuss during the lesson, but it is also evident that they must possess additional knowledge in order to be able to teach mathematics in an appropriate way to their students. However, what belongs to the professional knowledge of teachers, and how does it relate to their practical abilities?

There is a rather recent research tradition in the field of educational psychology that studies teachers as experts. The notion of "experts" expresses the programmatic reference to questions, research methods, and views of expert research in cognitive psychology. This approach analyzes the connection between the professional knowledge and professional activity of good performers within a certain field of activity. The expert approach provides a good starting position to approach such questions with empirical methods. When applying this approach to the study of teachers' cognitions, one is faced with the question of what shall be counted as professional knowledge. The concept of professional knowledge must be decomposed analytically. This is what this contribution is about.

2. A TOPOLOGY OF TEACHERS' PROFESSIONAL KNOWLEDGE

At first glance, professional knowledge seems to be sufficiently described by "subject matter," "pedagogy," and "specific didactics." These fields, however, have to be decomposed further if the intention is to understand the special characteristics of professional knowledge.

R. Biehler, R. W. Scholz, R. Sträßer, B. Winkelmann (Eds.),
Didactics of Mathematics as a Scientific Discipline, 73-88.

Shulman (1986) has presented a classification of teachers' knowledge. It comprises: "content knowledge," "curricular knowledge," "pedagogical knowledge," and "pedagogical content knowledge." These suggestions have proved to be very stimulating for research into teacher cognitions (Grossmann, 1990). In order to be able to describe qualitative features of professional knowledge, Shulman's categories must be differentiated further. This is why I take up his suggestion, but extended by both the concept of "philosophy of content knowledge" and a clear distinction between the knowledge of the academic discipline and that of the subject in school.

This section will provide a brief sketch of my topology of areas of teachers' professional knowledge. The following sections shall consider some areas of this topology in greater depth in order to cast light on the complex nature of professional knowledge.

2.1 Content Knowledge About Mathematics as a Discipline

This is what the teacher learns during his or her studies, and it contains, among other things, mathematical propositions, rules, mathematical modes of thinking, and methods.

2.2 School Mathematical Knowledge

The contents of teaching are not simply the propaedeutical basics of the respective science. Just as the contents to be learned in German lessons are not simplified German studies, but represent a canon of knowledge of their own, the contents of learning mathematics are not just simplifications of mathematics as it is taught in universities. The school subjects have a "life of their own" with their own logic; that is, the meaning of the concepts taught cannot be explained simply from the logic of the respective scientific disciplines. Or, in student terms: Mathematics and "math," theology and "religious studies" are not the same. Rather, goals about school (e.g., concepts of general education) are integrated into the meanings of the subject-specific concepts. For the psychological analysis of professional knowledge, this is important, as these aspects of meaning are, in part, implicit knowledge.

2.3 Philosophy of School Mathematics

These are ideas about the epistemological foundations of mathematics and mathematics learning and about the relationship between mathematics and other fields of human life and knowledge. The philosophy of the school subject is an implicit content of teaching as well, and it includes normative elements. Students, for instance, will learn whether the teacher adheres to the view that the "essential thing" in mathematics is operating with a clear, completely defined language, no emphasis being set on what the things used refer to, or whether the view is that mathematics is a tool to describe a reality, however it might be understood.

2.4 Pedagogical Knowledge

This means that part of knowledge that has a relatively independent validity separate from the school subjects. This includes how to introduce the behavior patterns necessary for handling a class (Kounin, 1970). It also concerns coping with parents in order to explain and influence student behavior. The pedagogical ethics of teachers with regard to treating their students justly is neatly interwoven with their pedagogical knowledge (Oser, in press). Pedagogical knowledge, of course, is very important for the teacher's professional activity; however, it shall not be treated extensively here, as I shall focus on those areas that are related to the subject matter.

2.5 Subject-Matter-Specific Pedagogical Knowledge

On the basis of the logical structure of the subject matter taken alone, no teaching decision can yet be made. Lesson observation shows still large interindividual differences in the didactical approach chosen, even if the subject matter and the textbook are the same (Leinhardt & Smith, 1985). To find suitable forms of presenting the subject matter, to determine the temporal order of treating the topics, and to assess which matters have to be treated more intensely requires subject-matter-specific pedagogical knowledge (Chevallard, 1985, chaps. 5, 6). This field of knowledge has a special character. It is integrated knowledge cross-referring both pedagogical knowledge and the teacher's own experience to the subject-matter knowledge. This integration is exhibited, for instance, when the logical structure of the subject matter is reshaped into a temporal sequence. Further, it consists in changing the structuring and relative weight of concepts and rules; something that is of central importance from the viewpoint of mathematical theory may be accorded less weight from the perspective of teaching.

2.6 The Cognitive Integration of Knowledge From Different Disciplines

The professional knowledge of teachers is not simply a conglomerate of various fields. Rather, an integration takes place during the course of practical training and professional experience, and the various fields of knowledge are related to practical experience. The fusing of knowledge coming from different origins is the particular feature of the professional knowledge of teachers as compared to the codified knowledge of the disciplines in which they have been educated.

In mathematics teachers, the subject-matter-specific pedagogical knowledge is to a large part tied to mathematical problems. In a way, it is "crystallized" in these problems, as research into everyday lesson planning has shown. In their lesson preparation, experienced mathematics teachers concentrate widely on the selection and sequence of mathematical problems. Both "thinking aloud" protocols (Bromme, 1981) and interviews with mathematics teachers, have provided hardly any indications of pedagogical considerations prior to the selection of problems. Nevertheless, pedagogical questions of shaping the lessons are also considered by teachers in their lesson planning, as these questions codetermine the decision about tasks. By choosing tasks with regard to their difficulty, their value for motivating students, or to illustrate difficult facts, and so forth, the logic of the subject matter is linked to teachers' assumptions about the logic of how the lesson will run and how the students will learn (for similar results, see, also, Tietze, 1986). Thus, the mathematical problems already contain the subject-matter core of the scenarios of activity that structure the teachers' categorical perception of the teaching process.

Teachers often do not even realize the integration they effect by linking subject-matter knowledge to pedagogical knowledge. One example of this is their (factually incorrect) assumption that the subject matter (mathematics) already determines the sequence, the order, and the emphasis given to teaching topics. The pedagogical knowledge that flows in remains, in a way, unobserved. To teachers who see themselves more as mathematicians than as pedagogues, their teaching decisions appear to be founded "in the subject matter," as Sträßer (1985) found in his interviews with teachers in vocational schools. In case studies with American teachers, Godmundsdottir and Shulman (1986) have reported an implicit integration of methodological and subject-matter ideas in teachers.

3. SUBJECT-MATTER KNOWLEDGE AND INSTRUCTIONAL OUTCOME

The subject-matter knowledge is not only an object of the professional activity of teachers but also, as a prerequisite of this activity, a major and extensive content of their professional training. But, how much knowledge of this type is necessary to be a successful teacher?

In the 1970s, some surprising empiricial studies were published. According to these, there was no measurable connection between the extent of teachers' subject-matter knowledge and instructional outcomes (Gage & Berliner, 1977, pp. 646-647). It seems to be immediately evident that teachers must have the subject-matter knowledge they are supposed to teach. This, however, does not permit the conclusion that there is a direct linkage between the extent of subject-matter knowledge and students' instructional outcomes measured by means of standardized tests.

Eisenberg (1977) tested the knowledge of 28 teachers in algebra, looking for connections to the growth of knowledge in their students. While student variables such as verbal competence and previous knowledge prior to the teaching unit contributed to the variance of the performance measured, this proved not to be true for teachers' amount of knowledge, confirming similar results obtained by Begle (1972). Both authors conclude that a relatively low stock of knowledge is sufficient to teach students. In a meta-analysis of 65 studies of teaching in the natural sciences, Druva and Anderson (1983) summarized the empirically established relationships between teacher variables (age, extent of education in the natural sciences) and both teacher behavior and student behavior as well as performance in class. The number of courses the teachers had taken in the natural sciences (as a measure of their knowledge) explained about 10% of the variance in student performance. Similar explanatory power was found for instructional quality variables, for instance, the posing of complex questions. The small (in absolute terms) share of variance explained by these variables is stressed by several authors and considered serious (Romberg, 1988). In contrast to this conclusion, it must be stated, however, that this indirect indicator of academic knowledge is even a good predictor of student performance, for individual variables in research on teaching, be they variables of teaching or so-called background variables in teachers or students, will always be able to explain only a relatively small percentage of variance, except for the variable of "pretest scores" (Brophy & Good, 1986; Dunkin & Biddle, 1974).

Nevertheless, a correlative connection between the extent of a teacher's training in the subject matter and student learning outcomes does not lend itself to causal interpretation as long as the process of mediation between these two variables is no topic. There are a few studies shedding light on some steps of these mediating processes. To give one example concerning the variable of clarity, a teacher's subject-matter knowledge contributes to his or her being able to stress important facts and ideas within the curriculum. This knowledge influences the quality of explanations given (Roehler et al., 1987) and the ability to integrate into their teaching student contributions that do not lie precisely on the teacher's intended level of meaning (Hashweh, 1986).

The effects of limited subject-matter knowledge were analyzed in a case study by Stein, Baxter, and Leinhardt (1990). They questioned a mathematics teacher extensively on his mathematical knowledge and educational ideas concerning the concept of function. Afterwards, they observed his teaching, looking for episodes in the videotape recordings in which a connection between subject-matter knowledge and teaching was recognizable. The teacher's ideas were limited to interpreting function as a calculating rule. He made no allowance for interpreting functions as mappings of quantities upon one another, nor for the possibility of one element being assigned to several corresponding elements. This limited idea of the function concept did not lead to classroom statements that were strictly false, but to the following three weaknesses in developing the subject matter in class: (a) Too much emphasis on special cases: The explanation of function given by the teacher was correct only for cases of one-to-one relations between the elements of the two quantities. (b) Too little profiting from teaching opportunities: Drawing function graphs was not referred back to defining functions, and hence appeared to the students as something entirely new. (c) Omission of preparation for an extended understanding of the concept:

While the examples had been chosen to solve the problems of this very class level, a more general understanding of the concept of function was more impeded than promoted.

Carlsen (1987) studied the connection between subject-matter knowledge and teachers' questioning in science teaching. He used interviews and sorting procedures to inquire into the knowledge of four student teachers. Classroom observations (9th to 12th grade) and analyses of lesson transcripts showed linkages between intraindividual differences in the extent of subject-matter knowledge and the teachers' questioning within their lessons. In teaching units on topics on which the teachers knew relatively little, they asked more direct questions, the questions having a low cognitive level. In topics on which the teachers knew their way better, the students talked more, offered more spontaneous contributions, and their contributions were longer; the teachers implicitly communicating how they expected the students to behave both by the manner of their questions and by the interest they showed in the subject matter (the variable of "enthusiasm"). Only teachers who possess good subject-matter knowledge are sufficiently sure of themselves to be able to direct classroom activities even in cases when the students take new paths of work (Dobey & Schafer, 1984).

Leinhard and Smith (1985) questioned teachers about their subject-matter knowledge on division (using interviews and sorting procedures) and subsequently observed their lessons. The teachers had different levels of knowledge about the properties of fractions. By strict confinement to algorithmic aspects of fractions, even those teachers with less conceptual knowledge were able to give lessons on this topic. In the classrooms, interindividual differences in the availability of various forms of representing fractions (e.g., as area sections, on the number line) were observed as well. The teachers who showed conceptual gaps in their knowledge also belonged to the expert group, having obtained good learning performance with their classes over years. The authors supposed that there is some kind of compensation between lack of subject-matter knowledge and more knowhow about techniques of organizing the teaching in class (but only within definite limits).

The partly disappointing results of the studies on the correlations between subject-matter knowledge and teaching success are rather more suited to point out the complexity of what belongs to a teacher's professional knowledge than to put in question the basic idea of investigating the relation between professional knowledge and successful teaching. The connection between a teacher's subject-matter knowledge and the students' learning performance is very complex. A large number of variables "interfere" with the effect the teacher's amount of subject-matter knowledge has on student performance. There is an interesting parallel to this in the history of educational psychology. With their Pygmalion effect, Rosenthal and Jacobson (1971) also described a connection between a cognitive teacher variable (anticipated student performance) and a product variable (actual student scores in tests). Only later studies (Brophy & Good, 1974; Cooper, 1979) were able to show how teacher expectations are communicated and how they are connected to student behavior, student cognitions, and, finally, student performance.

4. THE "PHILOSOPHY OF SCHOOL MATHEMATICS" IN TEACHERS

Structuring the problems to be worked on and evaluating goals and subgoals is a typical abilty for effective professionals in several professional fields (Schön, 1983). It requires normative components within the professional knowledge. Those professions that legitimize their daily activities by referring to a so-called scientific base often gloss over these normative elements in silence. Hence, such normative ideas will be treated here somewhat more extensively.

Only recently, normative ideas of teachers related to the subject matter and their effect on teaching (mostly called teachers' beliefs) have come under closer scrutiny (For the teaching of English: Grossmann, 1990; the natural sciences: Hollon & Anderson, 1987; mathematics: Cooney, 1985, this volume; Heymann, 1982; Kesler, 1985; McGalliard, 1983; Pfeiffer, 1981; Thompson, 1984; Tietze 1986; comparison of school subjects: Yaacobit & Sharan, 1985).

The concept of "philosophy" for this part of teachers' knowledge is intended to stress that this means an evaluating perspective on the content of teaching. It is not a matter of subjectively preferring this or that part of the curriculum. Therefore I prefer the notion of philosophy instead of the notion of belief in order to emphasize that it is a part of metaknowledge, soaked with implicit epistemology and ontology (see, also, Ernest, this volume) .

The effect of teachers' philosophy of school mathematics on their teaching is much more strongly verified empirically than the influence of the amount of subject-matter knowledge discussed above. A good example for studies on the philosophy of school mathematics is that of Thompson (1984). The author compared ideas about mathematics teaching in three woman teachers. Teacher J considered mathematics to be a logical system existing independent of whether it is acquired or not. She took her task to be clear and consistent presentation of the subject matter. She expected her students to learn, first of all, the connection between what they had already learned and what was new. In contrast, Teacher K had a more process-oriented conception of mathematics. Accordingly, her teaching was aligned to encourage students to discover for themselves. A third principle found was to listen attentively to and to take up and understand the ideas that students advanced. Thompson (1984) also found discrepancies between teachers' normative ideas and their teaching behavior. Thus, while Teacher J stressed how important mathematics is for solving practical problems, she had difficulties in introducing practical examples of this into her teaching. In two case studies, Cooney (1985) and Marks (1987) each examined a teacher's conception of problem-solving. Both teachers named "mathematical problem-solving" as their most important goal. They showed, however, rather different conceptions of what can be termed problem-solving in mathematics and can be encouraged by a teacher.

We compared the mathematics instruction on the topic of "stochastics" given by two teachers whose teaching obviously did not have the same degree of "smoothness" (Bromme & Steinbring, 1990). A group of teachers was observed across several lessons, and their behavior was judged according to scales listing their quality of teaching (providing guidance to the class, clearness in presenting the subject matter, etc.). This served to identify the two teachers. The next step was to investigate their difference in instructional quality. For this purpose, lesson transcripts were coded for two subsequent lessons for each teacher. The coding focused on the question of which aspects of mathematical meaning had been thematized by the teachers in class: the symbolic-formal side, the applications of formal calculus, or the relationship between formal calculus and the object to which it is applied. Both teachers were confronted with student contributions alternately thematizing these two aspects of mathematical meaning in an inconsistent way. The two teachers differed markedly in how they treated student contributions and in how they used what had been offered to develop the subject matter. The teacher whose teaching went more "smoothly" showed a more appropriate switching between the aspects of mathematical meaning and the establishment of explicit relationships between the levels of meaning. This suggests the assumption that normative views about school mathematical knowledge (i.e., about what is really worth knowing in a mathematical object) influence teacher behavior.

In the present empirical studies concerning the subject-matter knowledge of teachers, there is a partial overlapping of the above-mentioned conceptual distinction between "subject-matter-specific pedagogical knowledge" and "philosophy of school mathematics." A strict distinction may not be appropriate. Certain variants of the philosophy of school mathematics also require a more profound mathematical understanding as well as more and different subject-matter-specific pedagogical knowledge. The philosophy of school mathematics contains certain judgments about what are the central concepts and procedures that should be taught, and what characterizes mathematical thought. These values, however, are tied closely to the subject matter-specific pedagogical knowledge and to disciplinary knowledge of facts, and they are often implicit. It may well be possible for a teacher to belong to a certain school of thought without being aware of the fact that subject-matter knowledge also contains a set of values. A psychological theory of teachers' professional knowledge must take into account that normative elements are interwoven with all areas of knowledge (Bromme, 1992, chap. 8.2).

5. FORMING PROFESSIONAL KNOWLEDGE BY PRACTICAL EXPERIENCE: EVERYONE MUST LEARN BY EXPERIENCE

Teachers do not have to effect the integration of pedagogical knowledge and subject-matter knowledge alone. The education of teachers in most countries contains practical elements aiming at such a linkage. Nevertheless, the teacher is still obliged to adapt his or her general knowledge to the conditions of teaching with which he or she is confronted. In the following, some empirical results will be described supporting the hypothesis that teachers' professional knowledge is a quite particular mixture of the above-mentioned areas of knowledge (especially subject-matter knowledge, philosophy, and pedagogical knowledge), and that this mixture is structured by teachers' practical experience with their own classrooms.

The requirements of teaching compel teachers to modify their previously learned theories about the content and the ways of teaching it. This, however, must not be seen as a mere simplification of previously differentiated knowledge, but rather as an enrichment by information referring to situations. Empirical evidence can be found in studies examining whether teachers rely on psychological theories or make allowance for facts that have been proven to be relevant for learning processes in psychological studies. The question thus is not whether these teachers had explicitly heard about such results; this can be left aside. What matters is only whether they think and act in a way that seems reasonable to the interviewers according to psychological facts about student learning. Thus, some of the empirical studies inspired by Shulman's (1986) concept of "pedagogic content knowledge" examine the question whether teachers consider recent concepts of their subject's didactics and developmental psychological concepts of strategies of learning (Clift, Ghatala, & Naus, 1987; Shefelbine & Shiel, 1987). To the disappointment of their authors, these studies showed that the teachers studied did not rely on psychological theories, but used other knowledge referring to experience. These results must sometimes be read at odds with their authors' interpretations in order to note that the teachers studied do not simply show a deficit in subject-matter-specific pedagogical knowledge.

The following study provides an example of this: Carpenter, Fennema, Peterson, and Carey (1988) have analyzed teachers' concepts about student errors in arithmetic. The psychological basis of this analysis was developmental findings on 1st-grade children's addition strategies. According to how the task is formulated and to age group, several techniques of counting visible elements (fingers) can be observed (Carpenter & Moser, 1984). The task ($5 + ? = 13$): "How many marbles do you still need if you already have 5 marbles and want 13?" for instance, is solved in three steps: counting 5 objects, continuing to count from 5 to 13, and then counting the fingers that have been added. Later, the first of these steps is left out. The authors interviewed 40 experienced elementary school teachers (with an average of 11 years of experience) regarding what they knew of such strategies, then studied the connection between knowledge and both teaching behavior and teaching performance. For this, they used a collection of tasks containing the various task types. Subjects had to compare tasks as to their difficulty for 1st-grade students (in general, not for their own students). The degree of difficulty assumed was then compared to empirically found solution rates (Carpenter & Moser, 1984). For most of the task types, the majority of assessments were correct. The teachers, however, had difficulties in stating reasons for their assessments. Above all, they did not name the students' solving strategies, such as counting the concrete objects. Only eight of the teachers referred to student strategies at all in assessing the difficulty of the task. In the case of the above subtraction task, 18 teachers mentioned the difficulty that what is sought is at the beginning of the task description, but did not relate this to the counting strategy. Instead, the subjects gave the formulation of the problem or the occurrence of key terms as reasons for the task's difficulty, for example: "If the task says 'how many more marbles has . . . ' the children will at once think of a problem of addition." The teachers presumed that the students seek to establish whether it is a problem of addition or one of subtraction. They grouped the tasks according to whether the problem formulation in the text facilitates this search or makes it more difficult.

The next step of the study concerned the students' solving strategies. The teachers were shown videotapes of children using various strategies while working on tasks. Then the teachers were presented with tasks of the same kind and asked to predict whether the student observed would be able to solve this task, and how he or she would proceed. Using this method, the researchers intended to find out whether teachers recognize that the above subtraction and addition task differs for the students in the very fact that a direct representation by fingers is possible in one case and impossible in the other. The result was that, while teachers were able to describe the students' strategy, they obviously had no concept of it, and hence had difficulties in predicting the solution behavior in tasks in which they could not observe the student's actual work on them.

Subsequently, subjects were asked to predict solving strategies and success for students from their own class chosen at random, and to describe the strategy they expected. The students were tested independent of the teachers. On average, teachers were able to predict success correctly in 27 of 36 cases, and to predict the solving strategy correctly in almost half of the cases. In the strategy prediction, however, the differences between teachers were much larger than in their predictions about success. There was, however, no significant connection between general knowledge about strategies (which was measured in the second step) and the quality of the prediction with regard to their students, nor between this knowledge and student performance on the tasks themselves.

Carpenter, Fennema, Peterson, and Carey (1988) were disappointed at this lack of "pedagogical content knowledge." In the teachers, the authors missed the knowledge about individual solving strategies of the students working on the tasks. They said that the teachers looked to superficial task features to assess the difficulty, instead of at the strategies the students used in solving.

The teachers' way of proceeding, however, indicates rich knowledge from experience. Thus, it is a basic difficulty for students to find out which type of task they have to work on. In the classroom context, tasks are connected with the previous tasks. The student is called to recognize whether he or she may maintain his or her former strategy (i.e., adding, because adding problems were on), or whether a new strategy is required. Nesher and Teubal (1975) found that students use key terms in a problem text in order to identify the required operations. Establishing which part of mathematical knowledge is asked for at the moment is an important element of mathematical competence (Greeno, Riley, & Gelman, 1984). The teachers' assessments are thus very much an indication of experience-based professional knowledge about these facts. This knowledge is more realistic than the observations of research on strategies of adding, as the real student performance in class does not just depend on the individually available strategy of learning. Their certitude in this judgment, on the one hand, and their difficulties in giving reasons for it, on the other, are an indication that this is a case of intuitive knowledge from experience (Hoge & Coladarci, 1989; Leinhart & Smith, 1985; more evidence about expert teachers' abilities to assess the difficulty of mathematical tasks can be found in Schrader & Helmke, 1989).

6. ACCUMULATING PROFESSIONAL EXPERIENCE: THE EXAMPLE OF TEACHERS' KNOWLEDGE ABOUT THEIR STUDENTS' UNDERSTANDING

The previous sections described the professional knowledge that is acquired in teacher training and then changed by experience. The following will consider the collecting of experience more closely. Teachers' observations on their students during lessons shall serve as examples.

In educational psychology, there is a widespread normative idea that teaching should be adapted as individually as possible to the knowledge and abilities of individual students (Corno & Snow, 1986), and that, hence, the difficulties encountered by students during lessons should be perceived as accurately as possible. The categorical perception of student understanding is a good example for the application of professional knowledge. Studies presented up to now show a rather negative picture. They reveal that teachers notice very little of the understanding of their students (Jecker, Macoby, & Breitrose, 1965; Putnam, 1987). Shroyer (1981) interviewed teachers while they jointly viewed videotape recordings after lessons. The teachers were asked to recall instances in which students had experienced particular difficulties or in which they had shown unexpected progress. Shroyer carried out parallel observations of these lessons and found that only 3% of the difficulties and advances observed were actually perceived by the teachers.

The above studies, however, are based on an implicitly unrealistic idea of the requirements asked of a teacher during a lesson, which, again, has resulted in an underestimation of teachers' professional knowledge. The following study on mathematics teachers has yielded indications of this (Bromme, 1987).

Our question was which problems of, and which progress in, understanding do mathematics teachers perceive. Interviews were based on a brief list of mathematical tasks in the lesson. Interviews of nineteen 5th- to 7th-grade mathematics teachers, which referred to one lesson each, were analyzed with regard to their content. We intended to establish whether the teachers remembered advances of learning or problems of understanding, and who played the active part in an episode: the entire class, individual students known by name, or subgroups of the class. Per lesson, the teachers named only an average of two students, with a maximum of six by two teachers. Eight of the 19 teachers did not remember a student known by name having problems of understanding in the lesson just given. In the case of the advances in learning, an average of three students was named.

Hence, there was little perception as to the way the subject matter was understood individually. Instead, the teachers interviewed had observed the class as a whole. For "the class" as actor, observations could be found in all the teachers, whereas almost half of the teachers were unable to name a student having problems of understanding, as has been said. The number of student problems and learning advances remembered was thus, on the whole, surprisingly small. The result is – at first glance – just as negative as that obtained in Shroyer's study mentioned above (1981). Only few episodes in the teaching process containing problems and progress of understanding were remembered. These, however, were precisely those episodes in which new steps in working through the curriculum were initiated. From the teachers' view, these were the key episodes. Student contributions were remembered if they had been of strategical value for the flow of dialogue about the subject matter, for example: "Nobody was able to give an answer to my question, then Alexander came up with a good idea." The term "strategical value" means that these contributions occurred in situations during the lesson in which there was, according to the teachers' view, "a hitch" (as one of the woman teachers said), or in which the transition proper from the old to the new knowledge was intended.

The teachers' memory and, as may be assumed, their categorical perception as well, did not concentrate on the diagnosis of individual student errors, but rather on the Gestalt of the entire lesson's flow. The active subject of learning activities was not the individual learner, but rather an abstract, but psychologically real unit that I have labeled the "collective student" (Bromme, 1987; see, also, Putnam, 1987, for similar results obtained in a laboratory setting). These results show that teachers judge their students' problems and advances of understanding against the background of an intended activity structure. The way of talking most teachers use in saying that "the class" did good work today, or had more difficulties with fractional calculus than others, is not only a verbal simplification but also an indication that entire classes are categorical units of perception for teachers (see, also, the similar result in Rutter, Manghan, Mortimore, & Queston, 1980). The categorical unit “whole class“ is rather neglected in theories on mathematical education, the focus being more on the ”individual student” as a categorical unit of perceiving and thinking. Therefore teachers have to develop their own concepts about the class as a unit, and it is not by chance that the notion of ”the class” as an indvidual unit is an important element of teachers' professional slang.

7. SUMMARY AND CONCLUSIONS

In the 1970s, there were a number of studies according to which teachers with better curricular expertise did not perform better in their teaching. These studies, however, had two deficits: They compared subject-matter knowledge of facts (as measured by tests or by the number of university courses taken) directly with the learning performance of students, omitting to analyze the connection between subject-matter knowledge and teaching activity of teachers. Subsequent studies in which lessons were observed as well showed, among other things, an influence of the amount of subject-matter knowledge and of the philosophy of school mathematics on the flexibility of teachers in coping with unexpected student suggestions. In addition, there was, within certain limits, the possibility of mutual substitution between the richness of subject-matter knowledge and more pedagogical knowledge. A second deficit of these studies was their poor theoretical conception of subject-matter knowledge. The mere familiarity with the contents of teaching constitutes only a part of the conceptual tools necessary for teachers' daily work. For the mathematics teacher, we can distinguish between five such fields of knowledge that are needed for teaching: (a) knowledge about mathematics as a discipline; (b) knowledge about school mathematics; (c) the philosophy of school mathematics; (d) general pedagogical (and, by the way, psychological) knowledge; and (e) subject-matter-specific pedagogical knowledge. Two of these areas have been treated more extensively, as they are significant for further empirical research on the structure of teachers' professional knowledge.

One of these fields comprises evaluative views about school mathematics, for instance, about the value of certain concepts and techniques for what makes mathematics a content of education. Several empirical studies have shown a strong impact of the values and goals about the school subject matter on the teaching process. These have been termed "philosophy of school mathematics" here in order to emphasize that the normative elements are closely tied in with the subject's facts and procedures. Hence, this is not a case of purely subjective beliefs. While it would seem to be undisputed that professional activity also follows normative principles and requires value decisions, it is less self-evident that such value systems are in a way interwoven into the subject-matter knowledge about mathematics. The close linkage between normative and factual elements, however, must be taken into account in a psychological theory of professional knowledge.

The second field of professional knowledge that has been treated more extensively is that of subject-matter-specific pedagogical knowledge. The concern here is with the relationship between curricular content and teaching-learning process, and it must be developed by one's own experience. In mathematics teachers, it crystallizes predominantly in their ideas about mathematical tasks and their uses in the classroom. The teacher categories about scenarios of activity are another example of this. These are categories within which knowledge of different origins (here: mathematics and pedagogy) and personal experience have been fused. The integration of knowledge originating from various fields of knowledge, discussions with colleagues, and experience is an important feature of the professional knowledge of teachers, that has to be taken into account when thinking about any educational innovation that requires the teachers' cooperation.

REFERENCES

Begle, E. J. (1972). *Teacher knowledge and student achievement in algebra* (SMSG Reports No. 9). Stanford: SMSG.

Bromme, R. (1981). *Das Denken von Lehrern bei der Unterrichtsvorbereitung. Eine empirische Untersuchung zu kognitiven Prozessen von Mathematiklehrern.* Weinheim: Beltz.

Bromme, R. (1987). Teachers' assessment of students' difficulties and progress in understanding in the classroom. In J. Calderhead (Ed.), *Exploring teachers' thinking* (pp. 125-146). London: Cassell.

Bromme, R. (1992). *Der Lehrer als Experte. Zur Psychologie des professionellen Wissens.* Bern: Huber.

Bromme, R., & Steinbring, H. (1990). Eine graphische Analysetechnik für Unterrichtsverläufe. In K. Haussmann & M. Reiss (Eds.), *Mathematische Lehr-Lern-Denkprozesse* (pp. 55-81). Göttingen: Hogrefe.

Brophy, J., & Good, T. (1974). *Teacher-student relationships. Causes and consequences.* New York: Holt, Rinehart & Winston.

Brophy, J., & Good, T. (1986). Teacher behavior and student achievement. In M. Wittrock (Ed.), *Handbook of research on teaching* (pp. 328-375). New York: McMillan.

Carlsen, W. S. (1987, April). *Why do you ask? The effects of science teacher subject-matter knowledge on teacher questioning and classroom discourse.* Paper presented at the annual meeting of the American Educational Research Association, Washington, DC.

Carpenter, T. P., Fennema, E., Peterson, P. L., & Carey, D. A. (1988). Teachers' pedagogical content knowledge of students' problem solving in elementary arithmetic. *Journal for Research in Mathematics Education, 19,* 385-401.

Carpenter, T. P., & Moser, J. (1984). The acquisititon of addition and subtraction concepts in grades one through three. *Journal of Research in Mathematics Education, 15,* 179-202.

Chevallard, Y. (1985). *La transposition didactique*. Grenoble: La Pensée Sauvage.
Clift, R. T., Ghatala, E. S., & Naus, M. M. (1987, April). *Exploring teachers' knowledge of strategic study activity*. Paper presented at the annual meeting of the American Educational Research Association, Washington, DC.
Cooney, T. J. (1985). A beginning teacher's view of problem solving. *Journal for Research in Mathematics Education, 16*, 324-336.
Cooper, H. M. (1979). Pygmalion grows up: A model for teacher expectation, communication and performance influence. *Review of Educational Research, 49*, 389-410.
Corno, L., & Snow, R. (1986). Adapting teaching to individual differences among learners. In M. Wittrock (Ed.), *Handbook of research on teaching* (pp. 605-629). New York: McMillan.
Dobey, D. C., & Schäfer, L. E. (1984). The effects of knowledge on elementary science inquiry teaching. *Science Education, 68*, 39-51.
Druva, C. A., & Anderson, R. D. (1983). Science teacher characteristics by teacher behavior and by student outcome. A meta-analysis of research. *Journal of Research in Science Teaching, 20*, 467-479.
Dunkin, M. J., & Biddle, B. J. (1974). *The study of teaching*. New York: Rinehart & Winston.
Eisenberg, T. A. (1977). Begle revisted: Teacher knowledge and student achievement in algebra. *Journal for Research in Mathematics Education, 8*, 216-222.
Gage, N., & Berliner, D. (1977). *Pädagogische Psychologie*. München: Urban & Schwarzenberg.
Greeno, J. G., Riley, M. S., & Gelman, R. (1984). Conceptual competence and children's counting. *Cognitive Psychology, 16*, 94-134.
Grossman, P. L. (1990). *The making of a teacher: Teacher knowledge and teacher education*. New York: Teachers College Press.
Gudmundsdottir, S., & Shulman, L. (1986). Pedagogical content knowledge in social studies. In J. Lowyck (Ed.), *Teacher thinking and professional action*. Proceedings of the Third ISATT Conference (pp. 442-455). Leuven: University of Leuven.
Hashweh, M. Z. (1986, April). *Effects of subject-matter knowledge on the teaching of biology and physics*. Paper presented at the annual meeting of the American Educational Research Association, San Francisco.
Heymann, H. W. (1982). Didaktisches Handeln im Mathematikunterricht aus Lehrersicht. Bericht über zwei Fallstudien zu subjektiven Hintergründen des Lehrerhandelns. In H. Bauersfeld, H. W. Heymann, G. Krummheuer, J. H. Lorenz, & V. Reiß (Eds.), *Analysen zum Unterrichtshandeln* (pp. 142-167). Köln: Aulis.
Hoge, R. D., & Coladarci, T. (1989). Teacher-based judgments of academic achievement. *Review of Educational Research, 59*, 297-313.
Hollon, R. E., & Anderson, C. W. (1987, April). *Teachers' beliefs about students' learning processes in science: Self-reinforcing belief systems*. Paper presented at the annual meeting of the American Educational Research Association, Washington DC.
Jecker, J. D., Mackoby, W., & Breitrose, M. S. (1965). Improving accuracy in interpreting non-verbal cues of comprehension. *Psychology in the Schools, 2*, 239-244.
Kesler, R. J. (1985). *Teachers' instructional behavior related to their conceptions of teaching mathematics and their level of dogmatism: Four case studies*. Dissertation, University of Georgia. Ann Arbor: UMI.
Kounin, J. (1970). *Discipline and group managment in classrooms*. New York: Holt, Rinehart & Winston.
Leinhardt, G., & Smith, D. (1985). Expertise in mathematics instruction: Subject matter knowledge. *Journal of Educational Psychology, 77*, 247-271.
Marks, R. (1987, April). *Problem solving with a small "p": A teachers' view*. Paper presented at the annual meeting of the American Educational Research Association, Washington, DC.
McGalliard, W. A. (1983). *Selected factors in the conceptual system of geometry teachers: four case studies*. Dissertation, University of Georgia, Athens. Ann Arbor: UMI.
Nesher, P., & Teubal, E. (1975). Verbal cues as an interfering factor in verbal problem solving. *Educational Studies in Mathematics, 6*, 41-51.

Oser, F. (in press). Moral perspectives on teaching. *Review of Research in Education.*
Pfeiffer, H. (1981). *Zur sozialen Organisation von Wissen im Mathematikunterricht.* IDM Materialien und Studien 21. Bielefeld: Universität Bielefeld.
Putnam, R. T. (1987). Structuring and adjusting content for students: A study of live and simulated tutoring of addition. *American Educational Research Journal, 24,* 13-48.
Roehler, L. R., Duffy, G. G., Conley, M., Hermann, B. A., Johnson, J., & Michelson, S. (1987, April). *Exploring preservice teachers' knowledge structures.* Paper presented at the annual meeting of the American Educational Research Association, Washington DC.
Romberg, T. (1988). Can teachers be professionals? In D. Grouws & T. J. Cooney (Eds.), *Perspectives on research on effective mathematics Teaching* (Vol. 1, pp. 224-245). Reston: NCTM & Erlbaum.
Rosenthal, R., & Jacobsen, L. (1971). *Pygmalion im Unterricht. Lehrererwartungen und Intelligenzentwicklung der Schüler.* Weinheim: Beltz.
Rutter, M., Manghan, B., Mortimore, P., & Queston, J. (1980). *Fifteen thousand hours. Secondary schools and their effects on children.* London: Butler & Tanne.
Schön, D. (1983). *The reflective practitioner.* New York: Basic Books.
Schrader, F. W., & Helmke, A. (1990). Lassen sich Lehrer bei der Leistungsbeurteilung von sachfremden Gesichtspunkten leiten? Eine Untersuchung zu Determinanten diagnostischer Lehrerurteile. *Zeitschrift für Entwicklungspsychologie und Pädagogische Psychologie, 22,* 312-324.
Shefelbine, J. L., & Shiel, G. (1987, April). *Preservice teachers' schemata for a diagnostic framework in reading.* Paper presented at the annual meeting of the American Educational Research Association, Washington, DC.
Shroyer, J. C. (1981). *Critical moments in the teaching of mathematics: What makes teaching difficult?* Dissertation, Michigan State University, East Lansing, MI.
Shulman, L. S. (1986). Those who understand: Knowledge growth in teaching. *Educational Researcher, 15,* 4-14.
Stein, M., Baxter, J., & Leinhardt, G. (1990). Subject-matter knowledge and elementary instruction: A case from functions and graphing. *American Educational Research Journal, 27,* 639-663.
Steiner, H. G. (1987). Philosophical and epistemological aspects of mathematics and their interaction with theory and practice in mathematics education. *For the Learning of Mathematics, 7*(1), 7-13.
Sträßer, R. (1985). Anwendung der Mathematik - Ergebnisse von Lehrer-Interviews. *Mathematicia Didactica, 8,* 167-178.
Thompson, A. (1984). The relationship of teachers' conceptions of mathematics and mathematics teaching to instructional practice. *Educational Studies in Mathematics, 15,* 105-127.
Tietze, U.-P. (1986). *Der Mathematiklehrer in der Sekundarstufe II. Bericht aus einem For schungsprojekt* (Texte zur mathematisch-naturwissenschaftlich-technischen Forschung und Lehre Nr. 18). Bad Salzdetfurth: Franzbecker.
Yaacobi, D., & Sharan, S. (1985). Teacher beliefs and practices. The discipline carries the message. *Journal of Education for Teaching, 11,* 187-199.

Acknowledgements

Parts of this contribution are based on Bromme, R. (1992). Der Lehrer als Experte. Zur Psychologie des professionellen Wissens. Bern: Huber.

DIALOGUE BETWEEN THEORY AND PRACTICE IN MATHEMATICS EDUCATION

Heinz Steinbring
Bielefeld

1. NEW PERSPECTIVES ON THE RELATION BETWEEN THEORY AND PRACTICE

Traditionally, the central task of mathematics education has been to contribute in a more or less direct manner to improving the practice of teaching mathematics and to solve teaching problems. Accordingly, the didactics of mathematics is mainly conceived of as an auxiliary science, which has to transform the scientific mathematical knowledge into a suitable form of knowledge for teachers and students and which has to provide well-tested methodological procedures to teach this knowledge effectively. Mathematics education often is taken as a methodology for elementarizing, simplifying, and adapting scientific subject matter to the abilities of students.

Additionally, the role of the referential sciences, such as pedagogics, psychology, or the social sciences, is mostly understood as a further support for this central task of didactics: to improve everyday teaching practice. In particular, these sciences should help solve those educational, psychological, and social problems that go beyond the actual field of teaching mathematics.

Also with regard to the mathematics teacher and his or her pre- and in-service training, the didactics of mathematics primarily has the role of a servant: Didactics should prepare teacher students methodically for their future teaching practice and endow them with useful teaching strategies. And, in in-service seminars, experienced teachers expect more or less direct support for their everyday teaching practice from confirmed research results and reliable teaching materials.

Such an expectation toward didactics of mathematics seems to be dominant in the beliefs of many mathematics teachers and researchers: Useful research in mathematics education is characterised by a straightforward applicability of research findings to the problems of teaching practice. This ought to bring about direct improvements of practice. But, contrary to this widespread opinion about didactics of mathematics, there is agreement that

R. Biehler, R. W. Scholz, R. Sträßer, B. Winkelmann (Eds.),
Didactics of Mathematics as a Scientific Discipline, 89-102.

most teachers simply do not refer to research findings at all and do not use them in their professional activity. "... if teachers needed information to solve a problem, it is unlikely that they would search the research literature or ask the researcher to find an answer" (Romberg, 1985a, p. 2).

Are the results of didactical research much too far removed from the actual problems of teaching practice? Is it necessary to adjust scientific results even more strongly to the conditions of teaching practice? Or are teachers, for different reasons, unable to make professional use of research findings in their teaching profession (Romberg, 1985b, 1988)? Or is it even impossible to meet these implicit expectations addressed by practitioners to didactical theory and, vice versa, the expectations of educators addressed to practitioners, because they are unfounded and must be reconsidered? Could it be that scientific results cannot be applied to teaching practice in a direct and immediate way, on principle, but that the application of theory to practice is always very complex and depends on many premises (Kilpatrick, 1981)?

The dominant structure that is believed to control the relation between theory and practice could be described as a linear follow-up: Theory furnishes results that gain direct access to practice, improving and developing it. This linear pattern is not just found between didactical research and the practice of teaching; the relation between teacher and student in teaching/learning-processes is often interpreted as a linear connection, too: The teacher is the conveyor of the mathematical knowledge that he or she must prepare methodically and then hand over to the students in order to extend their comprehension and insights into mathematics.

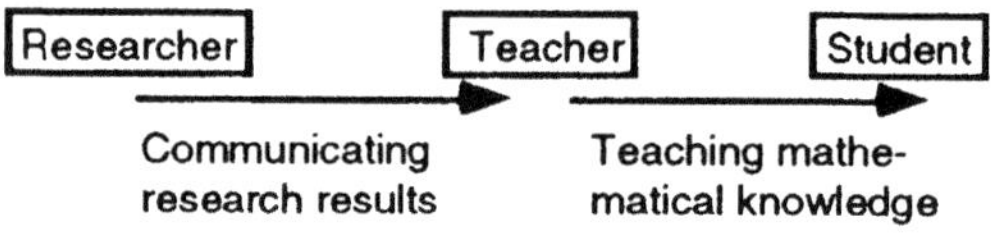

Figure 1. The linear relationships between researcher, teacher, and student.

This view is based on an interpretation of mathematical knowledge, as criticized by, for example, D. Wheeler (1985):

> In this model, the subject matter to be taught is already determined in content and form, the teacher knows this subject matter and passes it on, "as it is," to the students, and the students rehearse it until they can show they know it as well as, or nearly as well, as their teacher. What place can there possibly be for research if this is the state of affairs? (p. 10)

According to this model, research, at best, has to determine content and form of new mathematical subject matter for mathematics teaching.

This comparative analogy of the relation between research and practice of teaching to the relation between teacher and student seems to be helpful for many reasons. The assumed interpretation of the organizational structure of one of these relations implies a similar conception of the other relation (cf.

AG Mathematiklehrerbildung, 1981, p. 205; Rouchier & Steinbring, 1988). A linear model of the connection between theory and practice often is based on a similar linear model of the teaching/learning process of mathematics. Many research studies have criticized the perspective of the teacher as the conveyor of mathematical knowledge and the student as the receiver (Cooney, 1988; Mason, 1987). The teacher is viewed as providing learning situations in which students have to contribute their own potential for actively reconstructing knowledge, for establishing a personal relationship toward this knowledge.

The central perspective on the relation between theory and practice in the following is the forms of cooperation between didactical research and the mathematics teachers who already possess some professional experience; that is, an in-service training perspective and not university training. The reality of everyday teaching cannot be influenced in a direct way by didactical research, nor is it arbitrarily changeable and restructurable. In the framework of its socioinstitutional conditions and with regard to the specific epistemology of school mathematics, teaching practice is relatively autonomous of external influences; indeed, it has produced very effective provisions for maintaining this autonomy. Changing interventions into this complex practice have to reflect more carefully the hidden preconditions and mechanisms that are relevant in teaching practice.

This leads to consequences for both parts of the theory-practice relation: Didactical science has no direct possibility of controlling the everyday practice of the mathematics teacher, and the teacher has no straightforward possibility of controlling the students' process of either learning or comprehension. The partners participating in this process of mediation (necessarily) act relatively autonomously within the framework of the socioinstitutional conditions, a fact due to the difficult epistemological character of the knowledge under discussion, which can ultimately only be understood by means of personal reconstructions.

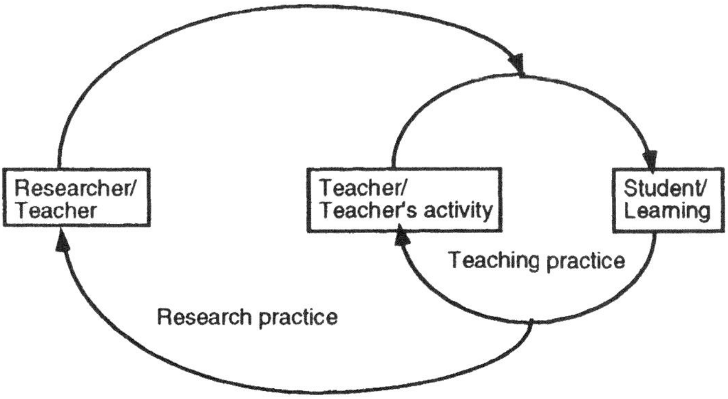

Figure 2. A model of the relation between theory and practice.

This requires a modified interpretation of the role and perspective of didactical theory in relation to practice. This could be expressed in the model shown in Figure 2.

This model tries to display the new fundamental paradigm shift in the theory-practice relation: There are no direct influences or hierarchical dependencies, but exchange and feedback between two relatively independent social domains of reflecting upon and mediating mathematical knowledge. Only such a structure could enhance a real dialogue: between teacher and students and between theory and practice, with all its ways of sharing, jointly observing, reflecting, and discussing, and its modes of communication that enable positive feedback that supports the subjective construction of mathematical meaning by means of integrating the fruitful ideas of different partners. The realization of such a dialogue can probably be established between researchers and teachers more easily if the teacher is not subjected to a "didactical contract" with the researcher. A dialogue between teachers and students under the usual conditions of the didactical contract is more difficult to establish. This model of cooperation between theory and practice must take into account the following three dimensions:

1. *Knowledge* (in very general terms about mathematics in teaching/learning situations): the relation between theoretical/scientific knowledge and practical/useful knowledge.

2. The *professional practice and social role* of persons involved in the theory-practice relationship, and the education of teachers.

3. *Forms and models of cooperation* between theory and practice in mathematics education.

Obviously, it is necessary for these three dimensions to overlap, but this analytic separation helps to get an adequate idea of the complex factors involved in the theory-practice relation. For 10 years, the international research project "Systematic Cooperation Between Theory and Practice in Mathematics Education (SCTP)" has been analyzing the problem of relating theory to practice from a broad perspective. A main basis has been a number of case studies from different countries reporting on diverse projects trying to improve the relation between didactical research and mathematics teaching practice (see Christiansen, 1985; Seeger & Steinbring, 1992a; Verstappen, 1988). Despite their examplary character, these cases in principle cover all the three dimensions developed here; some of the research papers reported below might be taken as an example of emphasis on some important aspect of the 3-dimensional network.

1. *Knowledge.* This is a complex dimension, because it not only contains the mathematical knowledge (the subject matter) to be learned by students or by teachers; it also refers to the related scientific and practical knowledge domains necessary to improve teachers' professional standards (epistemology, history of mathematics, psychology, pedagogics, curricular questions, etc.) and it has to deal with the difficult problems of mathematical

meaning and understanding (at the university and at school; cf. Bazzini, 1991; Ernest, 1992; Seeger & Steinbring, 1992b; Wittmann, 1989).

2. *Professional practice and social role.* This relates to the social framing factors influencing and supporting endeavors to mediate knowledge, be they in the classroom or in cooperation between researchers and teachers. The indirect ways of relating theory to practice require forms of social participation and sharing common experiences that belong to different professional practices and communicative situations (cf. Andelfinger, 1992; Brown & Cooney, 1991; Mason, 1992; Voigt, 1991; Wittmann, 1991).

3. *Forms and models of cooperation.* Cooperative efforts to implement this changed intention often take the form of case studies and applied projects, implicitly or explicity using attributes to describe the role of the partners involved and the status of the mathematical knowledge. Such practical case studies necessarily have their own "history," but a fruitful connection between the complex knowledge involved and the social embedment of cooperation between theory and practice can be organized only in concrete frameworks that then have to be investigated for general and universal insights. (cf. Bartolini Bussi, 1992; Bell, 1992; Burton, 1991; von Harten & Steinbring, 1991; Verstappen, 1991).

A major fundamental insight discussed and explored in the SCTP group is to more thoughtfully analyze the conditions of the "dialogical structure" of communication, cooperation, and materials (textbooks, reports, research papers) in the relation between theory and practice. Unlike a hierarchically structured conveyance of "context-free," absolute knowledge, a dialogical structure aims to be particularly aware of the specific contexts and conditions of application and interpretation for the mediated knowledge in which the partner of cooperation is involved. Scientific knowledge for mathematics teachers essentially has to refer to the circumstances of everyday teaching practice. A consequence is that neither a separate change of research nor of practice could improve cooperation, but that the relation between theory and practice has itself become a problem of research.

2. THE THEORETICAL NATURE OF MATHEMATICAL KNOWLEDGE: COMMUNICATING KNOWLEDGE AND CONSTRUCTING MEANING

In the framework of the range of important topics in the theory-practice relation, I shall concentrate on certain aspects of the mathematical knowledge negotiated and mediated in this relationship. The theoretical perspective will not be curricular, historical, or mathematical, but an attempt to use the *epistemological* basis of mathematics. If it is accepted that epistemology is the scientific enterprise of investigating the status, structure, and meaning of knowledge, then this perspective becomes indispensable for the analysis of such indirect modes of cooperation between scientific didactics and everyday teaching practice that aim at communication as a reciprocal dialogue searching for possibilities of constructing and enhancing meaning and not simply conveying knowledge matter. The intention is not to describe the mediation of a coherent didactical theory named "mathematical epistemology" to the practice of mathematics teaching, but to stress and to use epistemological considerations of mathematical knowledge, because this is an essential characteristic of every process of mediating knowledge between teacher and students as well as between researcher and teacher. This section presents an epistemological analysis. The next section discusses how classroom episodes can be interpreted along these lines and discussed with teachers as part of a theory-practice cooperation.

There is a fundamental *epistemological dilemma* in every mediation of mathematical knowledge: When teachers intend to provide new knowledge to their students, they have to use some specific (mathematical) signs and diagrams (carriers of the new knowledge), which are connected by some stringent rules, and they have to focus the students' attention on these knowledge carriers. However, the knowledge and its meaning is not contained in these carriers. These symbolic signs and diagrams are some kind of concrete substitute for the knowledge itself; they can only point to the knowledge and its meaning intentionally. This cannot be read off directly from these symbolic means, but has to be reconstructed from them actively.

Would it not be easier to communicate the mathematical meaning directly? But is this at all possible? This problem is the basis of the epistemological dilemma: Teachers have to use some kind of knowledge carrier, and have to cope with it strictly, and, at the same time, they have to be aware and to let their students know that the students themselves have to search for the meaning of the knowledge, which is not inherent to the symbolic means but is constituted in the relations students are able to construct between the symbols and some intended referential context.

An example may illustrate this epistemological dilemma. Consider the following problem from a textbook for 6th-grade students:

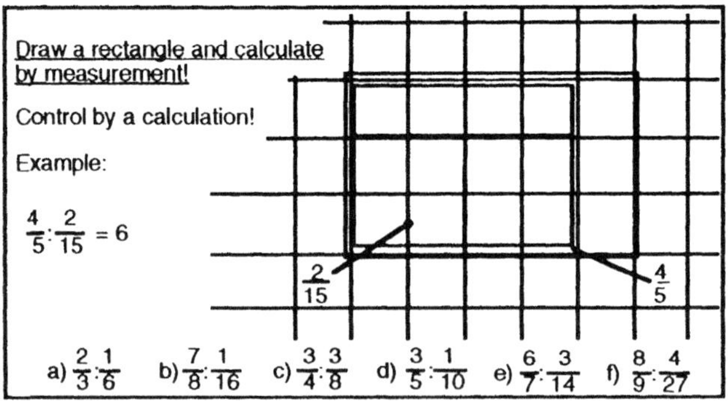

Figure 3. Visualizing and the division of fractions.

This problem deals with the division of fractions and tries to use a graphic diagram to mediate in a direct way the meaning of fraction division. This contrast between formula and graphic diagram is suitable to clarify some epistemological aspects between sign and object (or referent) in school mathematics. On the one side, there are mathematical signs connected by some operational symbols, functioning as a little system: $^4/_5 : {^2/_{15}} = 6$. On the other side, there is a geometrical reference context, intended to furnish meaning for the signs and operations. The diagram should support the process of constructing a meaning for the formula. The *relational structures* in the geometrical diagram and the formula are the important aspects and not the signs itself.

In which way can this diagram give meaning to the formula? Is it possible to deduce the idea of the division of fractions from it? Is it adequate to conceive of the elements in this diagram as concrete objects for directly showing the meaning of division?

First of all, one observes that all problems to be tackled have denominators that are a multiple of the denominator of the other fraction. Consequently, the intended explanation with the help of the diagram cannot be universal. A certain type of fractions seems to be presupposed, indicating a first reciprocal interplay between diagram and formula. There are more indications for this interplay: In this representation, a variable comprehension of 1 or the unit is necessary. The big rectangle with the 15 squares once is the unit, used to visualize the proportions of $^4/_5$ and $^2/_{15}$ as four rectangles (with 3 squares each) and as a rectangle of 2 squares respectively. The composition of three squares to a rectangle represents a new unit or 1. When interpreting the operation $^4/_5 : {^2/_{15}} = 6$, the epistemological meaning of the result "6" changes according to the changes of the unit. How is the 6 represented in the diagram? It cannot be the sextuple of the original rectangle, hence no pure empirical element.

The 6 could mean: In $^4/_5$ there are 6 times $^2/_{15}$ or there are 6 pairs of two squares in $^4/_5$. Or, interpreting $^4/_5$ as $^{12}/_{15}$ as implicitly suggested in the diagram itself, the operation modifies to: $^{12}/_{15} : {^2/_{15}} = 6$. But this is nothing other than the operation: $12 : 2 = 6$, because the denominator can be taken as a kind of "variable," that is, the 15 could also be 20, or 27, and so forth. In this division, in principle, the half is calculated, a division by 2 is made.

The analysis shows changing interpretations of the unit: First, the unit is represented by the big rectangle of 15 squares, then one single square also represents the unit. The epistemological reason is that a fraction like $^{12}/_{15}$ is not simply and exclusively the relation of the two concrete numbers 12 and 15, but a single representative of a lot of such relations: $^4/_5$, $^8/_{10}$, $^{16}/_{20}$, $^{20}/_{25}$, ... What is defined as the unit in the diagram is partly arbitrary and made by some convention, and, furthermore, the constraints of the geometrical diagram and of the given numerical sign structure determine partly the choice of the unit. For instance, for this arithmetical problem, it would not be an adequate choice to take the rectangle of 5 x 7 squares as the unit; whereas a rectangle of 6 x 10 squares, or subdivision of the squares into quarters, would be valid.

The intentional variability implicit in the numerical structure of a fraction is partly destroyed in the geometrical diagram used to represent the fraction; this variability has to be restored in the diagram by means of flexibly changing the unit. The concrete single diagram, with its parameters once chosen, has to be conceived of as a "general" diagram.

The relational structures in the object (referential system) and in the symbol system depend on each other. The relations have to be installed by the subject in accordance with structural necessities; a certain compatibility between the system of symbols and referents can be obtained only through the intended generalization of epistemological relations. This generalization is the objective to be learned and to be constructed by the learner.

Epistemological, didactical, and historical research has extended the perspective on the specific nature of mathematical knowledge (cf., e.g., Balacheff, 1987; Jahnke, 1978; Lakatos, 1976; Otte, 1984b; Steinbring, 1991a; Steinbring, 1993). The mathematical meaning results from relations within a system; knowledge is represented by a specific way of constructing relations. The most elementary relational form of theoretical mathematical knowledge can be characterized as the *epistemological triangle*:

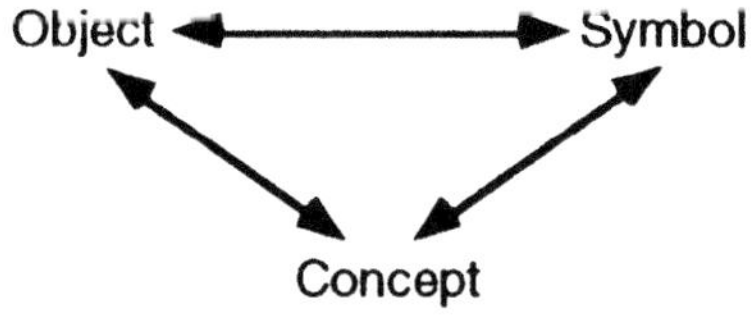

Figure 4. The epistemological triangle.

The meaning of theoretical knowledge emerges in the conflict between symbol/model on the one side and object/problem area on the other side (cf. Otte, 1984a; Steinbring, 1989). This epistemological triangle of mathematical knowledge is based on the characterization of "meaning" as the "triad of thoughts, words and things" (Odgen & Richards, 1923, p. 11) . With regard to this epistemological triangle of "object," "sign," and "concept," it is not assumed that the relations between the "corners" of the triangle are fixed a priori, but that they must continously be developed, installed, and eventually modified according to new prerequisites (cf. Bromme & Steinbring, 1990).

The peculiar aspect of mathematical concepts described by this epistemological triangle is the fact that the reference between object and symbol is not organized simply as a conventionalized name, but must be developed as a conceptual relationship. The ciphers 2 and 15 in the fraction $^2/_{15}$ given in this example are not an economic name for an object, like, for instance, the parts of a pie or the parts of a surface in a diagram, but they constitute a little "system of relations" that refers conceptually to the structure of a referential situation (cf. Steinbring, 1992). Mathematical symbols do not denote names, but display a system structure that relates variably to the referent structure.

The epistemological dilemma in every mathematical communication of the need to take symbolic carriers for the knowledge to be transported, and, at the same time, to go beyond these concrete carriers, requires a dualistic conception of mediating processes: In the classroom, mathematics teachers have to present the learning situations for their students in specific contexts, which can be shared in communication, and then, by means of generalization, they must initiate a process of decontextualization that helps students to subjectively reconstruct the meaning of the mathematical knowledge hidden in the context. Processes of decontextualization support the revelation of underlying structural relations in the object that make it possible to develop the conceptual relation between object and symbol in the epistemological triangle.

Fruitful dialogues between researchers and mathematics teachers also need contextualized situations representing examples of the teacher's object of professional activity to enable teachers and researchers to share a situation from which different decontextualizations can be created according to the objectives of different professional domains. An example will be discussed in the following.

3. ASPECTS OF THE COMPLEX NATURE OF THEORETICAL KNOWLEDGE IN SOCIAL PROCESSES

3.1 The Need for a Common Context

The implicit premise of much didactical research is to abstract all information for practice from seemingly superfluous context-dependent aspects. In return, this forces teachers to embed this abstract knowledge into their own context of experience. This implies a fundamental separation between the researcher's and the teacher's understanding of professional knowledge and its meaning: Both refer this scientific information to different reference contexts. Of course, there are necessary and even positive differences between the referential knowledge domains of the teacher and the researcher, but the crucial point for a fruitful dialogue is not to take divergent contexts of reference, but to look jointly at the same context of references, and develop a basis for shared views without supposing there could be identical interpretations in all aspects. Looking at the same context situation is a connecting element for different, contrasting, and complementary interpretations and applications of abstract, general professional knowledge.

The dialogue between theory and practice has to develop both levels. The decontextualized knowledge and examples of contextualized referential situations for the abstract information cannot be conveyed directly. However, in this way, they open a framework for reconstructing the meaning of this professional knowledge in relation to a common object of reference and in agreement with the different experiences from the teacher's or the researcher's professional activity.

Joint reading, interpretation, and analysis of lesson transcripts is an example of discussing a common object of interest and developing a dialogue between theory and practice (von Harten & Steinbring, 1991; Voigt, 1991).

> [Lesson transcripts] are well suited because they take classroom reality seriously, that is have teaching in its concrete form as their object, a fact which induces the participants to become aware of the conditions of this teaching and of the opportunities of change. Interpretation and evaluation of the actual immediate classroom reality indeed requires us to adopt a theoretical view. Insofar, the seemingly immediately empirical and real lesson transcripts are highly theoretical constructs. They must be understood as individual cases of a varying scope of possible classroom situations. (von Harten & Steinbring, 1991, p. 175)

Such cooperative work between teachers and researchers serves a twofold purpose: It is a means for researchers to communicate their theoretical ideas in a context of shared perspectives and it is used to explore exemplarily the teacher's practice, or better, to obtain feedback and to learn from the teachers.

3.2 A Classroom Episode

An example may illustrate the development of the two epistemological levels (contextualized and decontextualized) for the teacher's professional knowledge within the framework of a fruitful dialogue between theory and practice (for more details, see Steinbring, 1991b).

A short grade-6 teaching episode contains a sequence of exercises that the teacher poses for training the translation of fractions into decimal numbers. Despite this intended character of a phase of exercise, a shift to conceptual problems occurs very soon, which the teacher does not notice at all. The teacher starts with the first problem: to translate $^3/_{10}$ into the correct decimal. The solution comes immediately: 0.3. The three following problems are also solved more or less quickly, with the help of a brief reminder on the rules of the fraction calculus: $^4/_{50} = 0.08$, $^4/_5 = 0.6$, $^8/_{20} = 0.4$.

The next problem causes some productive confusion: What is the decimal for $^5/_8$? The students can no longer simply follow the teacher's explicit methodological intention to first enlarge the fraction given to one of the form: $^x/_{10}$, $^x/_{100}$, $^x/_{1000}$, ... When trying to solve the problem, the students propose the following transformation: $^5/_8 = {}^{25}/_{40} = {}^{2.5}/_4$. The teacher rejects this result, because it ignores the formal method he has proposed. In a second attempt, the students come up with a similar solution: $^5/_8 = {}^{25}/_{40} = {}^{62.5}/_{100}$. Now the methodological rule is fulfilled, but still the teacher is unsatisfied. There is a decimal number as numerator in this fraction, a nonadmitted combination of signs! In a kind of funnel pattern (Bauersfeld, 1978), the teacher forces the correct solution by first calculating the number of enlargement to the fraction $^{62.5}/_{100}$, that is, 125; the necessary arithmetical division of 1,000 : 8 =?, is more complex than the division of 5 : 8 =?, which would have given the solution directly.

Different intentions were interacting during this student-teacher episode: The teacher simply followed his methodological aim of training the fraction translation into decimals; and he relied on one rule, which he thought of as easy and universal: "Transform the fraction given into one of the form: $^{x}/_{10}$, $^{x}/_{100}$, $^{x}/_{1000}$, and so forth, and then read off the correct decimal number!" The students still have to cope with the unfamiliar new mathematical knowledge. They try to uncover the teacher's expectations and to follow his methodological rule as far as possible. The first four problems are solved; for the teacher, the fifth problem seems to be only technically more complex, but the students really encounter a new conceptual problem. In their attempts to give a solution, they offer (still unknowingly) an interesting conceptual generalization and, at the same time, an improved understanding of the connection between fractions and decimals. However, the teacher is not aware of this, because he is keeping strictly to his methodological plan.

Because of his strict goal of performing only some exercises, the teacher is not open to the conceptual ideas hidden in the students' proposals. He simply rejects the two fractions $^{2.5}/_{4}$ and $^{62.5}/_{100}$ for reasons of method and definition. The interpretation from our perspective is that the teacher was not sensitive to the epistemological dilemma of the mathematical symbols. He could not understand or accept the possible new meaning of these signs, the combination of decimals and fractions, which reflects the fundamental conceptual relation of decimals in a new way: the variable choice of the unit of measurement as a fraction with a denominator as a power of 10. Accepting the fraction $^{62.5}/_{100}$ would lead immediately to the answer $^{625}/_{1000}$ or $^{6.25}/_{10}$ or 0.625 by using the already known rule of shifting the position of the point. But being able to agree with this interpretation would require an epistemological vigilance toward the changing meaning of mathematical signs and their combinations, which is regulated within the framework of the epistemological triangle of object, symbol, and concept.

3.3 Analysis of Lesson Transcripts in a Dialogue Between Teachers and Researchers

This episode, and some of the epistemological issues presented here, can and have been taken as the common referential situation in a dialogue with a group of teachers together with the teacher of this episode. This common object served as a reference context to explain general epistemological ideas (i.e., the epistemological triangle, the epistemological dilemma, etc.) and, at the same time, to try to detect general constraints of the given concrete teaching situation.

The exemplary dialogue between theory and practice in this case included general and specific aspects. The discussion of the transcribed episode offered means for the teacher to detach himself from his subjective immersion in the teaching episode. This opened perspectives for a better comprehension of the students' remarks and intentions and for seeing some general features in the specific and particular teaching situation; a view that was supported by the different interpretations given by colleagues. Specific aspects concerned the interference of the teacher's methodological intentions with the epistemological constraints of the mathematical knowledge and its meaning as constituted in this interaction with the students. The seemingly unique mathematical signs and operations developed by the teacher entered a different context of interpretation in the students' understanding. How can the teacher become sensitive to such epistemological shifts of meaning? Here again, the very fundamental problem of the nature of (school) mathematical knowledge is questioned: The new knowledge cannot be "given" to the students; the teacher has to be aware of the way the students are trying to reconstruct the meaning of the mathematical signs and operations he has presented to the students. The shared discussion and dialogue between different practices enhanced the possibilities of becoming aware of underlying complementary perceptions and ways of integrating them.

This social situation of dialogue and sharing between theory and practice displayed the different paradigm of the theory-practice relation: to reconstruct from a common object one's own conceptual ideas and practical consequences by seeing the variable and general in the concrete, singular situation with the help of critics and the different perspectives of the participants.

4. CONCLUSIONS

Every productive dialogue between theory and practice in mathematics education has to unfold the dialectic between the concrete context and abstracting decontextualizations. This is not simply for reasons of presenting an illustrative example for abstract theoretical considerations. The concrete context has to play a basic role in the sense that it serves common and distinct roles for the different partners: It links different views, which are based on different professional activities, and it offers the establishment of referential connections and referential meaning with particular and comparable aspects.

In this respect, communication and mediating materials in the relation between theory and practice need to reveal different conceptual components:

1. a common referential object;
2. specific generalizations of the knowledge (mathematical, epistemological, professional) bound to the particular domain of experience;
3. means of social sharing, participating, and exchanging in communicative situations.

The dialogue between theory and practice in mathematics education cannot aim at a direct conveyance of ready knowledge, but can offer only occasions for a self-referential reconstructing of all aspects of professional knowledge necessary for the teacher. These productive occasions are based on the requirement for the teacher always to explore the conceivable relations between the complexity of an exemplary concrete situation and the intended, disguised, and variable generalizations and universal conceptions inherent in this situation. In a way, this paper has also tried to take this situation as a structuring lineament for mediating its theoretical message.

REFERENCES

A. G. Mathematiklehrerbildung. (1981). *Perspektiven für die Ausbildung des Mathematiklehrers.* Köln: Aulis.

Andelfinger, B. (1992). Softening the education of mathematics teachers. In F. Seeger & H. Steinbring (Eds.), (1992a), (pp. 225-230).

Balacheff, N. (1987). Processus de preuve et situations de validation. *Educational Studies in Mathematics, 18*(2), 147-176.

Bartolini Bussi, M. (1992). Mathematics knowledge as a collective enterprise. In F. Seeger & H. Steinbring (Eds.), (1992a), (pp. 121-151).

Bauersfeld, H. (1978). Kommunikationsmuster im Mathematikunterricht - Eine Analyse am Beispiel der Handlungsverengung durch Antworterwartung. In H. Bauersfeld (Ed.), *Fallstudien und Analysen zum Mathematikunterricht* (pp. 158-170). Hannover: Schroedel.

Bazzini, L. (1991). Curriculum development as a meeting point for research and practice. *Zentralblatt für Didaktik der Mathematik, 23*(4), 128-131.

Bell, A. (1992). Studying teaching. In F. Seeger & H. Steinbring (Eds.), (1992a), (pp. 153-163).

Bromme, R., & Steinbring, H. (1990). Die epistemologische Struktur mathematischen Wissens im Unterrichtsprozeß. In R. Bromme, F. Seeger, & H. Steinbring (Eds.), *Aufgaben als Anforderungen an Lehrer und Schüler* (pp. 151-229). Köln: Aulis.

Brown, S., & Cooney, T. J. (1991). Stalking the dualism between theory and practice. *Zentralblatt für Didaktik der Mathematik, 23*(4), 112-117.

Burton, L. (1991). Models of systematic co-operation betweeen theory and practice. *Zentralblatt für Didaktik der Mathematik, 23*(4), 118-121.

Christiansen, B. et. al. (1985). *Systematic co-operation between theory and practice in mathematics education.* Copenhagen: Royal Danish School of Educational Studies (ICME V).

Cooney, T. J. (1988). The issue of reform: What have we learned from yesteryear? *The Mathematics Teacher, 81*(5), 352-363.

Ernest, P. (1992). The relationship between the objective and subjective knowledge of mathematics. In F. Seeger & H. Steinbring (Eds.), (1992a), (pp. 33-48).

Harten, G. von, & Steinbring, H. (1991). Lesson transcripts and their role in the in-service training of mathematics teachers. *Zentralblatt für Didaktik der Mathematik, 23*(5), 169-177.

Jahnke, H. N. (1978). *Zum Verhältnis von Wissensentwicklung und Begründung in der Mathematik - Beweisen als didaktisches Problem.* IDM Materialien und Studien 10. Bielefeld: Universität Bielefeld.

Kilpatrick, J. (1981). Research on mathematical learning and thinking in the United States. *Recherches en Didactiques des Mathématiques, 2*(3), 363-380.

Lakatos, I. (1976). *Proofs and refutations. The logic of mathematical discovery.* Cambridge.

Mason, J. (1987). What do symbols represent? In C. Janvier (Ed.), *Problems of representation in the teaching and learning of mathematics* (pp. 73-81). Hillsdale, NJ: Erlbaum.

Mason, J. H. (1992). Reflections on dialogue between theory and practice, reconciled by awareness. In F. Seeger & H. Steinbring (Eds.), (1992a), (pp. 177-192).

Odgen, C. K., & Richards, F. A. (1923). *The meaning of meaning.* London: Routledge and Kegan Paul.

Otte, M. (1984a). *Komplementarität*. IDM Occasional Paper 42. Bielefeld: Universität Bielefeld.

Otte, M. (1984b). *Was ist Mathematik?* IDM Occasional Paper 43. Bielefeld: Universität Bielefeld.

Romberg, T. A. (1985a). Research and the job of teaching. In T. A. Romberg (Ed.), *Using research in the professional life of mathematics teachers (ICME 5)* (pp. 2-7). Madison, WI: Wisconsin Center for Education Research, School of Education, University of Wisconsin.

Romberg, T. A. (Ed.). (1985b). *Using research in the professional life of mathematics teachers (ICME 5)*. Madison, WI: Wisconsin Center for Education Research, School of Education, University of Wisconsin.

Romberg, T. A. (1988). Can teachers be professionals? In D. A. Grouws, T. J. Cooney, & D. Jones (Eds.), *Effective mathematics teaching* (pp. 224-244). Reston, VA: NCTM & Lawrence Erlbaum.

Rouchier, A., & Steinbring, H. (1988). The practice of teaching and research in didactics, *Recherches en Didactique des Mathématiques*, *9*(2), 189-220.

Seeger, F., & Steinbring, H. (Eds.). (1992a). *The dialogue between theory and practice in mathematics education: Overcoming the broadcast metaphor. Proceedings of the Fourth Conference on Systematic Cooperation between Theory and Practice in Mathematics Education (SCTP).* Brakel, Germany. IDM Materialien und Studien 38. Bielefeld: Universität Bielefeld.

Seeger, F., & Steinbring, H. (1992b). The myth of mathematics. In F. Seeger & H. Steinbring (Eds.), (1992a), (pp. 69-89).

Steinbring, H. (1989). Routine and meaning in the mathematics classroom. *For the Learning of Mathematics*, *9*(1), 24-33.

Steinbring, H. (1991a). The concept of chance in everyday teaching: Aspects of a social epistemology of mathematical knowledge. *Educational Studies in Mathematics*, *22*, 503–522.

Steinbring, H. (1991b). Eine andere Epistemologie der Schulmathematik - Kann der Lehrer von seinen Schülern lernen? *mathematica didactica*, *14*(2/3), 69-99.

Steinbring, H. (1992). The relation between social and conceptual conventions in everyday mathematics teaching. Unpublished manuscript. Bielefeld: IDM.

Steinbring, H. (in press). Epistemology of mathematical knowledge and teacher–learner interaction. *The Journal of Mathematical Behavior*.

Verstappen, P. (1991). Ten major issues concerning systematic cooperation between theory and practice in mathematics education. *Zentralblatt für Didaktik der Mathematik*, *23*(4), 122-127.

Verstappen, P. F. L. (Ed.). (1988). *Report of the Second Conference on Systematic Cooperation Between Theory and Practice in Mathematics Education*. Lochem/Netherlands. Enschede: S.L.O.

Voigt, J. (1991). Interaktionsanalysen in der Lehrerbildung. *Zentralblatt für Didaktik der Mathematik*, *23*(5), 161-168.

Wheeler, D. (1985). The utility of research. In T. A. Romberg (Ed.), *Using research in the professional life of mathematics teachers (ICME 5)* (pp. 8-15). Madison, WI: Wisconsin Center for Education Research, School of Education, University of Wisconsin.

Wittmann, E. C. (1989). The mathematical training of teachers from the point of view of education. *Journal für Mathematik-Didaktik*, *10*, 291-308.

Wittmann, E. C. (1991). From inservice-courses to systematic cooperation between teachers and researchers. *Zentralblatt für Didaktik der Mathematik*, *23*(5), 158-160.

ON THE APPLICATION OF SCIENCE TO TEACHING AND TEACHER EDUCATION

Thomas J. Cooney
Athens (Georgia)

1. INTRODUCTION

In this chapter, I will raise the issue of what it means to be scientific in the context of conducting research on teaching and teacher education. I will argue that our notion of being scientific is related to how we see change evolving in the teaching and learning of mathematics. The concepts of authority and adaptation will be considered as they are related to teacher education.

2. THE NOTION OF BEING SCIENTIFIC

The notion of being scientific has many connotations as it is applied to improving the teaching and learning of mathematics. A view of science that emphasizes regimented procedures yielding sweeping generalizations led Highet to conclude that science, so conceived, had little relevance to improving the art of teaching.

> I believe that teaching is an art, not a science. It seems to me very dangerous to apply the aims and methods of science to human beings as individuals, although a statistical principle can often be used to explain their behavior in large groups . . . . A scientific relationship between human beings is bound to be inadequate and perhaps distorted. (Highet, 1950, p. viii)

Davis (1967) echoed the same sentiment when he argued that teaching mathematics "is not the application of a science in any presently meaningful sense of such a phrase" (p. 38).

But some disagreed. Gage (1972), for example, argued that the objectivity of science could contribute to the improvement of education and could eventually provide a basis for constructing teacher education programs. This argument was echoed many times throughout the 1970s. Gallagher (1970) maintained that it was through science that the artistry of teaching can be revealed to those trying to master the art. Brophy put it quite bluntly.

> Teacher educators and educational researchers need to pay more attention to the accumulation of a data base that would allow truly prescriptive teacher education to emerge. Propounding ideas on the basis of commitments rather than supportive data is unscientific to say the least, and blowing with the wind by propounding

R. Biehler, R. W. Scholz, R. Sträßer, B. Winkelmann (Eds.),
Didactics of Mathematics as a Scientific Discipline, 103-116.

the latest educational fad is even worse. (Brophy, 1975, p. 15)

While the debate raged in the 1970s over the applicability of science to the art of teaching, what was obscured was the question of what constitutes science. A review of published research in the United States during this period suggests a view of science as an exercise in yielding statistical generalizations. Most of this research involved the process/product paradigm in which teacher behaviors were correlated with achievement – usually defined in terms of basic skills (see, e.g., Rosenshine & Furst, 1973). In the main, this research had little impact on the field of mathematics education.

By the late 1970s, the field was beginning to turn its head. Researchers, at least in the United States, began to study teachers' decision-making processes, thereby giving the impression that the questions were more cognitively oriented, yet holding tightly to the notion of "traditional" science. A study by Peterson and Clark (1978) is illustrative, as they traced the nature and types of decisions teachers made using correlational analyses. But there were other voices being heard, some inside and some outside the field of mathematics education, that raised more fundamental issues. From a methodological perspective, Mitroff and Kilmann (1978) concluded that "science is in serious need of methodological and epistemological reform" (p. 30). The authors maintained that "Even if there were no 'crises of belief' in science, there would still be good reasons for considering reform at this time, given the new cultural forces and streams of thought being articulated" (p. 3). Mitroff and Kilmann's (1978) analysis led them to identify four types of scientist. One type, the analytic scientist, believes in the value-free nature of science, that is, knowledge is separable from values. In contrast, the authors identified two other types, the conceptual humanist and the particular humanist, who focus on descriptions of human activity, raising the question of whether stories are an appropriate mechanism for communicating research findings.

Perhaps the most serious attack on the notion of "traditional science" came from Feyerabend (1988) who maintained that "the events, procedures, and results that constitute the sciences have no common structure" (p. 1). Feyerabend's (1988) orientation toward science supports an eclectic view of the way science should be conducted. According to Feyerabend, science, as defined by an allegiance to regimented procedures, runs the risk of undermining the value gained from human ingenuity, insight, and compassion. Similarly, Mitroff and Kilmann (1978) observed that, "The greatest scientists seem not only to combine the attributes of opposing types but to delight in doing so" (p. 12).

At one level, we can say that research on teaching has moved from what teachers *were* (i.e., their characteristics) in the 1950s and 1960s, to what teachers *did* in the 1970s, to what teachers *decided* in the early 1980s, to the more recent focus on what teachers *believe* (see Brown, Cooney, & Jones, 1990; Thompson, 1992). Such an analysis would miss, however, what was happening conceptually and methodologically in mathematics education. With the emerging prominence of the constructivist epistemology (in its many forms), a premium has been placed on meaning and context. This emphasis challenges us to reconsider what we mean by being scientific, including the notion of being objective. Von Glasersfeld addresses the issue of objectivity in the following way:

> In order to observe anything, in order to "collect data," one must have some notion – no matter how primitive and preliminary – of the particular experiences one intends to relate to one another. It is, obviously, these experiences that one will be looking for. In order to find them, one necessarily assimilates and disregards all sorts of differences in individual observations. The longer this goes on successfully and the more often the model one has constructed proves useful, the stronger becomes the belief that one has discovered a real connection, if not a *Law of Nature*. And once that belief has been established, there is a powerful resistance against any suggestion of change and – as Thomas Kuhn has so nicely shown with examples form the history of science – there will be powerful efforts to sweep any observed irregularity under the rug. (von Glasersfeld, 1987, p. 11)

What becomes obvious to anyone who has tried to understand why human beings behave as they do is that the lenses through which people see their world are intertwined with the context in which those lenses were created. Bauersfeld commented on this "fundamental relativism."

> Altogether, the subjective structures of knowledge, therefore, are subjective constructions functioning as viable models which have been formed through adaptations to the resistance of "the world" and through negotiations in social interactions. This *triadic nature of human knowledge* makes impossible an ascription of causes, which would dissect internal from external causations (Seiler, 1984; Seiler & Wannenmacher, 1983). The separation for analytical purposes may be necessary, but is helpful only provided the researcher does not lose sight of the fundamental inseparability. (Bauersfeld, 1988, p. 39)

While we are quick to use the word *theory* in discussing issues in mathematics education, we would be wise to view theory as something other than a monolithic concept rooted in a notion of objectivity defined by a sense of reality. Snow (1983) maintains that theory has many forms, ranging from a set of well-defined propositions as suggested by "traditional" science, to conceptual analyses, even to the inclusion of metaphors that reflect and influence our thinking. Given the nature of our field, it is difficult to imagine that theory in mathematics education is likely to result in a set of interdependent propositions. In fact, we might be wiser to conceptualize theory development as an exercise in revealing the human ingenuity, insight, and compassion of which Feyerabend (1988) speaks.

Consistent with the notion that theory in mathematics education is likely to be eclectic is the notion that stories (e.g., anecdotes, case studies) play an integral role in communicating what we learn from research. From this perspective of science, research is more akin to understanding the transformation of Van Gogh's beliefs and values as his paintings shifted from bright sunflowers to tortured landscapes, to understanding Goethe's motivation and needs as revealed in Eissler's insightful analysis of his psyche, to appreciating Janos Bolyai's mental state following his rebuff by Gauss and his ultimate rejection of mathematics as a field of inquiry than it is to describe and predict behavior through quantified generalizations. From such a perspective of science, the central issue of research on the teaching of mathematics and on teacher education becomes one of describing how teachers ascribe meaning to their lives in the classroom and how that meaning contributes to the selection of some teaching behaviors and the rejection of others. This is not to say that quantification does not play a role in coming to understand how teachers construct meaning. Indeed, the most enlightening research often consists of thick descriptions punctuated by statistical data. Lortie's (1975) classic study the *School Teacher* represents such a blending of qualitative and quantitative data that foreshadowed the blending of methodologies used in many of the case studies being conducted today. Nevertheless, the issues raised here do encourage us to consider that the notion of being scientific and developing theory may be much more problematic than it might at first appear to be.

3. WHAT WE HAVE LEARNED ABOUT TEACHING AND TEACHER EDUCATION

Teaching and teacher education are inherently practical matters, which is not to say that both cannot be improved through the practice of science, broadly interpreted. Consider, for example, a project conducted at the University of Wisconsin, called *Cognitively-Guided Instruction* (CGI), which has a teacher education component based on a research program that focuses on students' higher-order thinking skills. This project has generated an extensive body of research findings on young children's higher-order thinking skills, which have, in turn, been used as a basis for conducting in-service programs for 1st- and 2nd-grade teachers. Although the nature of the teacher education experience is not entirely clear, teachers were better able to adapt instruction to meet students' cognitive needs when given explicit information about how children learn mathematics (Peterson, 1988).

With respect to research in teacher education *per se*, Weiss, Boyd, and Hessling (1990) surveyed final reports from in-service projects to the National Science Foundation and interviewed project directors and found that in-service programs help teachers develop a richer knowledge base for teaching, which, in turn, seemed to promote a more open-ended teaching style. This was particularly true for teachers from largely minority or urban schools. The mostly anecdotal evidence indicates that teachers who participated in in-service programs were less likely to see the textbook as the sole determinant of the instructional program. Further, the teachers developed an increased sense of professionalism and became influential partners for other teachers in their schools and school districts. There is not much analysis of why these changes occur except that they seem related to the teachers' perceptions of themselves as professionals rather than any particular format for the in-service programs.

One of the intriguing notions embedded in teacher education programs is the relationship between teachers' knowledge of mathematics and their ability to teach mathematics. It is difficult to imagine a reasonable argument that a sound knowledge of mathematics is not related to developing a quality instructional program, albeit the documentation of this relationship remains elusive. (see Begle, 1968; Eisenberg, 1977). There is no shortage of evidence (e.g., Fisher, 1988; Graeber, Tirosh, & Glover, 1986; Mayberry, 1983; Wheeler & Feghali, 1983) that many elementary teachers lack the mathematical sophistication necessary to promote the kind of reform being called for by the National Council of Teachers of Mathematics (NCTM, 1989, 1991). While the documentation that elementary teachers lack an understanding of topics such as ratio and proportion, geometry, measurement, and number relationships is not unusual, it begs the question of how this lack of understanding influences instruction or inhibits reform. Although there is little evidence about the relationship of elementary teachers' knowledge of mathematics to the way mathematics is taught, such information seems critical to considering the means by which the problem can be addressed in teacher education programs. There can be little doubt that teacher education programs can increase a teachers' knowledge of mathematics. But, if the means of achieving this goal is inconsistent with the instructional process deemed necessary to impact on children, then what have we gained? Too often the medium belies the message as we try to "give" teachers mathematics, failing to realize that the teacher receives two messages: knowledge gained and the *means* by which it was gained. If teachers are asked to learn mathematics through a process of transmission, then there is an increased probability that they will come to believe that their students will also learn through the transmission process – a position counter to meaningful reform.

At the secondary level, there is virtually no research on the relationship between a teachers' knowledge of mathematics, other than the coarse method of defining one's knowledge of mathematics in terms of courses taken, and the teaching of mathematics. Indeed, it is highly doubtful that any meaningful statistical relationship will emerge between any reasonable measure of teachers' knowledge and the nature of instruction. There is evidence, however, that what a teacher thinks about mathematics *is* related to the way mathematics is taught. Hersh put it the following way:

> One's conception of what mathematics is affects one's conception of how it should be presented. One's manner of presenting it is an indication of what one believes to be most essential in it . . . . The issue, then, is not, What is the best way to teach? but What is mathematics really all about? (Hersh, 1986, p. 13)

A series of studies conducted at the University of Georgia by Thompson (1982), McGalliard (1983), Brown (1985), Kesler (1985), Henderson (1988), and Jones (1990) reveals that many teachers communicate a limited view of mathematics. Although it is not clear whether the teachers held a limited view of mathematics or whether the ethos of the classroom encouraged the communication of a limited view, the question seems moot when you consider the effect on students. Too, the issue is not just the mathematics that is taught, but the mathematics that is *assessed.* Cooney (1992) conducted a survey of 201 middle school and secondary school mathematics teachers' evaluation practices in which the teachers were asked to create an item that assessed a minimal understanding of mathematics and an item that assessed a deep and thorough understanding of mathematics. More than one-half (57%) of the teachers created computational items in response to a question about assessing a deep and thorough understanding of mathematics. The following items were typical of such responses:

1. $4\frac{1}{3} + 2\frac{2}{5} = ?$
2. Solve for x: $6x - 2(x + 3) = x - 10$
3. How much carpet would it take to cover a floor that is 12.5 ft by 16.2 ft?

These teachers conflated the notion of difficulty with the notion of assessing a deep and thorough understanding of mathematics. Teachers of below-average students were particularly likely to give computational items to assess what they considered a deep and thorough understanding of mathematics. Again, we can only conjecture whether this circumstance reflected the teachers' limited view of mathematics, or whether the conditions in the classroom mandated the use of computational items given the oft asked question by students, "Will this be on the next test?"

Studies by Helms (1989), Owens (1987), and Wilson (1991) suggest that beliefs about mathematics and the teaching of mathematics are rooted in experiences long before the teachers encounter formal training in mathematics education. Further, these beliefs do not change dramatically without significant intervention (Ball, 1988; Bush, 1983). Lappan et al. (1988) addressed the issue of changing teachers' style of teaching through an extensive in-service program. They found that a 2-week summer workshop was sufficient for the teachers' to learn the information presented, but clearly insufficient for them to transform that knowledge into viable teaching strategies. They concluded that this complex issue of transformation requires a sustained in-service program of at least 2 years duration in which teachers are provided not only technical assistance in using the project's materials but also intellectual and emotional support as well. When growth was exhibited, it seemed to involve the increased confidence that the teachers gained in dealing with more exploratory teaching situations.

Over a decade ago, Bauersfeld (1980) argued that teaching and teacher education are inherently social matters and, consequently, that change in the teaching of mathematics can only occur through the reflective act of conceptualizing and reconceptualizing teaching. In short, our beliefs about teaching are shaped by social situations and therefore can only be reshaped by social situations. Attending to this circumstance in a teacher education program involves far more than providing field experiences – the typical solution. It involves analysis and reflection, a coming to realize that learning – both the teachers' and the students' – is a function of context. This is not to say that the professional development of teachers is somehow based on generic notions about teaching and learning. Indeed, our ability to be reflective is necessarily rooted in what we understand about mathematics, psychology, and pedagogy.

Wittmann (1992) has argued that the formalism of mathematics itself encourages a broadcast metaphor of teaching in which the primary task of the teacher is to make the lectures clear and connected so that the student can absorb an appreciation and understanding of mathematical structure. A few years ago, I interviewed a mathematician who emphasized mathematical structure in his classes and maintained that his lectures could help students see mathematics come alive. Although he appreciated the formalistic nature of mathematics, he failed to realize the incongruity that exists in trying to make something come alive through a passive medium such as broadcasting information. One could argue that the question of what constitutes mathematics and where it resides (in the mind or on the paper) is largely philosophical. I maintain that, in terms of the teaching of mathematics, the real issue is what teachers believe about mathematics and how they envision their role as teachers of mathematics. Indeed, the "philosophical" debate plays itself out every day in classrooms around the world as teachers struggle to help kids learn mathematics. This suggests that considerable attention needs to be given to how beliefs are formed and how effective interventions can be created to help break the cycle of teaching by telling.

Somehow, as a profession, we seemed to lose sight of the importance of meaning that highlighted the work of such people as Brownell (1945) when we accepted the premise that science, narrowly defined, could reveal effective ways of teaching mathematics. More recently, we are again emphasizing meaning in research, particularly that involving classroom situations (see, e.g., Yackel, Cobb, Wood, Wheatley, & Merkel, 1990). Despite this apparent maturity in our profession and the fact that we seem to be asking questions that strike at the heart of what it means to teach and to learn mathematics, progress in teacher education is much less apparent. Nevertheless, we have at least come to realize that teachers are not tabula rasa, that a knowledge of mathematics alone is not sufficient to insure change in the classroom, and that change evolves over time.

4. THE NOTION OF AUTHORITY

An issue of importance to almost all beginning teachers, especially at the secondary level, and to many experienced teachers as well, is that of classroom management. While the authority of a teacher is a legitimate concern, there is, unfortunately, a certain conflation between interpreting teachers' authority as the responsibility for the physical well-being of students and as the legitimizing agent for the mathematics being taught. A teacher who encourages students to think creatively and who promotes a problem-oriented approach to the teaching of mathematics will encounter, by definition, a greater number of unpredictable moments in the classroom – thereby making the use of open-ended teaching methods somewhat risky. The difficulty is that when a teacher's authority is translated into defining the quality of mathematical thinking, the students' goals become defined in terms of social outcomes rather than cognitive ones (Bauersfeld, 1980; Cobb, 1986). In many classrooms, the teacher plays a dual role for students: the authority figure and the determiner of mathematical truth. This creates a certain blurring between social goals and mathematical goals; the better student is perceived as the one who produces answers the teacher desires.

Scholars such as Rokeach (1960) and Perry (1970) have addressed the role of authority as one defines his or her relationship to the world. Although differences exist, both take the position that when authority is defined external to the individual, a dogmatic state exists. This state accentuates the development of what Green (1971) calls nonevidentually held beliefs, that is, beliefs immune from rational criticism. The differences between nonevidentually and evidentually held beliefs and between dogmatism and rationality emphasize the distinction between indoctrination and teaching. Fundamentally, the issue is one of how a person comes to know something. In this sense, there is a certain inseparability between the mathematics that is taught and the means by which it is taught. This inseparability is often lost in our zeal to "train" or to "give" teachers whatever we deem their "deficiency" to be. It is a common trap for all teacher educators, as we fail to see the symmetry between what and how we teach teachers and what and how they teach their students.

In a recent methods course, we were doing an experiment in which we collected data, analyzed the data, generated an appropriate function to model the situation, and subsequently discussed the implication of this activity for teaching. At one point, a very enthusiastic preservice teacher proclaimed with both confidence and a sense of satisfaction, "I finally know the right way to teach mathematics!" It was a moment of both triumph and defeat. Triumph because she conveyed a sense of exuberance and understanding the function that modeled the data; defeat because she missed the more general point that the teaching of mathematics is problematic and cannot be reduced to any predetermined "right" way.

Our challenge as teacher educators is to create contexts in which teachers, at all levels of professional development, can envision teaching methods that reflect reasoning, problem-solving, communicating mathematics, and connecting mathematics to the real world (NCTM, 1989, 1991) and yet feel comfortable with their role as classroom managers. Given that some teachers expect a teacher education program to give them the "right way to teach," we face the difficult task of helping teachers realize the problematic nature of both mathematics and the teaching of mathematics, and that reliance on external authority encourages a passive view of teaching and learning that fails to honor the student's role in determining the validity of mathematical outcomes.

5. THE NOTION OF ADAPTATION

The notion of adaptation provides a means by which we can break the cycle of teaching by telling that permeates many classrooms. Von Glasersfeld's (1989) identification of the following two principles of constructivism: (a) Knowledge is not passively received but actively built up by the cognizing subject, and (b) the function of cognition is adaptive and serves the organization of the experiential world, not the discovery of ontological reality, focuses our attention on the importance of context in the creation of knowl edge. Von Glasersfeld's second principle, in particular, emphasizes the importance of context as individuals create their knowledge about either mathematics or the teaching of mathematics. As Kuhn (1970) has so persuasively argued, knowledge structures are necessarily contextual. The implication of this for teacher education is that acquiring new methods of teaching mathematics is necessarily and fundamentally connected to our conception of what it means to teach mathematics and what it is that we think mathematics is. For the preservice teacher, this may be the result of accumulated experiences as a student of mathematics; for the in-service teacher, conceptions are more likely rooted in what worked yesterday.

If we believe that teacher education should be an exercise in learning to be adaptive, then we can envision different kinds of teacher education programs than are typically the case. While the content of such programs may not differ, what does differ is the means by which this content is acquired. If we take seriously the notion that the way we learn is a significant factor in how we eventually teach, then we have the laid the groundwork for teachers becoming adaptive agents in the classroom. The shift being called for emphasizes the notion of "pedagogical power" as compared to "mathematical power" that is emphasized throughout the NCTM *Standards*. The notion of problem-solving involves identifying the conditions and constraints of a problem and subsequently considering ways of solving the problem. Pedagogical power also involves recognizing conditions and constraints (of a classroom situation), weighing the consequences of possible actions, and then deciding which course of action best addresses the situation in a particular classroom. Unlike solving a mathematical problem, however, pedagogical problem-solving results in a dynamic state – a process of searching for better classrooms.

Cooney (in press) has identified a number of activities that can move teachers along the continuum of reflection and adaptation. Suffice it to say here that any teacher education program interested in reflection and adaptation must begin with what teachers bring to the program and consider the means by which teachers can restructure what it is that they believe about mathematics and its teaching. This is not to diminish the importance of knowing mathematics, knowing how students learn, and being able to create different mathematical activities for students. It is, however, the orientation toward that knowledge that is of utmost importance. Further, it is unlikely that this orientation will be realized unless it is fostered and encouraged throughout the teacher education program.

6. CONCLUSION

Despite the fact that research is sometimes perceived by practitioners as being disjointed from the practice of schooling, it is often the case that research mirrors practice. This is particularly so for much of the research on teaching and teacher education. While such research may help us better understand some events, the strategy is inherently conservative. It tends to make practice better as we presently conceive it. On the other hand, if we think about the notion of being scientific as one of understanding how it is that teachers come to believe and behave as they do, then we have positioned ourselves for creating contexts in which teachers can consider the consequences of their teaching. From this perspective, we can encourage the teacher to become scientific in the sense that they, too, can engage in the process of understanding why their students behave as they do. This orientation casts the teacher as an adaptive agent, that is, as one who sees his or her task as one of adapting instruction to be consistent with their students' thinking and to enable students to provide their own rationale as to why certain mathematical generalizations are true or not. That is, the teacher plays the role of being the intellectual leader rather than the determiner of mathematical truth.

Currently, I am directing a project designed to help teachers develop and use alternate items and techniques in assessing their students' understanding of mathematics. One of the teachers provided the following analysis as she compared her former test questions with the current ones.

Interestingly, this change was affecting her teaching as well. She felt that she had "a responsibility to train the students to use these items in class so that they would be prepared for the tests." Hence, her teaching became punctuated with asking students to explain why something was or was not the case, to create examples to satisfy certain conditions, and to explore different ways of solving problems. What a marvelous testimony to a teacher becoming an adaptive agent using assessment as the vehicle for change.

Traditional Questions	Alternate Test Questions
1. Find the GCF of 12 and 15?	1.Why can't 12 be the GCF of 6 and 18?
2. Round each of the following to the nearest tenth: a. 48.128 b. 1399.92435 c. 4.2378	2. Create a number that would round to be each of the following: a. 54.32 b. 853.8 c. 91.998
3. Categorize each of the following numbers as prime or composite: 81, 2, 53, and 111.	3. Buddy thinks that 91 and 93 are prime because they end in an odd digit. Pete says that Buddy is wrong. Who is correct and why?

Another project teacher provided the following analysis with respect to the question:

Is it possible for an equilateral triangle to have a right angle? If so, give an example. If not, why not?

Level One: Yes. Sides are straight at a right angle.
Level Two: Yes, as long as all of the sides are the same length.
Level Three: No, because all sides must be equal.
Level Four: (a) No, because there must be one side of the triangle (hypotenuse) that is longer in a right triangle and equilateral has all sides the same.
(b) No, all the angles have to be the same and all three have to equal 180 degrees.
Level Five: (a) No, you can't have 3 right angles because the sum of the angles would be 270 degrees and it must equal 180. The angle measure are all the same in an equilateral triangle.
(b) No, because an equilateral triangle has all the same angles. If you had a triangle with 3 right angles, you would have 3/4 of a square of the sides would not connect.

Argue as we might about how the students' responses could have been categorized, what is indisputable is that the teacher had to make judgments about the quality of students' thinking. This is a far cry from judging the correctness of computational items as was typically the case in the survey cited earlier (Cooney, 1992).

What we need are descriptions, stories, about what influences teachers, how they can become adaptive agents, and what forms of teacher education facilitate an adaptive orientation toward teaching. As part of a research and development project, we have been conducting case studies about how preservice secondary teachers have interacted with materials on mathematical functions. Wilson (1991) has found, for example, that it is easier to impact on teachers' knowledge and beliefs about mathematics than it is to influence their knowledge and beliefs about the teaching of mathematics. We need a deeper understanding of the process by which teachers learn to teach so that we can have a better basis for developing teacher education programs. Appropriately defined and applied, science can enable us to develop this understanding and allow us to impact on the practical art of teaching and teacher education in a way not foreseen by Highet and many of our professional forefathers who ascribed to an analytical view of science.

REFERENCES

Ball, D. L. (1988, April). *Prospective teachers' understanding of mathematics: What do they bring with them to teacher education?* Paper presented at the annual meeting of the American Educational Research Association, New Orleans, LA.

Bauersfeld, H. (1988). Interaction, construction, and knowledge: Alternative perspectives for mathematics education. In D. Grouws, T. Cooney, & D. Jones (Eds.), *Perspectives on research on effective mathematics teaching* (pp. 27-46). Reston, VA: National Council of Teachers of Mathematics.

Bauersfeld, H. (1980). Hidden dimensions in the so-called reality of a mathematics classroom. *Educational Studies in Mathematics, 11*, 23-41.

Begle, E. G. (1968). Curriculum research in mathematics. In H. J. Klausmeier & G. T. O'Hearn (Eds.), *Research and development toward the improvement of education* (pp. 44-48). Madison, WI: Dembar Educational Research Services.

Brophy, J. E. (1975, November). *Reflections on research in elementary schools.* Paper presented at the conference on research on teacher effects: An examination by decision-makers and researchers, University of Texas, Austin, TX.

Brown, C. A. (1985). *A study of the socialization to teaching of beginning secondary mathematics teachers.* Unpublished doctoral dissertation, University of Georgia, Athens, GA.

Brown, S. I., Cooney, T. J., & Jones, D. (1990). Mathematics teacher education. In W. R. Houston, M. Haberman, & J. Sikula (Eds.), *Handbook of research on teacher education* (pp. 639-656). New York: Macmillan.

Brownell, W. A. (1945). When is arithmetic meaningful. *Journal of Educational Research, 38*, 481-498.

Bush, W. (1983). Preservice secondary mathematics teachers' knowledge about teaching mathematics and decision-making during teacher training (Doctoral dissertation, University of Georgia, 1982). *Dissertation Abstracts International, 43*, 2264A.

Cobb, P. (1986). Contexts, goals, beliefs, and learning mathematics. *For the learning of mathematics. 6*(2), 2-9.

Cooney, T. (in press) Teacher education as an exercise in adaptation. In D. Aichele (Ed.), *NCTM yearbook on teacher education*. Reston, VA: National Council of Teachers of Mathematics.

Cooney, T. (1992). *A survey of secondary teachers' evaluation practices in Georgia.* Athens, GA: University of Georgia.

Davis, R. B. (1967). *The changing curriculum: Mathematics.* Washington, DC: Association for Supervision and Curriculum Development, NEA.

Eisenberg, T. A. (1977). Begle revisited: Teacher knowledge and students achievement in algebra. *Journal for Research in Mathematics Education, 8*, 216-222.

Feyerabend, P. (1988). *Against method.* New York: Verso.

Fisher, L. C. (1988). Strategies used by secondary mathematics teachers to solve proportion problems. *Journal for Research in Mathematics Education, 19*, 157-168.

Gage, N. (1972). *Teacher effectiveness and teacher education: The search for a scientific basis.* Palo Alto, CA: Pacific Books.

Gallagher, J. J . (1970). Three studies of the classroom. In J. J. Gallagher, G. A. Nuthall, & B. Rosenshine (Eds.), *Classroom obsservation.* American Educational Research Association Monogaraph Series on Curriculum Evaluation, Monograph No. 6. Chicago: Rand McNally.

Glasersfeld, E. von (1987). *The construction of knowledge.* Seaside, CA: Intersystems Publications.

Glasersfeld, E. von (1989). Constructivism in education. In T. Husen & N. Postlethwaite (Eds.), *International encyclopedia of education* (pp. 162-163). (Supplementary Vol.). Oxford: Pergamon.
Graeber, A., Tirosh, D., & Glover, R. (1986). Preservice teachers' beliefs and performance on measurement and partitive division problems. In G. Lappan & R. Even (Eds.), *Proceedings of the Eighth Annual Meeting of the North American Chapter of the International Group for the Psychology of Mathematics Education* (pp. 262-267). East Lansing, MI: Michigan State University.
Green, T. (1971). *The activities of teaching*. New York: McGraw-Hill .
Helms, J. M. (1989). *Preservice secondary mathematics teachers' beliefs about mathematics and the teaching of mathematics: Two case studies*. Unpublished doctoral dissertation, University of Georgia, Athens, GA.
Henderson, E. M. (1988) *Preservice secondary mathematics teachers' geometric thinking and their flexibiltiy in teaching geometry*. Unpublished doctoral dissertation, University of Georgia, Athens, GA.
Hersh, R. (1986). Some proposals for revising the philosophy of mathematics. In T. Tymoczko (Ed.), *New directions in the philosophy of mathematics* (pp. 9-28). Boston: Birkhauser.
Highet, G. (1950). *The art of teaching*. New York: Vintage Books.
Jones, D. L. (1990). *A study of the belief systems of two beginning middle school mathematics teachers*. Unpublished doctoral dissertation, University of Georgia, Athens, GA.
Kesler, T. (1985). *Teachers' instructional behavior related to their conceptions of teaching and mathematics and their level of dogmatism: Four case studies*. Unpublished doctoral dissertation, University of Georgia, Athens, GA.
Kuhn, T. (1970). *The structure of scientific revolutions* (2nd ed.). Chicago: University of Chicago Press.
Lappan, G., Fitzgerald, W., Phillips, E., Winter, M. J., Lanier, P., Madsen-Nason, A., Even, R., Lee, B., Smith, J., & Weinberg, D. (1988). *The middle grades mathematics project. The challenge: Good mathematics – taught well* (Final report to the National Science Foundation for Grant #MDR8318218). East Lansing, MI: Michigan State University.
Lortie, D. C. (1975). *School teacher: A sociological study*. Chicago: University of Chicago Press.
Mayberry, J. (1983). The Van Hiele levels of geometric thought in undergraduate preservice teachers. *Journal for Research in Mathematics Education, 14*, 50-59.
McGalliard, W. (1983). Selected factors in the conceptual systems of geometry teachers: Four case studies (Doctoral Dissertation, University of Georgia, 1982). *Dissertation Abstracts International, 44*, 1364A.
Mitroff, I., & Kilmann, R. (1978). *Methodological approaches to social sciences*. San Francisco: Jossey-Bass.
National Council of Teachers of Mathematics. (1989). *Curriculum and evaluation standards for school mathematics teaching*. Reston, VA: National Council of Teachers of Mathematics.
National Council Of Teachers Of Mathematics. (1991). *Professional standards for the teaching of mathematics*. Reston, VA: National Council of Teachers of Mathematics.
Owens, J. (1987). *A sudy of four preservice secondary mathematics teachers' constructs of mathematics and mathematics teaching*. Unpublished doctoral dissertation, University of Georgia, Athens, GA.
Perry, W. (1970). *Forms of intellectual and ethical development in the college years: A scheme*. New York: Holt, Rinehart, & Winston.
Peterson, P. L. (1988). Teaching for higher-order thinking in mathematics: The challenge for the next decade. In D. Grouws, T. Cooney, & D. Jones (Eds.), *Perspectives on research on effective mathematics teaching* (pp. 2-26). Reston, VA: National Council of Teachers of Mathematics.
Peterson, P. L., & Clark, C. M. (1978) Teachers' reports of their cognitive processes during teaching. *American Educational Research Journal, 15*, 555-565.
Rokeach, M. (1960) *The open and closed mind*. New York: Basic Books.

Rosenshine, B., & Furst, N. (1973). The use of direct observation to study teaching. In R. Travers (Ed.), *Second handbook of research on teaching* (pp. 122-183). Chicago, IL: Rand McNally.
Seiler, T., & Wannenmacher, W. (Eds.). (1983). *Concept development and the development of word meaning.* New York: Springer.
Seiler, T. B. (1984). Was ist eine "konzeptuell akzeptable Kognitionstheorie"? Anmerkungen zu den Ausführungen von Theo Herrmann: Über begriffliche Schwächen kognitivistischer Kognitionstheorien. *Sprache & Kognition, 2,* 87-101.
Snow, R. E. (1983). Theory construction for research on teaching. In R. W. Travers (Ed.), *Second handbook of research on teaching* (pp. 77-112.). Chicago, IL: Rand McNally.
Thompson, A. (1982). *Teachers' conceptions of mathematics and mathematics teaching: Three case studies.* Unpublished doctoral dissertation. University of Georgia. Athens, GA.
Thompson, A. (1992). Teachers' beliefs and conceptions: A synthesis of the research. In D. Grouws (Ed.), *Handbook of research on mathematics teaching and learning* (pp. 127-146). New York: MacMillan.
Weiss, I. R., Boyd, S. E., & Hessling, P. A. (1990). *A look at exemplary NSF teacher enhancement projects.* Chapel Hill, NC: Horizon Research.
Wheeler, M. M., & Feghali, I. (1983). Much ado about nothing: Preservice elementary school teachers' concept of zero. *Journal for Research in Mathematics Education, 14,* 147-155.
Wilson, M. R. (1991). *A study of three preservice secondary mathematics teacher's knowledge and beliefs about mathematical functions.* Unpublished doctoral dissertation, University of Georgia, Athens, GA.
Wittmann, E. (1992). One source of the broadcast metaphor: Mathematical formalism. In F. Seeger & H. Steinbring (Eds.), *The dialogue between theory and practice in mathematics education: Overcoming the broadcast metaphor. Proceedings of the Fourth Conference on Systematic Cooperation between Theory and Practice in Mathematics Education (SCTP).* Brakel, Germany (pp. 111-119). IDM Materialien und Studien 38. Bielefeld: Universität Bielefeld.
Yackel, E., Cobb, P., Wood, T., Wheatley, G., & Merkel, G. (1990). The importance of social interaction in children's construction of mathematical knowledge. In T. J. Cooney & C. R. Hirsch (Eds.), *Teaching and learning in the 1990s* (pp. 12-21). Reston, VA: National Council of Teachers of Mathematics.

CHAPTER 3

INTERACTION IN THE CLASSROOM

edited and introduced
by
Rudolf Sträßer
Bielefeld

While Chapter 2 on *teacher education and research on teaching* took the principal agent inside the classroom – the teacher – as the focus of the papers and thus analyzed one pole of the "didactical triangle" (the teacher, the student, and the knowledge (to be) taught/learned, i.e., the didactical system in a narrow sense), chapter 5 on the *psychology of mathematical thinking* can be taken as an attempt to analyze the second human pole of this triangle. This chapter 3 on *interaction in the classroom* focuses on research concerned with communication and social interaction processes in mathematics teaching and learning. Concentrating on the interaction of the human agents does not just provide a link between chapter 3 on the teacher and chapter 5, which concentrates on the student, the learner. These perspectives also provide new insights into problems of teaching and learning that could not have been gained from the reduced perspectives. Research on teachers and teacher cognition already spread in the context of the modern mathematics reform movement in the late 1960s and early 1970s. Research on student's cognition has even a much longer tradition. Detailed studies on classroom

R. Biehler, R. W. Scholz, R. Sträßer, B. Winkelmann (Eds.),
Didactics of Mathematics as a Scientific Discipline, 117-120.

interaction, however, had to wait until the second half of the 1970s and were – at least partly – undertaken to understand and explain the "failure" of this movement in the so-called industrialized countries. In the 1980s, research on classroom interaction gained momentum with large research programs being funded and growing attention being gained in the research community. Because of the wealth of this field, some pertinent topics are not treated separately in this chapter. For example, the most important question of research methodology is discussed in each of the papers at least implicitly, but is not given a separate place. The first two papers of the chapter (Bartolini-Bussi and Bauersfeld) can serve as an illustration of a second most important distinction in the field: the complementarity of supporting innovations in mathematics teaching and of constituting a body of reliable knowledge on the teaching/learning process in the mathematics classroom. The two papers present two different research approaches and two different paradigm choices and by doing so throw light on the methodology issue.

In *Theoretical and empirical approaches to classroom interaction, Maria Bartolini-Bussi* starts by sharply marking two contrasting approaches: an approach called "recherches en didactique des mathématiques (RDM)" and "research on innovation (RI)." RDM is presented as an attempt to describe the functioning of didactical situations with the researcher acting as a detached observer of the didactical system. This approach aims at building a coherent theory of phenomena of mathematics teaching, with conditions of reproducibility in the teaching experiments as a major requirement on the research results. It is oriented toward *knowledge,* while "research on innovation (RI)" is oriented toward *action,* interested in the introduction of examples of good didactical transpositions and the analysis of the resulting processes. It aims at producing tools (either adapting them or constructing by itself) to transform directly the reality of mathematics teaching. Knowledge-oriented RDM is supposed to ignore the results of the action-oriented RI, while RI can borrow results from the former because of its intrinsic eclecticism. In her paper, Bartolini-Bussi explicitly describes research in support of innovation in mathematics teaching, while, implicitly, Bauersfeld writes from a perspective that takes knowledge production as the most important aim, and teaching innovations as desired and most welcome side effects. Bartolini-Bussi analyzes and compares Piagetian constructivism and Vygotskyan activity theory. She is searching for adequate theoretical tools for performing research in the RI tradition. She presents research examples from elementary mathematics education that were mainly based on an activity theoretical basis but in which conceptual elements from other theoretical traditions were also applied to cope with the complexity of an innovation – not hiding her preference for activity theory as the foundation of her work.

Heinrich Bauersfeld's contribution on *theoretical perspectives on interaction in the mathematics classroom* also starts with an overview of existing theoretical paradigms with activity theory and constructivism as two major strands, but then opts for a third perspective, analyzing the interaction in a mathematics classroom from an "interactionist" point of view. Presenting "interactionism" as a mediating approach, Bauersfeld clarifies the core convictions of this position on learning, meaning, languaging, knowing or remembering, and mathematizing. He shows consequences for the issues of understanding mathematics and language within elementary education in mathematics. As an outlook, Bauersfeld sketches how the recent transdisciplinary concern for "connectionism" may shed new light and explain some convictions of the interactionist perspective. However, connectionism is also taken as an example that theories in didactics of mathematics continually take advantage of new theoretical developments in other related sciences.

Nevertheless, an exhaustive discussion of the problems and potentials of the knowledge-versus-action controversy is still missing. Is it possible to follow a knowledge-oriented approach within the activity-theory paradigm, or can an action-oriented approach be founded on the constructivist research paradigm? Answers to these questions cannot be found in this volume.

The two other papers in this chapter analyze two special aspects of interaction in the mathematics classroom. In her paper, *Working in small groups: A learning situation?*, *Colette Laborde* starts from the perspective of the knowledge-oriented approach and analyzes the efficiency of a special learning situation: the case of students working together at a joint task of finding a common solution to a mathematical problem. The paper elicits the role of interpersonal processes in the construction of mathematical knowledge in mathematics classrooms and tries to determine some variables affecting these processes. The teacher (as a person) is only marginal in the learning situation, while special attention is given to joint work at the computer. Within this "ecology," she analyzes a learning situation that is of growing importance: Project work and home work often are done in small groups, and most computer-assisted learning takes place with two or three students in front of one computer. The role of the teacher may be taken over by a task to be fulfilled or a problem to be solved. Research on this arrangement is shown to produce contradictory results on its effectiveness as compared to a traditional classroom setting with three major factors for the effectiveness of cooperative work: choice of partners, choice of tasks, and length of the interaction process. A common feature in this research is the learner's charge to cope with the social situation as an additional demand to subject-matter learning in mathematics. The social complexity of the learning situation is shown as a problem as well as an additional potential for learning.

The paper *Mathematical classroom language: Form, function and force* by *David Pimm* concentrates on the most important means of the interaction in the classroom: language. Apart from other – and rarely used – physiological measures (e.g., eye movements) and test procedures (like multiple-choice testing), language seems to be the best analyzed set of "data" in didactics of mathematics. The paper first offers a survey of some recent work on mathematical classroom language in the context of work on language and mathematics in general. A few research results from the different linguistic aspects of classroom language (reading, writing, listening, and discussing) are presented, followed by research on the form of the mathematical communication in classrooms. Analysis of the almost incessant repetition of the sequence of initiation – response – feedback in teacher-student exchanges is taken as an example for discourse analysis techniques that ignore content and attend only to the form of the classroom language. Two alternative routes from informal spoken to formal written language are distinguished and commented on. Following this survey of research on language, Pimm discusses a more idiosyncratic and personal set of interests and emphases: meta-knowledge and meta-communication, modality, and "hedges" and "force," the inner purposes and intentions of the speaker. The paper finishes with some suggestions for future areas of important work yet to be done.

On the whole, the four papers of this chapter show the potential of concentrating on the interaction of teachers and students. The papers of C. Laborde and D. Pimm widen this perspective still further by commenting on special aspects of the "ecology" of this interaction: computers and language, by analyzing the most important means of representation and communication of mathematics. Chapter 4 on *technology and mathematics education* presents a complementary approach to questions raised in this chapter, in that it concentrates on means of teaching and learning.

THEORETICAL AND EMPIRICAL APPROACHES TO CLASSROOM INTERACTION

Maria G. Bartolini Bussi
Modena

1. INTRODUCTION

In recent years, the study of classroom interaction in the mathematics teaching-learning process has received more and more attention in the literature on didactics of mathematics: Whenever at least two persons are engaged (e.g., two students or a teacher and a student), factors depending on their mutual interaction are involved. It is opportune to attempt an overview of related literature: The whole spectrum of research is very broad and ranges from *analyses* of existing situations in standard classrooms (for a review of German literature, see Maier & Voigt, 1992) to studies of *transformation* of the teaching-learning process. I recognize the importance of the first kind of study to make both teachers and researchers aware of the existence of an implicit ideology of teaching as well as of the power of some hidden interaction rules. The above studies act, so to speak, as *demolishers of illusion* (ICMI, 1993) and are both a backdrop and an incentive for other studies. Yet, in my paper, I shall consider other kinds of study that are supposed to be more pragmatic (yet not at all atheoretical, as I shall argue in the following), because they are based on designing, implementing, and analyzing teaching experiments, in which the traditional implicit rules of interaction and the underlying ideology are voluntarily and systematically substituted by different explicit ones.

I shall be concerned with two issues, which need to be discussed before any tentative overview of literature: the function of theoretical assumptions (section 2) and the effects of choosing among different theoretical elaborations (section 3). The former is prior to any choice, while the latter concerns just the choice of a theory of learning. The aim of this paper is to elaborate Steiner's (1985) claim for *complementarity* on both issues from the perspective of my research on the relationship between social interaction and knowledge in the mathematics classroom (Bartolini Bussi, 1991).

R. Biehler, R. W. Scholz, R. Sträßer, B. Winkelmann (Eds.),
Didactics of Mathematics as a Scientific Discipline, 121-132.

2. PRIOR TO FRAMEWORKS: THEORY AND PRACTICE IN THE STUDY OF SOCIAL INTERACTION

2.1 Research For Knowing Versus Research For Acting Purposes

Two contrasting perspectives are represented by the so-called *Recherches en didactique des mathématiques* (Douady & Mercier, 1992), which are peculiar to the French community (referred to in the following as RDM), and by research on innovation (RI) developed in different countries (e.g., the *Purdue Problem Centered Mathematics Project,* Cobb, Wood, & Yackel, in press; the *Genoa Project,* Boero, 1988, 1992; the *Mathematical Discussion in Primary School Project,* Bartolini Bussi, 1991).

The purpose of RDM, at least as regards its core (the *theory of didactical situations* by Brousseau, 1986), is to describe the *functioning of didactical situations.* The researcher acts as a *detached observer* of the didactical system and looks for conditions of *reproducibility* in the teaching experiments. The possibility of *falsification* is a criterion to judge the acceptability of results.

Research for innovation (RI) is not framed (it cannot be framed, as I shall argue in the following) by such a coherent theoretical approach as RDM. Its main purpose is to introduce examples of good didactical transpositions and to analyze the resulting processes. As reproducibility cannot be assured by the mere description of the teaching setting, it is substituted by gradual *expansion* to larger and larger groups of teachers. The possibility of *verification* is a criterion for the relevance of results.

The main difference is in the underlying motive for research. RDM aims at building a coherent theory of phenomena of mathematics teaching; RI aims at producing tools (either adapting them or constructing by itself) to transform directly the reality of mathematics teaching. RDM is oriented toward *knowledge of classroom processes,* while RI is oriented toward *action in classroom processes.* RDM is supposed to ignore the results of the latter, as they usually do not meet its criteria, while RI can borrow results from the former, because of its intrinsic eclecticism.

2.2. Action and Knowledge Reconciled

The development of different conceptions of didactics of mathematics is surely dependent on social and historical factors. The analysis of this issue could be the subject matter of comparative studies in the social history of didactics of mathematics. References to some documents (e.g., Barra, Ferrari, Furinghetti, Malara, & Speranza, 1992; Douady & Mercier, 1992; Schupp, Blum, Keitel, Steiner, Straesser, & Vollrath, 1992) reveals that national conditions of development are very different. The image of didactics of mathematics seems to suffer from *local* conditioning (Boero, 1988). However, when an image is built or in construction, criteria to judge the relevance of problems and acceptance of methodologies within a scientific community are given.

Balacheff (1990a) calls for a confrontation and discussion of theoretical research and research for innovation. In my opinion, this sounds difficult: What is in question is not only the nonexistence of a universal language in which to execute the critical comparison (which is involved whenever competing theories are confronted) but also the existence of different meanings of didactical research. I shall adopt Raeithel's (1990) description of three models of relationships between actor and observer in the enquiring activity: (a) the *naive problem solver* who considers the symbolic structure inseparable from the perceived reality; (b) the *detached observer*, who represents reality by means of symbolic models, and (3) the *participant observer*, who develops the split between observing and observed subject into a dialogical relation. The first concerns radical realizations of *action-research projects*, which consider innovation as an ideological value and reject the development of progressive knowledge of classroom processes; as they are ideologically atheoretical, I have not considered them in this paper. The second is realized, for instance, by theoretical research programs such as the core of RDM; they share some methodological aspects with classical natural sciences and with experimental psychology in laboratory settings. The third is realized by RI, which aims at turning into reality some examples of anticipated classroom processes. The responsibility for choices is shared by a larger group that comprises *at least* researchers and teachers (it could also include administrators, parent representatives etc.). It is important to distinguish between action-research projects, in which action is a value and an end in itself (Model 1), from innovative projects (Model 3), in which action is both a means and a result of progressive knowledge of classroom processes. *The core of RDM and the core of RI address different problems, answer different questions, and refer to different models of enquiring activity.*

The human need to turn theoretical elaborations into reality is represented in the French community by so-called *didactical engineering* (Artigue, this volume). It shares some features with RI: for instance, the attention paid to long-term processes. Yet they cannot be confused. The teacher's role in the development of research acts as a litmus paper. In didactical engineering, the split between the time of designing/analyzing (which occurs *outside the classroom*, maybe with the participation of teachers, too) and the time of acting (when teachers are observed by *detached observers*) seems radical; in innovative research, teachers, as full members of the research team, are allowed to take part in the observation of their own classroom as *participant observers* (Eisenhart, 1988) and to make decisions even *in the course of action* (Davis, 1992; Steffe, 1991). In other words, didactical engineering derives from RDM, and shares the same model of enquiring activity. It is possible and even desirable to try to coordinate results with RI, but it is necessary to first take into account the basic difference of perspectives.

3. INSIDE FRAMEWORKS: CONSTRUCTIVISM VERSUS ACTIVITY THEORY OR PIAGET VERSUS VYGOTSKY

3.1 Foundation Aspects

In every research project, some basic assumptions about learning are supposed to be shared by the research team, even when they are not stated explicitly. In the following, I shall sketch some contrasting issues from two major perspectives on the role of social interaction in the process of learning: *constructivism*, in its more or less radical forms, and *activity theory*. The former refers to Piaget and the latter to Vygotsky, so that a distinction could be made between Piagetian and Vygotskyan frameworks. The above distinction, like every radical "either-or" classification, does not give full justice to the complex reality of research. For instance, the so-called *Geneva school* (e.g., Perret-Clermont, 1980) tries to coordinate Piaget and Vygotsky; the *ethnomethodological perspective* is introduced into radical constructivism to study the culture of mathematics classrooms (e.g., Bauersfeld, 1988). Besides, *connectionist* models of the human mind have entered the scene, even if their appearance is too recent to judge their relevance for and influence on didactical research (a meaningful exception is reported in Bauersfeld, this volume). Because of this complexity, I shall adopt the previous distinction, in spite of its limits, to keep the discussion at the level of the large community of mathematics educators.

The most important difference between Piagetian and Vygotskyan approaches concerns just foundation aspects and is still the same difference that divided Piaget and Vygotsky in the 1930s. Constructivism considers learning as the result of two inseparable complementary processes of interaction between the individual and the environment: *assimilation,* that is, the process of integration of either new objects or situations into the existing individual schemes; and *accomodation,* that is, the individual effort to adjust schemes to the environment (Piaget, 1936). Activity theory is centred upon *internalization* or *interiorization,* understood (in contrast to Piaget) as the transformation of an *interpsychological* (i.e., between individuals) into an *intrapsychological* process (i.e., within individuals). To put it in a few radical words, the Piagetian approach is based on *individual schemes*, while the Vygotskyan approach is based on *social relations;* for Piaget, the learning process is determined from *inside,* for Vygotsky, it is determined from *outside.*

It is no surprise that the Piagetian approach fits in with the Western tendency in psychological research to study *human mental functioning as if it exists in a cultural, institutional, and historical vacuum* (Wertsch, 1991, p. 2), even if it would be misleading to ascribe to Piaget the whole responsibility for this trend. In fact, the focus on the individual also fits in with some underlying ideas: Consider, for instance, the myth of genius, which is present in popular books on the history of mathematics (Bell, 1937) as well as in the professional education of mathematicians (Eisenberg, 1991). These facts, together with the scarce, late, and biased diffusion of the original papers of Vygotsky may give an early explanation of the evident hegemony of the Piagetian approach in Western literature on didactics of mathematics. Yet, outline presentations of activity theory exist (e.g., Christiansen & Walther, 1986; Mellin-Olsen, 1987), and quotations from Vygotsky are more and more frequent in the literature.

I shall not present a detailed comparison of the two approaches, as this would first require a reconstruction of the conceptual structure of both. Besides, such critical comparisons already exist from either competing perspective (Bauersfeld, 1990; Raeithel, 1990). Rather, I shall describe some implications for the development of didactical research. More space shall be devoted to the Vygotskyan perspective, as it is supposed to be less well-known.

3.2. Implications for Research on Didactics of Mathematics

Because of its focus on the learning subject, the Piagetian approach tends to neglect the role of cultural tradition represented by the teaching subject. Artigue (1992) attributes the influence of Piaget on the development of RDM to the need to contrast the empirical-sensory or behaviorist theory of learning, *to put the student back in the right position.* The same reason could apply to other Western RI projects as well: Being Piagetian was considered as *the way* to overcome the behaviorist theory of learning. However, it was only one of the existing opportunities.

Vygotsky could have offered a different one. For Vygotsky, the process of learning is not separated from the process of teaching: the Russian word *obuchenie,* which is used throughout Vygotsky's work, means literally the *process of transmission and appropriation of knowledge, capacities, abilities, and methods of humanity's knowing activity; it is a bilateral process, that is realized by both the teacher and the learner* (for a discussion by Mecacci, see Vygotsky, 1990). The social relation between teacher and learner cannot be avoided, as learning is not a relation between individuals and knowledge, but is rather the individual's introduction into an existing culture. The implications for didactical research are very strong, especially as far as the teacher's role is concerned. The metaphorical space in which to study the interaction between teacher and learners is the so-called *zone of proximal development.* One of the basic processes is *semiotic mediation* (Vygotsky, 1978, p. 40), determined when the direct impulse of the learner to react to a stimulus is inhibited through the intentional teacher's introduction of a sign. The very effect is that learners, by the aid of extrinsic stimuli drawn by the teacher, may *control their behavior from outside.*

Rigid applications seldom give full justice to the richness and complexity of the original ideas of founders. Piaget (1962) tried to coordinate his ideas to Vygotsky, while Vygotsky himself was more Piagetian than his followers (van der Veer & Valsiner, 1991, p. 392). If we look at recent developments, a greater separation is evident. A recent publication (Garnier, Bednarz, & Ulanovskaya, 1991) presents a collection of studies on didactical research (not limited to mathematics) from either Western countries or Russia. The provocative heading is *Après Vygotsky and Piaget. Perspectives sociales et constructiviste. Ecoles russe et occidentale.* Even if contributions are limited to researchers from French-speaking countries (Western school) and from the Moscow Institute for Psychology and Pedagogy (Russian school), the book is very stimulating. The same position on social interaction as a founding element of individual development is shared, by means of direct derivation from Vygotsky, as regards the Russian researchers, and by means of the Geneva school, as regards Western researchers. Apart from that, the two schools have developed in relative isolation from each other. Differences are relevant: For instance, when problem-solving is concerned, the starting point is given, on the one side, by a general model proposed by the teacher to solve a general class of problems (Moscow school) and, on the other side, by a collection of students' early conceptualizations to be modeled (Western school). In the former case, group work itself is often structured on the basis of the analysis of the item of knowledge. In the latter case, group work is often organized to provoke cognitive conflicts between learners. The purposes are different: internalization of interpsychological activity as such versus restructuring of early conceptualizations. I do not wish to assume personal responsibility for criticizing the development of the Vygotskyan school in Russia on the basis of the very scarce documents available to a Western researcher. Yet, according to Engestrom (1991), *concrete research and experimentation inspired by activity theory has been strongly dominated by the paradigm of internalization* with a scarce emphasis on the individual's *creation*, which was carefully studied by Vygotsky in *The Psychology of Art.* According to Davydov (1991), who was a student and a colleague of Vygotsky, the very difference between individual and collective activity is still an unsolved problem of activity theory.

3.3 The Problem of Choice

As I have argued above, there was a parallel destiny for Piagetian- and Vygotskyan-oriented research. With the relevant exception of the Geneva school, which is nevertheless engaged in psychological and not in didactical research (e.g., more attention has been focused on peer interaction than teacher-learner interaction), both seem to have led to extreme consequences for the individual and the social foundation. Later, because of the establishment of two competing schools with rigid membership to be defended, the reference to some radical slogans seems to have substituted the reference to the original papers. So, which framework to choose?

The problem does not seem so dramatic, in a sense, in theoretical research. As often happens in the development of science, the selection of "narrow" pieces of reality to be modeled can solve the problem of both acceptable modeling and theoretical coherence: "Narrowness" could result, in turn, in a limitation of either the number of subjects involved, the duration of observation, or the items of knowledge. A good example is the *theory of didactical situations* (Brousseau, 1986), which is successful for microdidactical studies, in which a given item of knowledge and a given problem situation is considered; the teacher has paid a price, but recent developments are going to fill the gap (Margolinas, 1992).

The situation is different in innovation projects in which the impact with complex reality is strong and unavoidable from the very beginning. No coherent theoretical framework is supposed to be sufficient to manage such complexity as a whole.

I can give an example by referring to the *Mathematical Discussion in Primary School Project* that is in progress in my research group (Bartolini Bussi, 1991). Background ideas came from Piaget, who still exerts the major influence on pedagogics in Italy. Later, more and more ideas from activity theory crept over the research group: Their adaptation for classroom work was (and still is) tested continuously. For instance, we used the concept of *semiotic mediation* to model (either design or analyze) the process of inhibiting the student's reaction by means of a cultural tool (Bartolini Bussi, in press a). The concept of *internalization* was used to model some special aspects of long-term teaching experiments on the coordination of spatial perspectives, when the teacher directly proposes a dialogical model for the solution of a drawing task that is gradually transferred from the interpsychological to the intrapsychological plane (Bartolini Bussi, in press b). Last, but not least, *activity theory* by Leont'ev (1977) offered a powerful tool to model long-term studies (Bartolini Bussi, in press a). *Our project is not an application of activity theory, but an example of progressive interaction between theory and practice, by means of appropriating existing theoretical tools.* Besides, the reference to original papers (rather than to subsequent applications) is a defence against radicalization. Yet, our work has also retained some ideas inherited from the Piagetian framework. Not only cooperation but also *cognitive conflicts* are focused. The concept of *epistemological obstacle,* inherited from Bachelard and Piaget via Brousseau (1986), has been used to model a teaching experiment on Cartesian graphs (Bartolini Bussi, 1992) and is the object of a permanent activity carried on with students (the reconstruction of a personal as well as a collective history of solution for a class of problems). Moreover, the collection of *students' conceptions* is always performed by teachers by means of collective discussions that act as the basis for the following activity.

Actually, if we had to decide whether to be considered Vygotskyan or Piagetian, we would say Vygotskyan, but our perspective could be better described by referring to *complementarity:* We allow ourselves to refer to approaches that are even theoretically incompatible. Maybe it is not possible to be simultaneously Piagetian and Vygotskyan, to encourage students to express their own conceptions while introducing a sign for semiotic mediation. Yet, in the design of long-term studies, it is possible to alternate phases influenced by either a Piagetian or Vygotskyan perspective. The acceptance of alternating phases does not result in an equidistant position from Piagetian and Vygotskyan perspectives: *The will to renounce theoretical coherence in favor of relevance to problems of action is deeply Vygotskyan*, as Vygotsky, unlike Piaget, was not a theoretician, but a protagonist of the great social and cultural struggles of the 1920s and the 1930s in Russia (Mecacci, in Vygotsky, 1990 p. ix). A similar (even if not identical) position on complementarity seems to be shared by the teams of other innovation projects (see Bartolini Bussi 1991).

4. TWO EXAMPLES TO THINK OVER

4.1. When the Child Is Speechless

Teacher: *That's fine! What is it?*
(on the table, there is a three-dimensional small cat of folded and stapled paper, built by the teacher in advance)
Child: . . . (silence)
Teacher: *Do you know what it is?*
Child: *Paper.*
Teacher: *Look at it well, what is it?*
Child: *Eyes. . . . that's an eye.*
Teacher: *An eye.*
Child: *Nose, mouth.*
Teacher: *And what is this?*
Child: *The other small eye, that is whiskers.*
Teacher: *And that* . . . (she points at the body)
Child: *Legs.*
Teacher: *This part all together, what is it?*
Child: . . . (silence)
Teacher: *Okay. There are the eyes, the nose, the mouth, the whiskers.*
(she points at each one)
All together, what is it?
Child: *It is paper.*
Teacher: *What is it? not what is it made of? What's its name?*
Child: *It's written?*
Teacher: *No.*
Child: . . . (silence)
Teacher: *You have said that it has eyes, a nose, and so on. What is it?*
Child: . . . (silence)

Teacher: *Is it a child?*
Child: . . . (silence)

This episode is taken from the observation protocol of a one-to-one interaction between an elementary school teacher (Bondesan, personal communication) and a low achiever (1st grade): The child already knows the teacher and the climate is very relaxed. This special interaction (a remedial workshop) was designed for low achievers in order to foster the development of planning and designing strategies by means of verbal language as a prerequisite for mathematical problem-solving (Boero, 1992). The goal of this session is to build a copy of the puppet while verbalizing the process. The child is a 1st grader with learning disabilities; she is not handicapped, but she has lacked family experiences of joint activity in which action is systematically accompanied by speech. As the protocol shows, she can name the different parts of the object, but cannot name the whole. The teacher feels responsible for unblocking the child, because of institutional needs (the very purpose of that remedial workshop) and for personal needs (the "revolutionary" will to offer equal opportunities to every child). What has theory to offer her? Two radical competing positions are offered by Piagetian versus Vygotskyan researchers: act as a clinical interviewer, encourage the child to express herself and to build her own knowledge; act as a guide, help the child, lend her the right gestures and words. Actually, the teacher behaved as a Vygotskyan and successfully offered the child *actions and utterances to be imitated;* maybe, being Piagetian, in this radical sense, could have resulted in abandoning the child to her destiny.

4.2 When Mathematical Behavior is Against Everyday Behavior

The problem of mathematical proof seems to be one of the crucial issues of didactics where advanced thinking is concerned. Balacheff (1990b) studied the students' treatment of a refutation by means of social interaction. His work confirmed the usefulness of social interaction, but enlightened its limits too, because of the major role played by argumentation. In a specific study on deductive thinking, Duval (1991) showed that the rules of deductive reasoning are very different from the rules of argumentative reasoning. The strategy that the same author experimented successfully to make the students (aged 13-14) distinguish between argumentative and deductive reasoning is supposed to be more Vygotskyan than Piagetian (actually, in the paper, disagreement with Piaget is explicitly stated even if Vygotsky is not referred to): They were given the rules for building an oriented propositional graph, to connect hypotheses to conclusions (a good example of *semiotic mediation).* We could even be critical about such an introduction of rules to be followed if they are perceived by students as rules of classroom contract only. Yet what seems to me unquestionable is that deductive reasoning depends on social factors: When students are approaching mathematical proof, they are entering a flow of thought that was (and still is) developed outside school by mathematicians, together with a related system of values as well as of acceptable behaviors. To cope with this problem, it is not sufficient to consider mathematics as an individual subjective construction, it is necessary to consider *mathematics as a collective cultural and social process.*

5. CONCLUSION

The examples in the last section show that the Vygotskyan perspective is useful for studies on both low attainers and advanced learners. They have not been proposed to deny the usefulness of Piagetian analysis, but only to recall situations that seem to fit the Vygotskyan perspective. Maybe they can also be managed in a Piagetian framework, but the burden of proof rests on Piagetian researchers. Nevertheless I am not so sure that the game is worth the candle. As history of science teaches us, the exclusive long-term adhesion to one system could result in either ignoring relevant aspects of reality, if theoretical coherence gets the upper hand, or introducing into the system such complications as to make it no longer manageable, if the modeling of increasingly complex events is pursued.

It seems to me that the only solution is to accept *complementarity* as a necessary feature of theoretical and empirical research in didactics of mathematics and look for conceptual tools to cope with it successfully, as Steiner (1985) suggests in the developmental program of the international study group on *Theory of Mathematics Education*.

REFERENCES

Artigue, M. (1992). Didactical engineering. In R. Douady & A. Rouchier (Eds.), *Research in Didactique of Mathematics* (pp. 41-66). Grenoble: La Pensée Sauvage.

Balacheff, N. (1990a). Towards a problématique for research on mathematics teaching. *Journal for Research in Mathematics Education, 21*(4), 258-272.

Balacheff, N. (1990b). Beyond a psychological approach of the psychology of mathematics education. *For the Learning of Mathematics, 10*(3), 2-8.

Barra M., Ferrari M., Furinghetti F., Malara N. A., & Speranza F. (Eds.). (1992). The Italian research in mathematics education: Common roots and present trends. *Progetto Strategico del C.N.R. - Tecnologie e Innovazioni Didattiche, 12.*

Bartolini Bussi, M. (1991). Social interaction and mathematical knowledge. In F. Furinghetti (Ed.), *Proceedings of the 15th PME Conference, 1*, 1-16.

Bartolini Bussi, M. (1992). Mathematics knowledge as a collective enterprise. In F. Seeger & H. Steinbring (Eds.), *The dialogue between theory and practice in mathematics education: Overcoming the broadcast metaphor* (pp. 121-151). Materialien und Studien Band 38, IDM Bielefeld.

Bartolini Bussi, M. (in press a). The mathematical discussion in primary school project: Analysis of long term processes. In L. Bazzini & H.-G. Steiner (Eds.), *Proceedings of the Second Italian-German Bilateral Symposium on Didactics of Mathematics.*

Bartolini Bussi M. (in press b). Coordination of spatial perspectives: An illustrative example of internalization of strategies in real life drawing, *The Journal of Mathematical Behavior.*

Bauersfeld, H. (1988). Interaction, construction and knowledge: Alternative perspectives for mathematics education. In T. A. Grouws & T. J. Cooney (Eds.), *Perspectives on research on effective mathematics teaching* (Vol. 1, pp. 27-46). Hillsdale NJ: Erlbaum.

Bauersfeld, H. (1990). *Activity theory and radical constructivism: What do they have in common and how do they differ?* Occasional Paper 121, IDM Bielefeld.

Bell, E. T. (1937). *Men of mathematics.* New York: Simon & Schuster.

Boero, P. (1988). An innovative curriculum: Changes in didactic phenomena and related problems. In H.-G. Steiner & A. Vermandel (Eds.), *Proceedings of the Second TME Conference* (pp. 280-296). Bielefeld-Antwerpen.

Boero, P. (1992). The crucial role of semantic fields in the development of problem solving skills in the school environment. In J. P. Ponte, J. F. Matos, J. M. Matos, & D. Fernandes (Eds.), *Mathematical problem solving and new information technologies* (pp. 77-91). Berlin: Springer.

Brousseau, G. (1986). *Théorisation des phénomenes d'enseignement des mathématiques.* Postdoctoral dissertation, University of Bordeaux.

Christiansen, B., & Walther, G. (1986). Task and activity. In B. Christiansen, A. G. Howson, & M. Otte (Eds.), *Perspectives on mathematics education* (pp. 243-307). Dordrecht: Reidel.

Cobb, P., Wood T., & Yackel E. (in press). Discourse, mathematical thinking and classroom practice. In E. Forman, N. Minick & A. Stone (Eds.), *Contexts for learning: Sociocultural dynamics in children development.* Oxford: Oxford University Press.

Davis, J. (1992). The role of the participant observer in the discipline of noticing. In F. Seeger & H. Steinbring (Eds.), *The dialogue between theory and practice in mathematics education: Overcoming the broadcast metaphor* (pp. 167-176). Materialien und Studien Band 38, IDM Bielefeld.

Davydov, V. V. (1991). The content and unsolved problems of activity theory. *Multidisciplinary Newsletter for Activity Theory, 7/8*, 30-35.

Douady, R. & Mercier, A. (Eds.). (1992). *Research in didactique of mathematics.* Grenoble: La Pensée Sauvage.

Duval, R. (1991). Structure du raisonnement deductif et apprentissage de la démonstration. *Educational Studies in Mathematics, 22*, 233-251.

Eisemberg, T. (1991). On building self-confidence in mathematics. In F. Furinghetti (Ed.), *Proceedings of the 15th PME Conference, 2*, 9-16.

Eisenhart, M.A. (1988). The ethnographic research tradition and the mathematics education research. *Journal for Research in Mathematics Education, 19*(2), 99-114.

Engestrom, Y. (1991). Activity theory and individual and social transformations. *Multidisciplinary Newsletter for Activity Theory, 7/8*, 6-17.

Garnier, C., Bednarz, N., & Ulanovskaya, I. (Eds.). (1991). *Après Vygotsky et Piaget: Perspectives sociale et constructiviste. Ecoles russe et occidentale.* Bruxelles: De Boeck - Wesmael.

ICMI (1993). What is research in mathematics education, and what are its results? - Discussion document for an ICMI study. *Zentralblatt für Didaktik der Mathematik, 23*(3), 114-116.

Leont'ev, A.N. (1977). *Attività, coscienza, personalità,* Firenze: Giunti Barbéra. (Original work published in 1975)

Maier, H., & Voigt, J. (1992). Teaching styles in mathematics education. In H. Schupp, W. Blum, C. Keitel, H.-G. Steiner, R. Straesser, & H.-J. Vollrath (Eds.), Mathematics education in the Federal Republic of Germany. *Zentralblatt für Didaktik der Mathematik, 24*(7), 248-252.

Margolinas, C. (1992). Elements pour l'analyse du rôle du maître: Les phases de conclusion. *Recherches en Didactique des Mathématiques, 12*(1), 113-158.

Mellin-Olsen, S. (1987). *The politics of mathematics education.* Dordrecht: Reidel

Perret-Clermont, A. N. (1980). *Social interaction and cognitive development in children.* London: Academic Press.

Piaget, J. (1936). *La naissance de l'intelligence chez l'enfant.* Neuchatel: Delachaux et Niestlé.

Piaget, J. (1962). *Comments on Vygotsky's critical remarks concerning The Language and Thought of the Child, and Judgment and Reasoning in the Child.* Boston, MA: M.I.T. Press.
Raeithel, A. (1990). Production of reality and construction of possibilities: Activity theoretical answers to the challenge of radical constructivism. *Multidisciplinary Newsletter for Activity Theory, 5/6,* 30-43.
Schupp, H., Blum, W., Keitel, C., Steiner, H.-G., Straesser, R., & Vollrath, H.-J. (Eds.). (1992). Mathematics education in the Federal Republic of Germany. *Zentralblatt für Didaktik der Mathematik, 24*(7).
Steffe, L. P. (1991). The constructivist teaching experiment: Illustrations and implications. In E. von Glasersfeld (Ed.), *Radical constructivism in mathematics education* (pp. 177-194). Dordrecht: Kluwer.
Steiner, H.-G. (1985). Theory of mathematics education: An introduction. *For the Learning of Mathematics, 5*(2), 11-17.
Veer, R. van der & Valsiner, J. (1991). *Understanding Vygotsky: A quest for synthesis.* Oxford: Blackwell.
Vygotsky, L. S. (1978). *Mind in society: The development of higher phychological processes.* Cambridge, MA: Harvard University Press.
Vygotsky, L. S. (1990). Pensiero e linguaggio. Bari: Laterza.
Wertsch, J. V. (1991). *Voices of the mind: A sociocultural approach to mediated action.* London: Harvester Wheatsheaf.

Acknowledgements

This paper was prepared with the financial support of C.N.R. and M.U.R.S.T.; I wish to thank Paolo Boero for helpful discussions and for comments on a previous version of this paper.

THEORETICAL PERSPECTIVES ON INTERACTION IN THE MATHEMATICS CLASSROOM

Heinrich Bauersfeld
Bielefeld

First they tell you you're wrong, and they can prove it.
Then they tell you you're right, but it's not important.
Then they tell you it's important, but they've known it for years.
(Charles F. Kettering, the inventor of the first successful electric automobile self-starter, citation from *TIME*, 1969, July 11, p. 45)

There is a growing interest in the theoretical foundations for mathematics education. But there is also a confusing plurality of deliberate labels in use for different positions. Since theories "in use" are always theories developing, related discussions suffer from the difficulty in identifying the status or branch of theory one refers to. The following attempt, therefore, aims at identifying basal backgrounds and orientations behind the special theoretical views under discussion. But the leading interest for this is of a pragmatic rather than theoretical or philosophical nature: it is with the developing of clearer consequences for the field of mathematical teaching and learning, clarifying the related impacts on practice.

1. THE PSYCHOLOGICAL TRADITION AND THE INDIVIDUAL

From a connectionist standpoint, this family of instructional theories has produced an abundance of technology on an illusionary psychological foundation. (Carl Bereiter, 1991, p. 15)

From the beginning of this century, we find a strong psychological line of research work focusing on learners, their intelligence, their abilities, and their thinking (for an overview, see, e.g., Ausubel, 1968; Hilgard & Bower, 1975). For decades, little educational research work was done outside this line. It was much later that educational research also began to include the issue of teaching. Still in 1974, Dunkin and Biddle in their *Study of Teaching* state: "Research on teaching is as yet a very young science" (p. vii). What remained the same was the focus on the *individual*, on the single learner as well as on the single teacher, isolated in his or her classroom.

Up into the 1980s, "interaction" was understood mainly as an interaction between variables, for example, as "Aptitude x Treatment interaction" (see Snow & Farr, 1987; Snow, Federico & Montague, 1980) rather than as so-

R. Biehler, R. W. Scholz, R. Sträßer, B. Winkelmann (Eds.),
Didactics of Mathematics as a Scientific Discipline, 133-146.

cial interaction. Only the very recent developments of cognitive science have begun to open into the social dimension (see the remarkable change of book titles from *Knowing, Learning, and Instruction* to *Perspectives on Socially Shared Cognition*; Resnick, 1989; Resnick, Levine, & Teasley 1991). In parallel, categories like "instruction" or "training" have nearly disappeared, not least because of the negative connotations that have developed with the growing insight and acceptance of the social dimension. "Information" and "intelligence" seem to follow (see Varela, 1990).

But, for a long time, the basal characteristic, common with many other approaches, was the focus on the individual. We may call this the *individualistic* stream of educational theories. The historical background clearly is the fascination with the individual, identifiable throughout the 19th century and well into the 20th century. Nietzsche's statement that the highest goal of humanity does not lie in its end, but in its highest exemplars marks a peak of this individualistic tradition.

2. THE SOCIAL CONCERN AND THE COLLECTIVE

> Verbal expression is never just a reflection of something existent beyond it which is given and "finished off." It always creates something absolutely new and unique, something which is always related to life values such as truth, goodness, beauty, etc. (Mikhail Bakhtin, citation written in 1920, first publication of the Russian original 1979 in Moscow; cited by Kozulin, 1990, p. 54).

During the same period, Soviet psychology developed quite differently. The 1917 revolution turned Marx' and Engels' texts to the rank of bibles. From the very beginning, this forced Soviet psychologists to take their theory of society into account. Typical is Vygotsky's program, dated from 1925, for developing a "general psychology" based on dialectical materialism:

> It is the theory of psychological materialism or the dialectic of psychology which I describe as general psychology. . . . One has to explore the *essence* of the given area of phenomena, the laws of their alteration, their qualitative and quantitative characteristics, their causality, one has to create related categories and concepts, in one word – a "capital" of its *own*. (Vygotsky 1985, pp. 251-252, referring to Marx' "capital")

Characteristic for the psychological movement in the Soviet Union at that time is also the separation from behaviorism as well as from Gestalt or holistic psychologies. In 1929, about 600 books on psychological themes appeared in the Soviet Union (Jaroschewski, 1975, p. 406), giving proof of the vivid discussion. Basov, a scholar of Bechterev, was the first to stress the importance of "activity" (instead of "behavior") for human mental development (Métraux, in Vygotsky, 1992, p. 9). And, nearly contemporary, Vygotsky was the first to analyze activity and consciousness from the perspective of dialectic materialism's doctrine of societal practice.

In a transient phase of his thinking about 1930, Vygotsky discriminated higher from lower mental functions through their genesis. The lower mental functions follow stimulus-response constellations; they develop through maturation. In contrast, higher mental functions are mediated through the use of tools and signs, and are open to conscious and deliberate training. The higher functions develop only within societal relations, "through the internalisation of selfregulatory pattern pre-given in society" (Métraux, in Vygotsky, 1992, p. 19).

It was in 1932 that Vygotsky changed his mind dramatically, as he noted in his diary (published in 1977 in Russian), in which he marked "the analysis of the meaning of signs as to be the only adequate access for an investigation of conscious human activities" (Métraux, in Vygotsky, 1992, p. 15). Reading Engels' *Dialectic of Nature*, he "abruptly was led to the issue of the relation not only between man and nature, but also between man and others, and man and himself, as mediated through tools." (p. 16). He, apparently, had arrived at what he was searching for so intensively: the instrument for bridging between the lower and the higher mental functions as well as for describing the interrelation between the psychological and the social. During the last two years of his life, he dealt with the key concept of "mediating activity" (adopted from Hegel's concept "vermittelnde Tätigkeit"), which he split into "use of tools" and "use of signs" (Vygotsky, 1992, pp. 152-153). Thus his last two years can be interpreted as the offspring of activity theory.

But it was as late as 1979, about half a century later, that:

> addressing a symposium on Vygotsky's theoretical legacy, Moscow philosopher and psychologist G. P. Schedrovitsky resolutely challenged the myth of succession and suggested that the activity theory substantially derivated from Vygotsky's original program. Schedrovitsky emphasised that the principle of semiotic mediation is the cornerstone of cultural-historical theory, representing its primary focus. (Kozulin 1990, p. 254).

It is remarkable that in his attempt to describe the development of sign use, Vygotsky turns to quote from special experimental work with children, whilst his more scholastic followers had (and still have) endless debates about the meaning of certain concepts and where their boundaries should be drawn. Some even deny whether Vygotsky can be named an activity theorist at all. Typical is Brushlinsky, who speaks of "the activity approach (of S. L. Rubinstein and A. N. Leont'ev, as we mentioned earlier) and non-activity approach (of, among others, L. S. Vygotsky)" (Lektorsky, 1990, p. 72).

Late in 1932, Vygotsky quotes Engels: "The tool means the specific human activity, the forming impact of man onto nature, the production," knowing that the impact is reciprocal: Man changes with the use of tools as well (Vygotsky, 1992, p. 102). Vygotsky understood "tool" primarily as the laborer's tool for his working activities:

> The tool is the mediator of the external activity of man, directed at the subjection of nature. But the sign does not alter the object of psychic operation. Rather it is a

> medium for the psychic influencing of behaviour – of the own or that of others. (Vygotsky, 1992, p. 154)

Thus ruling the nature and ruling the behavior of others is the function of "mediating activities." The fascination of his last two years of life was with function and use of signs, which, in his understanding, include language "use:"

> According to the cultural-historical theory evolved by L. S. Vygotsky in the last years of his life, it is speech or to be more exact, speech and other cultural signs social in origin and thus distinguishing men from animals that serve as the "producing cause" (his own expression) of the child's psychic development. (Brushlinsky, cited in Lektorsky, 1990, p. 72)

Comparing Vygotsky's late texts with the related production of his followers – particularly Rubinstein, Leont'ev, and Davydov on "activity theory" – produces the impression that he seemed to be much more sensitive, more empirically oriented, and less scholastic. (There is an interesting parallel, at least for German readers, with the famous educator Herbart [1776-1841], whose writings were almost forgotten under the sweeping success of his scholars Ziller, Dörpfeld, and Rhein. They turned his very reflected ideas into handy recipes, teachable concepts, and a scholastic system of "formal steps," but missed his reflectedness and sensitivity through simplification and formalized representations.) The followers generalized Vygotsky's key concept and spoke of "mediator objects" (sometimes directly in German "gegenständliche Mittel"), which, as *objects,* include even language (see Lektorsky 1984, 1990), and they identify mediator objects as "carriers of meaning:" "Mediator objects used in the process of cognition do not have a value as such but merely as carriers of knowledge about other objects" (Lektorsky, 1984, pp. 142-143).

Recently they also introduced the notion of "collective subject" (Davydov, 1991; Lektorsky, 1984, pp. 232-233), which incorporates the individual: "The individual subject, his consciousness and cognition must be understood in terms of their incorporation in different systems of collective practical and cognitive activity" (Lektorsky, 1984, p. 240).

Such shifts of meaning absolutize the social – or better: the collective – dimension. And it is no remedy to modify this by stating "the collective subject itself does not exist outside concrete persons" (Lektorsky, 1984, p. 240). The crucial points are the stated dominance of the social and the related objectifying of language – making an object of something, what Engels called "Mythos der Verdinglichung," the myth of objectification. Lektorsky accuses Vygotsky of being "one-sided," because of his "exaggerated" identification of egocentric speech with thinking: ". . . if speech fulfills the function of planning and even that of solving problems, what is thought supposed to do?" (Lektorsky, 1984, p. 240; Lektorsky uses scientifically quite dubious arguments for this, like: "It is common knowledge that speaking does not yet mean thinking, although it is impossible to think without speaking at all." Lektorsky, 1984, p. 240). But just this presumed separating of languaging and thinking carries the temptation for an objectivation of language (see Bauersfeld, 1992a). Likewise Brushlinsky states "speech . . . *cannot be activity*" (cited in Lektorsky, 1990, p. 72), because "word-sign" does not have the same importance as activity (in his sense). But what – if not language as an objective body of meaning is meant – will be left with a word-sign, once it becomes separated from its use? Vygotsky, obviously, was much more careful with related descriptions.

Taking the followers' activity theory as a prototype, I will call related theoretical views the *collectivist* stream of educational theories. There are interesting attempts toward the development of "social theories" for learning and teaching (see, e.g., Markowitz, 1986; Miller, 1986).

3. A MEDIATING POSITION – INTERACTIONISM

> With primitive means the child tries to react upon a complicated structure. (Vygotsky, 1992, p. 252)

Following both Paul Feyerabend's advice: "All you can do, if you really want to be truthful, is *to tell a story*" (1991, p. 141) and Gregory Bateson's conviction that stories can be very "informative" in research and in education, allow me to give a brief personal account of how I arrived at somewhat different positions. In the early 1960s, our empirical work with students in Grades 1 through 6, especially related to the changes from elementary into secondary education (Grades 5 and 6 are the transition levels in Germany), appeared to produce quite weak outcomes, because little was known at that time about the relations between teacher and student(s). There was no sufficient answer to questions like: How does a teacher identify a student's mistake? How do both teacher and student arrive at somewhat viable agreements and meanings for continuing? How does a student understand the teacher's inventions?

The availability of video recorders then elicited fundamental changes in our approaches. When videotaped classroom scenes could be played back on and on, applying different foci of attention from passage to passage, a tremendous need for the theoretical orientation of such interpretative procedures became evident. Psychological theories, as helpful as they are, did not cover the complicated reflexive relations among teachers and students. But well developed means for describing the interaction among human beings were available in special wings of sociology and linguistics: Ethnomethodology, Social Interactionism, and Discourse Analysis, the branch of linguistics investigating language pragmatics (initially, we found most help in Berger & Luckmann, 1966; Blumer, 1969; Mehan & Wood, 1975; later, also Cazden & Hymes, 1972; Goffman, 1974; and many more).

Since sociologists are interested in social structures only, but not in learning and teaching subject matter issues, we had to transfer concepts and relations into our field of concern. Early products were the identification of "patterns of interaction" (Bauersfeld, 1978; Voigt, 1984), of "domains of subjective experiences" (Bauersfeld 1983), and, more generally, of a specific "hidden grammar" for the activities in mathematical classrooms, which – from an observer's view – students and teacher often seem to follow, though not consciously (Krummheuer, 1992). We abandoned simple cause/effect ascriptions and favored an "abductive" hypothesis formation (Pierce, 1965). In order to understand sufficiently the individual gains and the social regularities emerging from certain classroom cultures, it was necessary to switch between both views, the psychological and the sociological, *without giving preference* to either one.

Across the years, the reactions of the wider community, particularly from both the extreme positions, were very much like the Kettering motto describes it (see above). On the other hand, the insight into the *reciprocity* of (a) *individual change* and development through participation in social interaction, including the insuperable subjectivity of personal constructions; and (b) the permanent *accomplishment* and change *of social regularities* through the individual members of the classroom culture made it very easy to adopt the radical constructivist principle when I came to meet Ernst von Glasersfeld. We, the research group in Bielefeld (Bauersfeld, Krummheuer, Voigt), had arrived at quite similar consequences, mainly from sociological reasons rather than from psychological and philosophical bases, which seem to have formed the basis for the genesis of the radical constructivist principle (via Vico, Kant, and others; for more details about our position, see Bauersfeld, 1988, 1991, 1992b; Krummheuer & Voigt, 1991).

The core convictions of our interactionist position are, in brief, as follows:

1. *Learning* describes a process of personal life formation, a process of an interactive adapting to a culture through active participation (which, in parallel, reversely constitutes the culture itself) rather than a transmission of norms, knowledge, and objectified items.
2. *Meaning* is with the use of words, sentences, or signs and symbols rather than in the related sounds, signs, or representations.
3. *Languaging* describes a social practice (the French *parole*), serving in communication for pointing at shared experiences and for orientation in the same culture, rather than an instrument for the direct transportation of sense or as a carrier of attached meanings.
4. *Knowing or remembering something* denotes the momentary activation of options from experienced actions (in their totality) rather than a storable, deliberately treatable, and retrievable object-like item, called knowledge, from a loft, called memory.
5. *Mathematizing* describes a practice based on social conventions rather than the applying of a universally applicable set of eternal truths; according to Davis and Hersh (1980), this holds for mathematics itself.
6. (Internal) *representations* are taken as individual constructs, emerging through social interaction as a viable balance between the person's actual interests and re-

alized constraints, rather than an internal one-to-one mapping of pregiven realities or a fitting reconstruction of "the" world.
7. Using *visualizations and embodiments* with the related intention of using them as didactical means depends on taken-as-shared social conventions rather than on a plain reading or the discovering of inherent or inbuilt mathematical structures and meanings.
8. *Teaching* describes the attempt to organize an interactive and reflexive process, with the teacher engaging in a constantly continuing and mutually differentiating and actualizing of activities with the students, and thus the establishing and maintaining of a classroom culture, rather than the transmission, introduction, or even rediscovery of pregiven and objectively codified knowledge. (Bauersfeld 1992b)

4. A SIMPLIFIED OVERVIEW

We now can arrange the identified basal positions into a simple schema (following an idea from Jörg Voigt):

Individualistic Perspectives	*Collectivist Perspectives*
Learning is individual change, according to steps of cognitive development and to context. Prototype: Cognitive Psychology.	Learning is enculturation into pre-existing societal structures, supported by mediator means or adequate representations. Prototype: Activity Theory.

Interactionist Perspectives

Teacher and students interactively constitute the culture of the classroom, conventions both for subject matter and social regulations emerge, communication lives from negotiation and taken-as-shared meanings.
Prototypes: Ethnomethodology, Symbolic Interactionism, Discourse Analysis (Pragmalinguistics).

The middle position is meant for and acts (at least for us) as a link between the two extremes. Many of the recent US reinterpretations of Vygotsky will fall under the collectivist perspectives, insofar as these usually neglect the social interactionist insights. In contrast, early applications of the radical constructivist principle will more likely belong to the individualistic views. Surely, there is an abundance of different perspectives in between and overlapping the extremes. Thus the scheme can mark poles only.

5. CONSEQUENCES FOR ELEMENTARY EDUCATIONAL PRACTICE

> Theorists often divide over the choice of guiding principles while maintaining a consensus on the rules specifying legitimate inferences from them. (Peter Galison, 1987, p. 244)

Both extremes, the individualistic and the collectivist stream, have their convincing practices in general education: The perhaps most famous case of an individualistically oriented educational practice is Pestalozzi's work in Stans, where he collected and educated the orphans left from the Swiss liberation war with France, reported in his *Letter from Stans* (1799). However,

Pestalozzi (1946) also pointed to the social function of labor. The most famous case of a collectivist-oriented practice is Makarenko's work near Poltava, Ukrainia, where he collected and educated dead-end youth (besprisorniks) right after the revolution (1920-1928), reported in his *Pedagogical Poem* (1940). In these two cases, quite different fundamental convictions have led to very similar – and very successful – practices, and both with severely damaged youth.

In mathematics education, things seem to be more complicated than in general education. According to my recent work, I will limit these remarks to elementary education in mathematics and, within this area, to the issues of the understanding of mathematics itself and of language. The contrast tried here contradicts the consequences from both the two extreme traditions with the consequences drawn from the intermediate interactionist position. On this level of discussion, it is clear that only quite general inferences are possible.

5.1 Understanding Mathematics

Fundamentally different practices arise from whether mathematics is taken as an objective truth, as a societal treasure, as something existing and documented objectively, or as a practice of shared mathematizing, guided by rules and conventions emerging from this practice.

The first conviction will lead teachers to "introduce" children, to use "embodiments" and "representations," which are structurally as "near to the structure mathematically meant" and as little misleading or distracting as possible. Children's errors will find corrections toward the expected correct answer and so forth. Objectively existing structures and properties also give space for "discovery" activities, given that the expected findings are in reach of the present cognitive aptitudes (e.g., "zone of proximal development").

The latter conviction will lead teachers to organize their activities as part of a practice of mathematizing, that is, as a challenging and supportive "subculture" specific to *this* teacher and *these* children in *this* classroom, which functions toward developing the students' "constructive abilities," their related self-concept, and self-organization, rather than as a management through product control and permanent external assessments. The diversity of subjective constructions of meaning and the necessity to arrive at viable adaptations – "taken-as-shared meanings" and "taken-as-shared regulations" – requires optimal chances for discussions based on intensive experiences and aiming at the negotiation of meanings. There is no discovery in the classical sense, there is subjective construction of meaning only, since "what is observed are not things, properties, or relations of a world that exists as such, but rather the results of distinctions made by the observer himself" (von Glasersfeld, 1991, pp. 60-61).

5.2 Language

Related to language, again, we arrive at very different practices depending on whether languaging is taken as the use of an objectively existing body of language, of the storehouse of societal knowledge and prepared meanings, or whether languaging is understood as a social practice of orienting.

Once we separate "language" and "activity," the primacy is given to activity (see Brushlinsky, above), and learning will have to begin with activities in which language is used as a pregiven "tool." The "collective subject" becomes "enculturated" into an already existing culture. The learning subject's creative inventions appear to be deviant moves, which have to undergo correction toward the standardized use of the "mediating tools."

> So long as language is considered to be denotative it will be necessary to look at it as a means for the transmission of information, as if something were transmitted from organism to organism . . . . when it is recognised that language is connotative and not denotative, and that its function is to orient the orientee without regard for the cognitive domain of the orienter, it becomes apparent that there is no transmission of information through language. (Maturana & Varela, 1980, p. 32)

In the latter case, again, we arrive at the necessity for an ongoing negotiation of meaning in the classroom, aiming not only at a viable adapting to taken-as-shared meanings of the subject matter pointed at but also at a related clarifying of the taken-as-shared meanings of the signs and words in use, and, particularly, at furthering the reflection of the underlying subjective constructive processes.

It is remarkable how far Vygotsky has pointed out the need to analyze higher mental functions as *processes*. Thinking of everyday classroom practices, the product orientation is still found to dominate the majority of classrooms everywhere: Teachers' inventions follow their subjective image of the *product* to be taught rather than ideas for developing useful constructive and descriptive *processes* with students. It is only in a much later state of rooted habits, conventions, and norms that a person's mathematizing can develop the properties, so much beloved by mathematicians, of curtailment and elegance, of forcing power, of precision and sharpness in thinking and presenting – "since there is no other way of thinking it" (as Jaspers, 1947, p. 467 enthusiastically said). The product illusion, perhaps, is the most devastating force in education, because it usually blinds the more knowledgeable and (in terms of subject matter) better prepared teachers.

6. OUTLOOK: THE NEXT CATCHWORD – CONNECTIONISM

> We should say: *it is thinking*, just as we say: *it is thundering*. To speak of *cogito* is too much already, if we translate it into *I am thinking*. (Georg Christoph Lichtenberg, 1971, in: Sudelbücher, K 76, p. 412. By the way, Vygotsky, 1992, p. 147, already has quoted the very same aphorism. He used it to introduce his excellent analysis of tying a knot in one's handkerchief and the related functioning for remembering.)

More than 200 years ago, Lichtenberg already pointed at a crucial fact that presently characterizes consequences from connectionist models. Indeed, across the last years, computer models for human brain functioning have come into favor under labels like "connectionism," "dynamic networks," and "parallel distributed processing," or "neural net" models. I am not interested in the technical realizations. But the interpretation of such models in our field of mathematics education opens quite fascinating perspectives. "The 'new connectionism' is causing a great stir in cognitive science and artificial intelligence" says Bereiter (1991, p. 10), himself a well-known cognitivist before. Clearly, these models are simpler, more powerful, and allow more convincing interpretations of educational experience and research outcomes than cognitive psychology has produced so far (see Varela, 1990, 1992; also, Hiebert & Carpenter, 1992; Ramsey, Stich, & Rumelhart, 1991; Rueckl & Kosslyn, 1992).

Common to all of these models is the interpretation of the human brain as a huge network consisting of nodes and connections, with many specialized sets of nodes and connections as part of it. The brief reinterpretation of a few key concepts from this perspective may enable the reader to assess the persuasive power her or himself:

Rule generation. Hebb's rule, fundamental in connectionism, states a reinforcement of the connection between two nodes once they are both in resonance (activated). Frequent activation, therefore, will lead to a preference for this connection, once one of the two nodes becomes activated. The same holds with chains or trees of connections. Once any part of such connective patterns becomes activated, as part of the global state of the whole network ("mind"), the related connections will work without further release (due to the increased "weight" of the connections).

No wonder that we experience children as perfect creators of regularities and rules: What has functioned twice already has good chances to undergo preferenced activation in case of the third appearance. Also the genesis of subjective routines and habits, emerging through participation and often without conscious notice, finds a simple explanation in this model. What is learned in the classroom is *co-learned* in its majority, it emerges by the way. The overtly and consciously learned issues probably would never function without these obscured co-learned backgrounds.

The *totality of experiencing.* Besides Hebb's rule, the brain connections follow the reciprocity rule: Connections between two different regions of the brain, layers, or patterns of nodes are reciprocal (with very few exceptions). Since practically every part of the brain is connected with every other part, there are *global states* of the mind only. Thus, not only all senses are involved but also emotions and even the position or movement of remote extremities of the body (kinesthetics). The brain is understood as a highly "cooperative system." "In the end all processes depend functionally upon the status of single elements," as Varela and Thompson (1991) have pointed out, and these depend upon their related global states (distribution of activations all over the network).

The globality of the states of the mind appears for us as the totality of experiencing. A smell can elicit a whole reminiscence in all details. In the classroom, even minor changes in the presentation of a task can evoke quite deviant interpretations from the students. The totality of our experiencing, however, unveils the secret of our creativity: A global state of mind can become activated just from any of its single parts, enabling us to combine elements from quite different domains of subjective experience by passing through a series of different global states.

Students' errors. If a network produces inadequate reactions, there are many options for interpretations. In a new situation, the reaction will be given tentatively, using partly available and partly new (weak) connections. In a routine situation, the reaction can come from a preferentially available (strong) but inadequately formed pattern of connections. Or, two likewise current alternatives can compete. In any case, the adequate definition of the situation can fail, which makes it impossible to activate the adequate pattern for the expected reaction, and so forth.

In a mathematics classroom, related to calculations, for example, the four different interpretations would require different help and inventions. In the new situation, encouraging the parts that are already functioning adequately will be a useful strategy, whereas the usual product correction would end in confusion. Product correction in the routine situation will leave the preferentially available connections almost untouched; in the very next similar situation, the inadequate pattern will "fire" again – if other and more intensive inventions have not enabled a comparably strong replacement. For many students, text problems produce the case of two strong options competing: "I don't know whether to multiply or to divide!" (The pursuit of this problem here would require a more intimate discussion of text problems.) In case of a miss of an adequate situational definition (adequate global state), metacommunication may form a helpful strategy, that is, negotiating about what we are talking about.

Forgetting. Connections, once ready for use but not active over a longer period, will fade away. Within larger layers or patterns of connections, this fading will hurt the weakest (the least or latest activated) parts first. Clearly, like a person's biography, such patterns have a "history" of activations and changes, and this, on the other hand, makes every reaction of the network a new and unique one.

Forgetting as a "fading away," often with a desperate search for the missing links or key parts, particularly when these had been "weak" all over, is a well-known feature.

Consciousness and control. There is no central agent in the brain steering or supervising ongoing activities. The brain is self-organizing, a "society of mind" (Minsky, 1987). The processual regularities, which an observer may describe, "emerge," they are global properties. The instant flow of global states controls itself through similarities and differences between global states, which require decisions between alternatives. Also, there is no issue like "knowledge" stored at any locations; "all knowledge is in the connections" (Rumelhart, 1989, p. 135).

Consequently, there is no arbitrary "retrieving" from "memory," as we know. And very little of the brain's processing is open to conscious control. There is no direct teaching of concepts, strategies, or "metaknowledge," since these are properties of (subjective) global states, which emerge from intensive experiences only (related to the culture of the classroom, to negotiating of meaning, and the active participation of the learner). And nobody can make up another person's internal global states. In particular, "if the world we live in is brought about or shaped rather than pregiven, the notion of representation cannot have a central role any longer" (Varela, 1990, p. 90).

Apparently, the way our brain is functioning is nearer to practices of pragmatical adaptation like "tinkering" or "bricolage" (the French equivalent) than to ideals of abstract thinking, rule-guided inferencing and reflecting, or rational production, as a mathematician would like to see it. As Bereiter (1991, p. 13) says, "[Networks] do best what people do best – recognize pattern and similarities. They work in the messy, bottom-up way that nature seems bound to. They approximate rather than embody rationality." We are left to rethink our usual convictions concerning teaching and learning.

REFERENCES

Ausubel, D. P. (1968). *Educational psychology: A cognitive view.* New York: Holt, Rinehart & Winston.

Bauersfeld, H. (1978). Kommunikationsmuster im Mathematikunterricht: Eine Analyse am Beispiel der Handlungsverengung durch Antworterwartung. In H. Bauersfeld (Ed.), *Fallstudien und Analysen zum Mathematikunterricht* (pp. 158-170). Hannover: Schroedel.

Bauersfeld, H. (1983). Subjektive Erfahrungsbereiche als Grundlage einer Interaktionstheorie des Mathematiklernens und -lehrens. In H. Bauersfeld, H. Bussmann, G. Krummheuer, J. H. Lorenz, & J. Voigt (Eds.), *Lernen und Lehren von Mathematik.* IDM-series Untersuchungen zum Mathematikunterricht, Vol. 6 (pp. 1-56). Köln: Aulis Verlag Deubner.

Bauersfeld, H. (1988). Interaction, construction, and knowledge: Alternative perspectives for mathematics education. In D. A. Grouws & T. J. Cooney (Eds.), *Perspectives on research on effective mathematics teaching* (pp. 27-46). Reston, VA: Erlbaum.

Bauersfeld, H. (1991). Structuring the structures. In L. P. Steffe (Ed.), *Constructivism and education.* Hillsdale, NJ: Erlbaum.

Bauersfeld, H. (1992a). Activity theory and radical constructivism - What do they have in common and how do they differ? *Cybernetics and Human Knowing, 1,* 15-25.

Bauersfeld, H. (1992b). Integrating theories for mathematics education. *For the Learning of Mathematics, 12*(2), 19-28.

Bereiter, C. (1991). Implications of connectionism for thinking about rules. *Educational researcher, 20*(3), 10-16.

Berger, P. L., & Luckmann, T. (1966). *The social construction of reality*. Garden City, NY: Doubleday.

Blumer, H. (1969). *Symbolic interactionism*. Englewood Cliffs, NJ: Prentice-Hall.

Cazden, C. B., John, V. P., & Hymes, D. (1972). *Functions of language in the classroom*. New York: Columbia University, Teachers College Press.

Davis, P. J., & Hersh, R. (1980). *The mathematical experience*. Basel, Switzerland: Birkhäuser.

Davydov, V. V. (1991). The content and unsolved problems of activity theory. *The Multidisciplinary Newsletter for Activity Theory*, 7/8, 30-35.

Dunkin, M. J., & Biddle, B. J. (1974). *The study of teaching*. New York: Holt, Rinehart & Winston.

Feyerabend, P. (1991). *Three dialogues on knowledge*. Oxford: Blackwell.

Galison, P. L. (1987). *How experiments end*. Chicago, IL: University of Chicago Press.

Glasersfeld, E. von (1991). Distinguishing the observer: An attempt at interpreting Maturana. *Methodologia, V*(8), 57-68.

Goffman, E. (1974). *Frame analysis - An essay on the organization of experience*. Cambridge, MA: Harvard University Press.

Hiebert, J., & Carpenter, T. P. (1992). Learning and teaching with understanding. In D. A. Grouws (Ed.), *Handbook of research on mathematics teaching and learning* (pp. 65-97). New York: MacMillan.

Hilgard, E. R., & Bower, G. H. (1975). *Theories of learning*. Englewood Cliffs, NJ: Prentice-Hall.

Jaroschewski, M. (1975). *Psychologie im 20. Jahrhundert*. Berlin: Volk und Wissen VEB.

Jaspers, K. (1947). *Von der Wahrheit*. München: Piper

Kozulin, A. (1990). *Vygotsky's psychology: A biography of ideas*. London: Harvester Wheatsheaf.

Krummheuer, G. (1992). Lernen mit "Format": Elemente einer interaktionistischen Lerntheorie. Weinheim: Deutscher Studien Verlag.

Krummheuer, G., & Voigt, J. (1991). Interaktionsanalysen von Mathematikunterricht: Ein Überblick über einige Bielefelder Arbeiten. In H. Maier & J. Voigt (Eds.), *Interpretative Unterrichtsforschung*. IDM-series Untersuchungen zum Mathematikunterricht, Vol. 17 (pp. 17-32). Köln: Aulis Verlag Deubner.

Lektorskij, V. A. (1984). *Subject - Object - Cognition*. Moscow: Progress Publ.

Lektorskij, V. A., & Engeström, Y. (Eds.). (1990). *Activity: The theory, methodology and problems*. Issues in Contemporary Soviet Psychology Series. Orlando, FL: Deutsch.

Lichtenberg, G. Ch. (1971). *Schriften und Briefe* (Vol. 2). München: Hanser Verlag.

Makarenko, A.S. (1954). *Der Weg ins Leben - Ein pädagogisches Poem*. Berlin: Aufbau-Verlag. [Original work published 1940]

Markowitz, J. (1986). *Verhalten im Systemkontext: Zum Begriff des sozialen Epigramms*. Frankfurt/Main: Suhrkamp.

Maturana, H. R., & Varela, F. J. (1980). Autopoiesis and cognition: The realization of the living. In R. S. Cohen & M. W. Wartofsky (Eds.), *Boston studies in the philosophy of science* (Vol. 42). Dordrecht, Netherlands: Reidel.

Mehan, H., & Wood, H. (1975). *The reality of ethnomethodology*. New York: Wiley.

Miller, M. (1986). *Kollektive Lernprozesse. Studien zur Grundlegung einer soziologischen Lerntheorie*. Frankfurt/Main: Suhrkamp.

Minsky, M. (1987). *The society of mind*. London: Heinemann

Pierce, C. S. (1965): *Collected papers II - Elements of logic*. Cambridge, MA: Harvard University Press.

Pestalozzi, J. H. (1946). Stanser Brief. In P. Baumgartner (Ed.), *Werke in 8 Bänden. Band 3: Schriften 1798-1804* (pp. 91-124). Erlenbach-Zürich, Switzerland: Rotapfel-Verlag. [Original work published 1799]

Ramsey, W., Stich, S. P., & Rumelhart, D. E. (1991). Philosophy and connectionist theory. In D. E. Rumelhart (Ed.), *Developments in connectionist theory*. Hillsdale, NJ: Erlbaum.

Resnick, L. B. (1989). *Knowing, learning, and instruction*. Hillsdale, NJ: Erlbaum.

Resnick, L., Levine, J. M., & Teasley, S. D. (Eds.). (1991). *Perspectives on socially shared cognition*. Washington, DC: American Psychological Association.

Rueckl, J. G., & Kosslyn, S. M. (1992). What good is connectionist modelling? A dialogue. In A. F. Healy, S. M. Kosslyn, & R. M. Shiffrin (Eds.), *From learning theory to connectionist theory: Essays in honor of William K. Estes* (Vol.1). Hillsdale, NJ: Erlbaum.

Rumelhart, D. E. (1989). The architecture of mind: A connectionist approach. In M. I. Posner (Ed.), *Foundations of cognitive science* (pp. 133-159). Cambridge, MA: MIT Press.

Snow, R. E., & Farr, M. J. (Eds.). (1987). *Aptitude, learning, and instruction. Vol. 2: Cognitive process analyses of learning and problem solving.* Hillsdale, NJ: Erlbaum

Snow, R. E., Federico, P.-A., & Montague, W. E. (Eds.). (1980). *Aptitude, learning, and instruction.* Vol. 1*: Cognitive process analyses of aptitude.* Hillsdale, NJ: Erlbaum.

Varela, F. J., & Thompson, E. (1991). *The embodied mind.* Newton, MA: MIT Press.

Veer, R. van der, & Valsiner, J. (1991). *Understanding Vygotsky: A quest for synthesis.* Oxford: Blackwell.

Voigt, J. (1984). *Interaktionsmuster und Routinen im Mathematikunterricht.* Weinheim: Beltz.

Vygotsky, L. S. (1985). *Die Krise der Psychologie in ihrer historischen Bedeutung.* In J. Lompscher (Ed.), *Ausgewählte Schriften* (Vol. 1, pp. 57-277). Köln: Pahl-Rugenstein.

Vygotsky, L. S. (1992). *Geschichte der höheren psychischen Funktionen.* W. Jantzen, J. Lompscher, A. Métraux, & M. Stadler (Eds.), *Fortschritte der Psychologie* (Vol. 5). Münster: Lit Verlag. [Original work published 1960]

WORKING IN SMALL GROUPS: A LEARNING SITUATION?

Colette Laborde
Grenoble

1. THEORETICAL FRAMEWORK AND QUESTIONS

In a widespread approach in "didactique des mathématiques," learning is considered as an adaptation to a new situation. In mathematics, this new situation is a problem students cannot solve with their available knowledge but for which they can develop new solution tools. These new tools are starting points for new knowledge. In this approach, it is also commonly assumed that this process of adaptation is not spontaneous, and conditions must be organized to allow it. Learning situations must be designed by the teacher. One of the main aims of didactique des mathématiques is to characterize these learning situations.

This approach seems to consider learning only as an individual interaction process between knowledge and student, whereas it is obvious that classroom situations are essentially social:

1. the choices about knowledge to be taught meet some social and cultural expectations;
2. the students are involved as cognitive and social subjects (in particular, even their representations of mathematical contents are partially of a social nature);
3. the progress of a class is based on social interactions between partners (teacher-students and student-student).

Vygotsky (1934), who distinguished the development of spontaneous concepts and of scientific concepts (but recognized the links between them), claimed the following thesis:

1. knowledge coming from the social environment plays an important role in the representations of scientific concepts by the child;
2. but the child does not assimilate the scientific concepts as such and reconstructs these concepts on his or her own. In this thesis, intrapersonal and interpersonal processes seem to interact in the construction of scientific knowledge by the child.

This presentation is an attempt

1. to elicit the role of interpersonal processes in the construction of mathematical knowledge in mathematics classrooms in the specific case of stu-

R. Biehler, R. W. Scholz, R. Sträßer, B. Winkelmann (Eds.),
Didactics of Mathematics as a Scientific Discipline, 147-158.

dents working together at a joint task of finding a common solution to a mathematical problem;

2. to determine some variables affecting these processes.

These group work situations are systematically used by some teachers in their class; they are also being developed in curricula that provide opportunities for project work (like in the UK), or recently in France in so-called "modules" (grade 10), in which mathematical activities not necessarily linked to the curricula can be organized in an open way. The introduction of computers in the classrooms also gives rise to joint work at the computer since very often the number of machines is limited.

In group work situations, students are faced with two kinds of problem: They must solve a mathematical problem, but they have to achieve this through a social activity. Thus, they are additionally confronted with a social problem. In order to know more about the role of interpersonal processes in the individual construction of mathematical knowledge, I will focus my study on the interrelations between these two kinds of problem.

Students must jointly solve a problem and agree on a common solution. The problem given to them does not depend on the fact that the solution must be found by one student alone or by a group of students (except in organized situations of task division like in some Russian experiments quoted in section 5). The respective roles of the partners are not determined by the situation: A student may agree to everything that is proposed by his or her partner or may systematically be against the partner's proposal. The "devolution" of the mathematical problem is not linked to the a priori social organization of the situation, but the development of the situation seems to depend on the partners.

I propose to distinguish two kinds of processes involved in the group work situation: the conflicting processes and the cooperation processes between the partners. A huge literature on the topic seems to point out some positive effects of such work, be it based on conflicting processes or on cooperation processes: The solution produced by the group is generally better than that produced by an individual, and group work seems to provide a positive impact on long-term learning. But, in some instances of collaborative work, some children seem to regress (Tudge, 1992). Some psychologists try to understand why some collaborations are more successful than others. Rogoff (1990) suggests, for example, that different social contexts may promote different aspects of intellectual functioning: A peer collaboration would facilitate a shift in perspective. These studies manipulated several factors in specifically defined experimental settings, but they did not deal with learning in a broader context such as the school context and they did not necessarily analyze the learning of complex knowledge like mathematical content. In what follows, I will focus on the role of social interaction in the mathematical aspects of solving processes and in the construction of knowledge.

2. SOCIOCOGNITIVE CONFLICTS IN THE SOLVING PROCESSES

The role of the sociocognitive conflict is presented in several studies as possibly producing a positive outcome on (a) the elaboration of the solution of the problem, and (b) the learning.

This claim is based on the theory of sociocognitive conflict, developed in particular by social psychologists at Geneva (cf. the collective book edited by Mugny, 1985). According to this theory, the contradiction coming from two opposite points of view is more readily perceived and cannot be refuted so easily as the contradiction coming from facts for an individual. The latter may either not perceive the contradiction or not take it into account when wavering between two opposite points of view and finally choosing one of them. In order to master a task, students working jointly are committed to overcoming conflict. When attempting to solve the contradiction, they may manage to coordinate the two points of view into a third one overcoming both initial points of view and corresponding to a higher level of knowledge. This is the starting point for learning. The above-mentioned social psychologists have tested the theory on the construction of general schemas studied by Piaget, like the schema concerning the conservation of liquids or of lengths. When we organized group work situations with students solving mathematical problems, we could also observe the construction of a new solution of higher conceptual level and the overcoming of the contradiction between the partners. Let me give two examples:

In a situation in which two students had to describe a geometrical diagram in a written message meant for two other students who did not know the diagram, the labeling of some elements of the diagram by the producers of the message often appeared as a solution overcoming the partners' disagreement about their mutual formulations in natural language: Each proposal was judged as erroneous or too complex by each partner and as possibly leading the receivers to a misunderstanding. Labeling some elements provided a means that was accepted as an unambiguous and economical way when describing elements depending on the labeled elements: Instead of writing "the line joining the point we made to the other point we have just drawn," they could write: "join Point A to Point B" (Laborde, 1982).

The example of a situation of ordering decimal numbers also illustrates how students can construct a new correct strategy when they have to decide between two strategies giving different results (Coulibaly, 1987). Leonard and Grisvard (1981, 1983) have shown that sorting a sequence of decimal numbers may pose a problem even for older students, and that with striking regularity, two erroneous rules often underlie the students ' solutions:

1. A rule R1 according to which among two decimal numbers having the same whole part, the bigger one is the number with the bigger decimal part, this latter being considered as a whole number; for example:

0.514 > 0.6 because 514 > 6 or 0.71 > 0.006 because 71 > 6.

2. A rule R2 according to which among two decimal numbers having the same whole part, the bigger one is the number with the decimal part having the smaller number of digits; for example:

0.6 > 0.514 because 0.514 has three digits after the decimal point, while 0.6 has only one digit after the decimal point, but 0.5 > 0.514 or 0.71 > 0.006.

One may be convinced of the strength of these rules insofar as, in some cases, they provide correct results. Teachers are very often not aware of these erroneous rules followed by their students, because they have access only to their final answers and not to the reasoning leading to them. Students are thus reinforced in their erroneous strategies. I leave to the reader the pleasure to check that, when R1 and R2 give the same answer, they are correct, while, when the results are contradictory, obviously only one of them is false. But the consequence of this observation is important from a didactical point of view. It implies that well-chosen numbers may allow the teacher or the experimenter to find which rule is followed by the student in the task of sorting decimal numbers. We must indeed note that it has very often been observed that a student's answers can be described by only one rule.

The experiment carried out by Coulibaly determined the rules underlying 8th-grade students' answers to a written test. Four pairs of students were formed by putting together students following different rules. Each pair then had to jointly order five sequences of decimal numbers and to elaborate a written explanation meant for other younger students on how to compare decimal numbers. The sequences were carefully chosen in order to provoke contradictions between R1 and R2. The first question gave rise to a conflict for three pairs, and for two of them, the conflict led to a new rule R'1 overcoming the contradiction: This rule consists in giving the same length to the decimal parts by adding the adequate number of zeros to the right of the shorter decimal part.

So Chrystel thought that 7.5 is less than 7.55, while Cecile argued for the reversed order; Chrystel convinced Cecile by proposing that she puts the same number of digits to both decimal parts: 7.5 equals 7.50 and 7.50 was recognized by Cecile as less than 7.55.

This new rule, which is adapted from R1, avoids the application of R2 and overcomes the conflict. It never occurred in the prior written test. It is noteworthy that these pairs elaborating the rule R'1 applied it in the next questions and could formulate it in the explanation meant for younger students.

Three consequences can be drawn from this example:

1. A social interaction could lead to a conflict, because of the choice of the numbers to be compared and of the composition of the pairs (students operating according to two different rules).

2. A conflict did not systematically appear in all cases in which it could have appeared.

3. Conflicts were not necessarily solved by the construction of a new rule.

This brings me to claim that the outcome of such social contradiction depends on several factors, some of which can be more or less controlled, such as the choice of the task variables of the problem given to the students. (By *task variables*, I mean features of the problem whose variations imply changes in the students' solving strategies; these variables, when used to promote learning, are also called "variables didactiques," didactical variables, in France.) The effect of the other ones linked to the individuals involved in the interaction is more uncertain: A social negotiation between two individuals is not predetermined, and all the past experience of each partner may play a role. So, in a study about group work at computers, Hoyles, Healy, and Pozzi (1993) were able to find a link between an initial antagonism between some group members and the emergence of a competitive organizational style within the group.

Several reports mention that conflicts are not always solved by rational arguments but also by authority arguments. Arrangements can be found among partners that are external to the mathematical problem. And if a conflict is solved by rational arguments, neither the solution nor the reason is necessarily correct from a mathematical point of view. Balacheff (1991, pp. 188-189) concludes from an experiment on proving processes developed by students working in pairs that social interaction may give rise to argumentative behaviors leading to a resolution of the conflict on a nonscientific basis. Balacheff claims that these behaviors may even become obstacles to the elaboration of a proof by students. They can, for instance, favor naive empiricism or the use of a crucial experiment instead of a higher-level proof.

Coming back to my initial interpretation, I interpret this claim as the social problem overtaking the mathematical one: Students are more eager to win socially than solve the problem. In this sense, it is possible to consider social interaction as a potential obstacle to the "devolution" (i.e., the appropriation) of the mathematical problem and thus to the development of mathematical processes.

3. COOPERATION IN THE SOLVING PROCESS

It has also been observed that, even when students are not in conflict, cooperative work may lead to a better solution than individual work (Vandenplas-Holper, cited in Beaudichon & Vandenplas-Holper, 1985). Uyemura-Stevenson (cited in De Avila, 1988, p. 113) found significant relationships between student-student consultation and performances or even math conceptualization, more than when student-student consultation was replaced with teacher-student consultation or when both consultations were combined.

Cooperative work is more widespread in classrooms than conflicting situations. Grevsmühl (1991) analyzed the verbal exchanges between students in pairs solving mathematical word problems and observed that the major part of the speech acts (70%) indicated a cooperation with the partner. In group work in the classrooms, in which groups are not constituted in a strict way like in research experiments, social interactions are less frequently based on well-delineated conflicts. Collaborative processes may take place. Proposals made by one student may be improved by the partner and transformed into more sophisticated solutions. New approaches toward a solution may be elaborated from proposals made by whatever students and overcome the simple addition of ideas. Robert and Tenaud (1989) could confirm this claim in a long-term teaching-learning experiment on geometry for 17- to 18-year-old students in which students regularly worked in groups. The group work was organized in interaction with systematic institutionalization phases made (after one or several sessions of group work) by the teacher not only about the mathematical solving strategies linked to the problem but also about the generalization of methodological points.

What are the features of group work favoring this phenomenon of a social construction of a higher-level solution than the individual proposals? I would like to refer to the notion of "zone of proximal development" proposed by Vygotsky (1985, p. 269), the zone of possible conceptual states reached by the student when interacting with an adult or a more advanced partner. It seems that it is possible to extend some characteristics of this notion to the case in which a group of peers is collaborating on a joint task. The two main characteristics in which scientific concepts differ from everyday life concepts, are (according to Vygotsky, 1985, p. 287) "the awareness and the voluntary aspects" of their genesis. Cooperating with others contributes to the development of these two characteristics through the explaining and refuting processes social interaction requires: Coming to an agreement on a common solution with others requires at least making one's own approach explicit, possibly comparing it with the approach of the partner, and even arguing against it (this is the extreme case of a conflicting situation). Robert and Tenaud (1989) assume that this phase of elicitation of the method is more widespread in group work than in individual work, and they consider it as supporting the development of an improvement of the solving process. Yackel (1991) develops a further argument, namely, that the discussion should involve several students (more than two), and supports her claim by an example of peer questioning in a 2nd-grade class, which fostered sophisticated forms of explanation and argumentation that were not present when students worked alone or in pairs.

Group work may also allow the exteriorization of various strategies and lead students to a decentration of their point of view, because it pushes them to situate their solution among the various other ones. Moving from one solving strategy to another one is a second feature that may also be the origin of conceptual progress: Knowing how to consider a problem under various points of view, how to move from one strategy to another one with regard to the problem to be solved, contributes to a more flexible use of knowledge and to a decontextualization of mathematical ideas.

It should be noted that this ability of moving from one strategy to another one is particularly efficient for complex problems, which cannot be solved by routines or algorithms but require the combination of several approaches. This was exactly the case in the geometry problems used by Robert and Tenaud. It means that the possible superiority of group work is strengthened in complex situations, allowing a multiple approach and not a single routine solution.

This interpretation of the role of the diversity of points of view is supported by research findings from Hoyles, Healy, and Pozzi (1993). They identified four organizational styles in the group work they observed on various tasks at computers and noticed that in the "competitive" style (the group splits into competitive subgroups without communication), the opportunity for exchanging and being confronted with alternative perspectives or different modes of representing the same problem space was reduced. These authors related this to the fact that this competitive style turned out to provide both less productivity (quality of the group outcome in the task) and less effectiveness on the learning of new knowledge than a "collaborative" style in which students shared their local and global targets on the tasks in common discussions.

However, the positive influence of peer discussion is questioned by some studies (Pimm, 1987, Pirie & Schwarzenberg, 1988). Fine-grained studies on episodes of collaborative small group activity (Cobb, Yackel, & Wood, 1992) focus on the construction of a shared meaning in social interaction (a meaning that is neither the intersection nor the addition of the individual meanings but arises out of the interaction), and state that this shared meaning emerges from a circular, self referential sequence of events rather than a linear cause-effect chain: "the students can be said to have participated in the establishment of the situations in which they learned" (Cobb, Yackel, & Wood, p. 99). This stresses the complexity of such social interaction situations and may explain the diversity of research results.

3.1 Group Work at Computers

Group work is enhanced in the mathematics classroom through the introduction of computers. Students very often work in small groups at the computer (2, 3, or 4 students). It has been observed that students are likely to subdivide the task into subtasks more often than in a paper-and-pencil task (Gallou, 1988, pp. 31-32; Hoyles & Sutherland, 1990): One student is in charge of manipulating on the computer (programming, typing, handling the mouse, etc.) while the other(s) propose(s) or even dictate(s) what is to be done, like in the case study of Janet and Sally (Hoyles & Sutherland, 1990, pp. 328-329). The necessity of material manipulation may be a cause of organization of work and "division of labor" hindering discussion.

In the analysis of structures of interaction between several students solving a joint task together at a computer, Krummheuer (1993) was able to observe a form of interaction that he calls "automatisiertes Trichtermuster" ("automatized funnel pattern"). This is very close to a common structure of interaction in traditional teaching between teacher and students: The "Trichtermuster" accounts for a communication that is established between the teacher and the students, in which, by narrower questions, the teacher manages to obtain the expected local answer from the students; this kind of interaction prevents students from constructing a global meaning of the situation. In computer tasks, a similar communication may be established between students dealing only with short actions to be done on a computer in order to obtain as rapidly as possible an expected effect on the screen instead of trying to carry out a shared reflection on a possible strategy for the whole mathematical problem. The device, through the material effects it can produce, absorbs all the interaction content, offering another kind of obstacle to the development of a solution. It must be stressed that it is difficult to escape the attraction of a narrow focusing on the computer, because the computer offers visible feedback to every action (effect of the action produced on the screen). Hoyles, Healy, and Pozzi (1993) also observed a better group outcome when students could have discussions away from the computer during global target episodes. This group work at computer needs to be investigated more closely, especially since the introduction of *direct manipulation*, which may reduce the discussion about local syntax problems of programming. But new problems may arise from the meaning students give to this direct manipulation (cf. Hölzl, 1992).

4. LIMITATIONS OF THE FUNCTIONING OF COOPERATIVE WORK

It has been mentioned that various immediate outcomes of a group work are possible even if the students agree on a common solution: (a) a better solution is found than a single student would have produced; (b) the agreement on a solution is based on authority arguments; and (c) the agreement is based on cognitive grounds, but not mathematically satisfying ones even in the case of a right solution.

We suggest that three categories of conditions play a role in the positive immediate outcome: choice of the partners, choice of the task, and length of interaction.

4.1 Choice of the Partners

In their experiments, the Genevian social psychologists stress that the problem posed to the children is essentially of social nature, that the sociocognitive tools are developed only for re-establishing an equilibrium of social nature (Carugaty & Mugny, cited in Mugny, 1985, p. 66). For them, the social problems precede the cognitive problem. The problem situations we organize involve a mathematical problem and complex contents; the solution processes require the use of mathematical knowledge, and we do not follow these researchers concerning the priority of the social problem. We consider that the overtaking of the situation by the social problem is a misdevelopment that must be avoided, and the equilibrium is, in our opinion, a conceptual equilibrium related to mathematical conceptions of students. That is why the "cognitive distance" between the partners must have an optimal size: not too big (they cannot understand each other) and not too small (they have identical points of view).

4.2 Choice of the Task

Researchers in mathematics education have stressed the influence of the task on the behavior of students in group work and on the content of their exchanges (Hoyles, Healy, & Pozzi, 1993; Robert & Tenaud, 1989). The task must provide a new situation for the students that they cannot solve immediately (a discussion in this case would be useless), but in which they can start with their previous knowledge, although it is not enough to achieve the task. The task must favor verbalizing and communicating between students: That is the reason why it can occur, when students have to do something without justification, that they do not really exchange arguments on performing the task (e.g., procedural tasks on a computer). Cooperative work is enhanced when students have to describe or justify their solutions. Immediate objective feedback may also prevent a discussion between partners.

4.3 Length of the Interaction Process

A too small period of time does not allow interaction to take place; the interaction process is not a sequential one. Time is needed to internalize what the partner is proposing, to relate the proposals to previous approaches, and to understand the consequences of the proposal. In many experiments (e.g., Cobb, Yackel, & Wood, 1992; Laborde, 1982), we could observe that a proposal made by a student is not adopted immediately by the partner, but may be taken into consideration when the latter has experienced some difficulties with his or her own approach. The complexity of the progress of the solving processes in group work is higher than in an individual situation (see below).

5. SOCIAL INTERACTION AS A COMPONENT OF THE "MILIEU"

A constructivist perspective pays attention to situations in which the student must evolve on his or her own and not with the help of the teacher. For these situations, Brousseau (1986, p. 49) has stressed the role played by the interactions of the student with a given "milieu," that is, all elements of the environment of the task on which students can act and which gives them feedback of various kinds on what they are doing. Offered by the situation itself, the feedback to the actions of the students must enable them to have access to information about what they have done, to infer some conclusions about the validity of their work, and to make other trials resulting in an adapted solution. Such feedback may give evidence to the students to what extent their solution is not pertinent, it may make contradictions apparent. These contradictions provoke an imbalance that can give rise to new attempts of equilibration: Knowledge can originate from this dynamical process of imbalance and re-equilibration. This feedback is not only of a material nature but can also be of an intellectual nature when it provokes some contradiction between what the student expects thanks to his or her previous knowledge and what he or she can observe in the situation. According to Margolinas (1993), the previous knowledge of the student takes the role of *validity criteria*. One can recognize the underlying Piagetian notions of equilibration and cognitive conflict.

In this theoretical framework, social interactions between students are part of the milieu. Because of their social nature and their dependence on elements related to human behavior and ideas, they are not so certain and do not work in such a deterministic way as feedback coming from the physical environment. In one sense, the complexity of the milieu is increased.

The Russian research trend can be interpreted as a way of organizing the "milieu" in relation to the content of the task. In some experiments (Rivina, 1991; Polivanova, 1991; Roubtsov, 1991), group work was organized by giving different subtasks to each partner – but these tasks were not independent, and students had to coordinate their solutions in order to achieve the whole task. The subdivision of the task was based on a content analysis of the task. This research may be perceived as an attempt to reduce the uncertainty of the social interaction while relating it to the conceptual nature of the task. It was done on tasks in physics and in mathematics.

6. CONCLUSION: COMPLEXITY

As a conclusion, I would like to stress the common flavor in all work on social interaction: In these studies, the focus is on the complexity of social interaction situations. Introducing a social dimension into a learning situation contributes to an increase in the complexity of the situation by introducing an additional problem to the mathematical one. My analysis shows that several elements may play a crucial role in the quality of the group work and in the subsequent learning outcome.

1. When working in small groups, students must be aware of the social demands of the task and of what these demands imply. They must attempt to meet these demands, and this awareness does not result in a spontaneous

adaptation but has to be learned. That is why a positive outcome of such situations requires long-term experience.

2. Working in small groups involves a multiplicity of approaches and points of view, and thus a greater conceptual work of coordination.

These elements may not easily be controlled – and this fact may be one of the reasons why some teachers avoid using group work in their classes.

We believe that the positive outcome of introducing a social dimension into learning situations in mathematics is related to the increased complexity of these situations due to social aspects: Perhaps the greater complexity is a major reason for more learning.

REFERENCES

Balacheff, N. (1991). The benefits and limits of social interaction: The case of mathematical proof. In A. Bishop, S. van Dormolen, S. Mellin-Olsen (Eds.), *Mathematical knowledge: Its growth through teaching* (pp. 175-92). Dordrecht, Netherlands: Kluwer.

Beaudichon, J., & Vandenplas- Holper, C. (1985). Analyse des interactions et de leurs effets dans la communication référentielle et la maîtrise de notions, In G. Mugny (Ed.), *Psychologie sociale du développement cognitif* (pp. 125-49). Bern: Lang.

Brousseau, G. (1986). Fondements et méthodes de la didactique des mathématiques. *Recherches en didactique des mathématiques*, 7(2), 33-115.

Carugati, F., & Mugny, G. (1985). La théorie du conflit socio-cognitif. In G. Mugny (Ed.), *Psychologie sociale du développement cognitif* (pp. 45-70). Bern: Lang.

Cobb, P., Yackel, E., & Wood, T. (1992). Interaction and learning in mathematics classroom situations. *Educational Studies in Mathematics*, *23*(1), 99-122.

Coulibaly, M. (1987). Les décimaux en quatrième: Analyse des conceptions. *Mémoire de DEA*. Université Joseph Fourier, Grenoble 1, Laboratoire LSD2-IMAG.

De Avila, E. (1988). Bilingualism, cognition and minorities. In R. Cocking & J. Mestre (Eds), *Linguistic and cultural influences on learning mathematics* (pp. 101-22). Hillsdale, NJ: Erlbaum.

Gallou-Dumiel, E. (1988). Symétrie orthogonale et micro-ordinateur. *Recherches en didactique des mathématiques, 8*, 5- 59.

Garnier, C., Bednarz, N., & Ulanovskaya, I. (1991). *Après Vygotsky et Piaget - Perpectives sociale et constructiviste. Ecoles russe et occidentale.* Bruxelles: De Boeck Wesmael.

Grevsmühl, U. (1991). Children's verbal communication in problem solving activities. In F. Furinghetti (Ed.), *Proceedings of the Fifteenth PME Conference* (Vol. 2, pp. 88-95). Dipartimento di Matematica dell'Universita di Genova.

Grisvard, C., & Leonard, F. (1983). Comparaison de nombres décimaux, *Bulletin de l'APMEP* No. 340, September 1983, pp. 450-459.

Hölzl, R. (1992). Interpretative Analyse eines Problemlöseversuchs im Zugmodus der Cabri-Geometrie. *Zentralblatt für Didaktik der Mathematik, 4*, 128-34.

Hoyles, C., Healy, L. & Pozzi, S. (1993, February). *Telling a story about computers, groups and learning mathematics.* Paper presented at the ESRC InterSeminar, Collaborative Learning. Oxford.

Hoyles, C., & Sutherland, R. (1990). Pupil collaboration and teacher intervention in the Logo environment. *Journal für Mathematik-Didaktik, 11*(4), 323-343.

Krummheuer, G. (1993). Orientierungen für eine mathematikdidaktische Forschung zum Computereinsatz im Unterricht. *Journal für Mathematik-Didaktik, 14*(1), 59-92.

Laborde, C. (1982). *Langue naturelle et écriture symbolique: Deux codes en interaction dans l'enseignement mathématique*. Unpublished postdoctoral dissertation, IMAG, Grenoble.

Leonard, F., & Grisvard, C. (1981). Sur deux règles implicites utilisées dans la comparaison de nombres décimaux positifs. *Bulletin de l'APMEP*, No. 327, February 1981, pp. 47-60.

Margolinas, C. (1993). *De l'importance du vrai et du faux en mathématiques*. Grenoble: La Pensée suavage.
Mugny, G. (1985). *Psychologie sociale du développement cognitif*. Bern: Lang.
Pimm, D. (1987). *Speaking mathematically: Communication in mathematics classrooms*. London: Routledge & Kegan Paul.
Pirie S., & Schwarzenberg R. (1988). Mathematical discussion and mathematical understanding. *Educational Studies in Mathematics, 19*(4), 459-470.
Polivanova, N. (1991). Particularités de la solution d'un problème combinatoire par des élèves en situation de coopération. In C. Garnier, N. Bednarz, & I. Ulanovskaya (Eds.), *Après Vygotsky et Piaget - Perpectives sociale et constructiviste. Ecoles russe et occidentale*. Bruxelles: De Boeck Wesmael.
Rivina, I. (1991). L'organisation des activités en commun et le développement cognitif des jeunes élèves. In C. Garnier, N. Bednarz, & I. Ulanovskaya (Eds.), *Après Vygotsky et Piaget - Perpectives sociale et constructiviste. Ecoles russe et occidentale* (pp. 163-178). Bruxelles: De Boeck Wesmael.
Robert, A., & Tenaud, I. (1989). Une expérience d'enseignement de la géométrie en Terminale C. *Recherches en Didactique des Mathématiques, 9*(1), 31-70.
Rogoff, B. (1990). *Apprenticeship in thinking*. Oxford: Oxford University Press.
Roubtsov, V. (1991). Activité en commun et acquisition de concepts théoriques par les écoliers sur le matériel physique. In C. Garnier, N. Bednarz, & I. Ulanovskaya (Eds.), *Après Vygotsky et Piaget - Perpectives sociale et constructiviste. Ecoles russe et occidentale*. Bruxelles: De Boeck Wesmael.
Tudge, J. (1992). Processes and consequences of peer collaboration. *Child Development, 63* (6), 1364-1379.
Vygotsky, L. (1985). *Pensée et langage* (Sève, F., Trans.). Paris: Editions Sociales. [Original work published 1934]
Yackel, E. (1991). The role of peer questioning during class discussion in second grade mathematics. In F. Furinghetti (Ed.), *Proceedings of the Fifteenth PME Conference*. (Vol. 3, pp. 364 - 371). Dipartimento di Matematica dell'Universita di Genova.

MATHEMATICS CLASSROOM LANGUAGE: FORM, FUNCTION AND FORCE

David Pimm
Milton Keynes

1. INTRODUCTION

The expression "the state of the art" has two main senses. The first refers to a domain as a whole and usually involves a broad survey of the current field, perhaps discussing how it came to be so. The second sense invokes a single, particular view located out on the rim. In this chapter, I shall endeavour to address both senses, firstly by offering a necessarily brief survey of some recent work on mathematical classroom language, in the context of work on language and mathematics in general, before discussing a more idiosyncratic and personal set of interests and emphases, finishing with some suggestions for future areas of important work yet to be done.

There are many different relationships that can be highlighted between language and mathematics. Such considerations can frequently be found under the heading of "the language *of* mathematics," though this latter phrase can be interpreted in a number of senses. It can variously mean:

1. the spoken language of the mathematics classroom (including both teacher and student talk);
2. the use of particular words for mathematical ends (often referred to as the *mathematics register*);
3. the language of texts (conventional word problems or textbooks as a whole, including graphic material and other modes of representation);
4. the language of written symbolic forms.

General collections on the area of language and mathematics include Cocking and Mestre (1988), Durkin and Shire (1991), Ellerton and Clements (1991), and a review of the area from a psychological research perspective is offered by Laborde (1990).

It is important to note that the phrase "the language of mathematics" can also refer to language used in aid of an individual doing mathematics alone (and therefore include, e.g., "inner speech"), as well as language employed with the intent of communicating with others. Language can be used both to conjure and control mental images in the service of mathematics. As Douglas Barnes (1976) has insightfully commented: "Communication is not the only function of language." And the Canadian literary critic Northrop

R. Biehler, R. W. Scholz, R. Sträßer, B. Winkelmann (Eds.),
Didactics of Mathematics as a Scientific Discipline, 159-169.

Frye (1963) has written: ". . . the language of mathematics, which is really one of the languages of the imagination, along with literature and music." However, in this chapter, after a few broader illustrations of the area in general, I shall focus particularly on issues of mathematics classroom language – though it is an interesting open question concerning how the fact that it is *mathematics* that is under discussion shapes and influences all of the language forms and functions that are customarily employed.

2. CHANNELS OF MATHEMATICAL COMMUNICATION

The teaching and learning of mathematics involves the activities of reading and writing, listening and discussing. Each of these linguistic aspects of classrooms has engendered considerable work. A few items in each activity are mentioned here.

Since the early 1980s, discussion in mathematics classrooms and teacher gambits to promote and facilitate it have moved onto the educational agendas in some Western countries (e.g., in the UK, with the Cockcroft report, DES, 1982; in the US, with the publication of the NCTM Standards document, 1989). Various attempts to specify which parts of classroom talk are to count as mathematical discussion have been proposed. For instance, Pirie and Schwarzenberger (1988, p. 461) offer: "It [mathematical discussion] is purposeful talk on a mathematical subject in which there are genuine pupil contributions and interaction."

However, there is still the vexed question of the particular contributions that talk of this kind (and whose "purposes" and whose decisions about the "genuine" nature of the interactions) can make to the specific learning of mathematics. I indicate below an example of teachers choosing to ignore the meaning in favour of attending to the form of an utterance: One possibility here is to find situations in which the teacher is making such judgements and endeavour to study them.

I have looked at the situation of reporting back on a range of open-ended or problem-solving activity and explored a number of questions about active listening, as well as the linguistic demands placed on all participants when engaged in reporting back to the rest of the class. More specific questions include: How can students develop the linguistic skills of reflection and selection of what to report? How can they work on acquiring a sense of audience? To whom is the reporter talking? (For more on this topic, see Pimm, 1992.)

Finally, a current general orientation to classroom talk (arising from ethnographic research) invokes the notions of representativeness and voice: Who gets to speak? Whom do we hear from in classrooms, and how? And about what? Who remains silent, how and why? (Are they silenced or do they silence themselves?) One result of the disciplined ways of looking that many fields develop may be that the same voice (or voices) gets replicated over and over. These questions are worth asking of mathematics classrooms. One focus might be on representativeness of the voices of the two genders, or the various ethnic or social groups, while another might be more on the form and structure of spoken interactions between mathematics teachers and students in general.

There are important differences between speech and writing, not least with regard to relative permanence and linear or non-linear flow in time, as well as being able to see the whole discourse when written down (an aid for reflection). There has been much research on reading in mathematics. Early work focused primarily on the problematic notion of "readability indices," which objectified the phenomena of interest and located it as a property of the text alone. Subsequently, some more interesting work has been done, in particular on the strategies and skills of what can be read (Borasi & Siegel, 1990). Part of reading facility involves constructing meaning from written texts, a task that becomes increasingly central as students progress through the educational system (see Laborde, 1991, for an interesting account).

One current theme of research on writing involves looking at the issue of student journal writing as an aid to learning mathematics. Some discussion of this issue can be found in Borasi and Rose (1989). Waywood (1990, 1992) has formulated an initial classification of types of secondary school mathematics journal writing as a framework for analysing how journals might provide a vehicle for student learning. His proposed triple, sequential categorization of use is: *recount* (narrative), *summary description*, *dialogue* (between ideas). His aim is one of reflection on learning, and from this work he has generated the hypothesis that the mode of journal writing reflects the stance towards learning on the part of the student.

3. COMMUNICATING MATHEMATICALLY IN CLASSROOMS

Since Aiken's seminal research review in 1972, entitled *Language factors in learning mathematics*, the area of mathematical classroom language has exploded dramatically in the subsequent 20 years, and a comprehensive bibliography would now run to hundreds of entries. In part, this phenomenal growth has paralleled the increasing interest in the role of language and social factors in schooling in general, after decades of relative under-emphasis during what might be called "the Piagetian years." A contemporary Western revival of interest in the work of Vygotsky on the one hand (see, e.g., Edwards & Mercer, 1988, discussed further below) and the specific examination of the classroom as a discourse context by linguists on the other (see Sinclair & Coulthard, 1975; Stubbs, 1983) have altered the research arena considerably.

In general, since the mid-1970s, techniques of discourse analysis have been used to examine aspects of classroom discourse, among other linguistic contexts, and to highlight certain normative aspects of language use in these particular speech settings. One early "finding" by Sinclair and Coulthard (1975) was the almost incessant repetition of the sequence

I(nitiation) – *R*(esponse) – *F*(eedback) in teacher-student exchanges. (In the excerpt that follows (Yates, 1978), T is the teacher, P refers to any pupil (student), and they are discussing the problem of finding a means of communicating what is on the blackboard (a route map of major cities and motorway links in England) to someone in the next room. I have added my suggested codings.

> P: Morse Code. (R)
> T: Morse code, well that is not necessary. We can speak to him – he is only the other side of the door. (F)
> P: Coordinates. (R)
> T: Coordinates would be one way of doing it. That would be a very good way of doing it. What do you mean by coordinates? (F then I)
> P: Say five across and down this way. (R)
> T: Well that is a very good idea, it is one I had certainly not thought of. Any other bright ideas? (F then I)
> P: Hold up a mirror. (R)
> T: Hold up a mirror – it cannot go through a solid door. . . . (F)

A more detailed, analytic account of this I–R–F sequence and some transcripts from lessons in which mathematics teachers have found ways of escaping from it is given in Pimm (1987).

However, there has been some concern about discourse analysis' technique of ignoring content and attending only to the form of an utterance in terms of classifying and analysing classroom language. Observations about what discourse analysis *cannot* offer are made by Edwards and Mercer (1988) in their book *Common Knowledge*. They comment:

> It may be thought that a concern with the content of the talk rather than with its form, and with interpreting people's meanings rather than coding their turns at speaking, is an altogether less rigourous and objective way of dealing with discourse. (p. 10)

But they then go on to offer three justifications for so doing. These are: formal discourse analysis does not allow them to answer the questions they want to ask; their analyses are offered in terms of the data themselves, not data already coded; discourse analysis itself also needs an interpretative framework in order to make judgements about coding. One interesting area of work that I shall mention later involves situations in which mathematics *teachers* themselves opt to ignore the content in favour of the form of what a student has said as part of their teaching strategies.

By means of a detailed study of some science and mathematics teaching in a classroom (particularly a set of lessons involving exploring relations among various features of a simple pendulum), Edwards and Mercer examine the rhetoric of "progressive" education in English elementary schools. They focus on the disparity between the level of freedom accorded the students at the level of action and that at the level of discourse and "generation" of the knowledge. They also detail various indirect teacher devices for constructing the "common knowledge" in the classroom, identifying: controlling the flow of conversation; determining who is allowed to speak, when and about what; use of silence to mark non-acceptance of a student's offering; reconstructing (and reformulating) "recaps" of what has been said, done or ostensibly discovered. All of these devices are common features of mathematics lessons.

One key focus they allude to is that of teacher questions. "Teachers may all be obliged to control classes and lessons, but they choose particular strategies for doing so . . . . However, there has been hardly any research on teachers' purposes in asking questions" (Edwards & Mercer, 1988, p. 30).

One researcher who has undertaken an exploration of this topic in the context of mathematics classrooms is Janet Ainley (1987, 1988). She has insightfully explored the varied functions of questions and how they are interpreted by students. She discovered many mismatches in interpretation of videotaped extracts of teaching. These excerpts were shown to elementary students and their teachers separately and they were asked about *why* they thought the teacher asked a particular question. The notion of the purpose of "focusing question" is of particular appositeness in mathematics, due to the problem of indicating where a student's attention should be (see, also, Love & Mason, 1991).

As I mentioned earlier, within natural language there are conventionally two main channels, those of speech and writing. (It is important, however, not to ignore the particular nature of working with mathematics and either the deaf – e.g., Barham & Bishop, 1991 – or the visually impaired.) One difficulty facing all teachers of mathematics is how to facilitate their students' moving from the predominantly informal spoken language, with which they are all pretty fluent (Brown, 1982), to the formal written language, which is frequently seen as the hallmark of much mathematical activity. The diagram below (Figure 1) indicates two alternative routes, and highlights different classroom practices in terms of working with students.

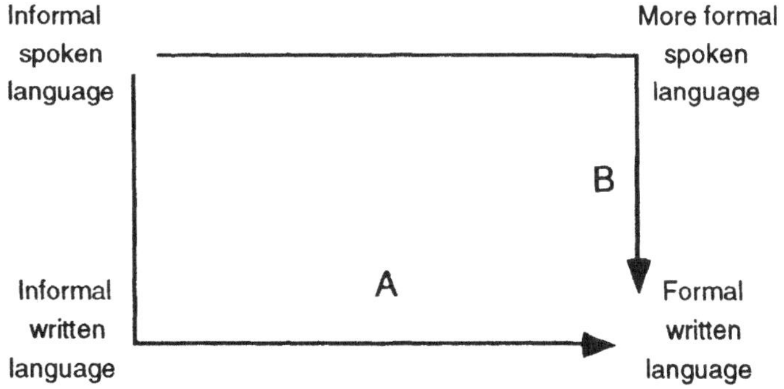

Figure 1. From informal spoken to formal written language.

Route A encourages students to write down their informal utterances, which are then worked on in terms of increasing the adequacy of the written form to stand on its own (e.g., by use of brackets or other written devices to convey similar information to that which is conveyed orally by emphasis or intonation). Route B involves work on the formality and self-sufficiency of the spoken language *prior* to its being written down. This usually involves constraints being placed on the communicative situation, in order to highlight attention to the language used. Reporting back, mentioned earlier, offers one such instance of this latter route.

Students learning mathematics in school in part are attempting to acquire communicative competence in both spoken and written mathematical language. Educational linguist Michael Stubbs claims (1980, p. 115): "A general principle in teaching any kind of communicative competence, spoken or written, is that the speaking, listening, writing or reading should have some genuine communicative purpose." Is this at odds with viewing a mathematics classroom as an avowedly, deliberately, *un-natural*, artificial setting, one constructed and controlled with particular aims in mind, one in which the structure and organization of the discourse by the teacher has some quite unusual features?

4. LOOKING TOWARDS THE FUTURE: FORM, FUNCTION AND FORCE

In the second half of this chapter, I turn to some of these particular features of mathematical classroom discourse that I am currently emphasizing in my own work. I focus in particular on the nature of teacher-student spoken interactions and forms of speech.

4.1 Meta-Knowledge and Meta-Commenting

My first proposed area for work involves examining the knowledge and levels of awareness students have (whether tacit or explicitly available) of some of the teacher's *forms* of utterance, and the extent to which they are identified as part of the role of being a teacher – as opposed to forming aspects of the idiolect of that individual (who happens to be their teacher). Explorations might be carried out where, for instance, students are recorded working in groups to see whether certain students "take on" some of the teacher's functions (acting *in loco domini* whether by agreement or assertion) and what language forms they use in so doing. By continuing to record student conversations when the teacher arrives at such a group, transitions to and from "teacher discourse" may be recorded.

Some teachers explicitly attempt to "hand over" some of their functionings to groups. If a teacher arrives at a table asking "What question am I about to ask you?", a different interpretative task is being offered from the one initiated by the request "Tell me what you are doing." The teacher question "What question am I about to ask you?" is a meta-question designed to encourage students to notice the teacher's interventions as regular and systematic. It also carries with it the implicit suggestion that the student might take on the particular function that the teacher has been carrying out up until now by asking the same question of herself.

In a paper entitled *Organizing classroom talk*, Stubbs (1975) offers the notion that one of the characterizing aspects of teaching discourse as a speech event is that it is constantly organized by meta-comments, namely that the utterances made by students are seen as appropriate items for comment themselves, and, in addition, that many of the meta-remarks are evaluative.

He comments:

> The phenomenon that I have discussed here under the label of meta-communication, has also been pointed out by Garfinkel and Sacks (1970). They talk of "formulating" a conversation as a feature of that conversation.
>
> > A member may treat some part of the conversation as an occasion to describe that conversation, to explain it, or characterise it, or explicate, or translate, or summarise, or furnish the gist of it, or take note of its accordance with rules, or remark on its departure from rules. That is to say, a member may use some part of the conversation as an occasion to *formulate* the conversation.
>
> I have given examples of these different kinds of "formulating" in teacher-talk. However, Garfinkel and Sacks go on to point out that to explicitly describe what one is about in a conversation, during that conversation, is generally regarded as boring, incongruous, inappropriate, pedantic, devious, etc. But in teacher-talk, "formulating" is appropriate; features of speech do provide occasions for stories worth the telling. I have shown that teachers do regard as matters for competent remarks such matters as: the fact that somebody is speaking, the fact that another can hear, and whether another can understand. (Stubbs, 1975, pp. 23-24)

A glance at any mathematics lesson transcript bears out Stubbs' claim – the language students use is more often in focus by the teacher than what they are trying to say with it. In addition to the general categories mentioned above, here is a more interesting "example" of more particular relevance to mathematics.

> Zena: Can I just rub it out?
> Teacher: Yes, do. [With slight irony, as she has already rubbed out the final 3 with her finger and changed it to a 4.] You can even use a board rubber if you want to.
> Zena: [Looks at the teacher who is standing at the back of the class] Is that all right?
> *Pause (2 secs)*
> Teacher: Zena asked a question.
> [Chorus of yesses from the class.]

In relation to my earlier mention of Ainley's work on questioning, I am led to wonder whether Zena appreciated any difference between her two questions that were dealt with very differently by the teacher. Here, his meta-comment "Zena asked a question" offers a deflection that allows him apparently to take a turn in the conversation yet without having to respond to Zena's request for evaluation directly.

Pragmatics is an area of linguistics dealing with how words can be used to do things, to achieve one's ends. The philosopher Paul Grice (1989) has proposed a co-operative principle and a series of very general maxims to try to account for how and why discourse works and coheres. He cites the example of the book review, which, in its entirety, runs: "This book has narrow margins and small type." What implicatures must be made in order to construe this as a book review? One of Grice's suggestions enjoins us to behave so as to "avoid obscurity of expression, avoid ambiguity."

1. The maxim of *Quality* (be truthful, according to the evidence you have).
2. The maxim of *Quantity* (be informative, but not over-informative).
3. The maxim of *Relevance* (be relevant to the conversation).
4. The maxim of *Manner* (say things clearly, unambiguously, briefly).

I have yet to look at the notion of meta-commenting in relation to violations of Grice's maxims. But it is an interesting observation that many of Grice's maxims of conversation are regularly and systematically violated in classroom discourse.

> A teacher and a student are putting up posters and having to take out many old staples:
> *Student*: Do we have to take them all out?
> *Teacher*: You can sweep dust under the carpet too.

4.2 Modality and Hedges[1]

A second general area arises under the general heading of "modality," which initially referred to the use of modal verbs (see Stubbs, 1986) to mark the degree of speaker certainty or uncertainty (e.g., "that *might* be true"), but now has a more general meaning. One discussion of the notion in relation to mathematics learning can be found in Anne Chapman's (1993) doctoral dissertation *Language practices in school mathematics: A social semiotic perspective.* She writes:

> Hodge and Kress (1988) use the semiotic term *modality* to describe the social construction or contestation of knowledge. Modality refers to the degree of certainty embedded in a statement . . . . In any school subject, the weighting attached to what is said is important. Mathematics, in particular, is typically regarded as a factual subject and thus is likely to have a high modality structure. (p. 57)

What other linguistic means are commonly available and used in mathematics classrooms for indicating the speaker's relation to or stance taken with respect to some knowledge claim uttered? In John Wyndham's novel *The Kraken Wakes*, for instance, one of the characters reports:

---

[1] A general term in this area is "hedge" (see, e.g., Lakoff, 1972), though Prince, Frader, & Bosk (1982) have usefully distinguished between "hedges" and "shields." An example of a *shield* is "I *think* that X is true," where the uncertainty is in relation to the speaker's level of confidence in the truth of the assertion, while a *hedge*, such as "the cost is approximately £20," has the uncertainty marker inside the proposition itself.

> "For present purposes the danger area is being reckoned as anything over four thousand", said Dr Matet . . . .
> "And what depth did you advise as marking the danger area, Doctor?"
> "How do you know I did not advise four thousand fathoms, Mrs Watson?"
> "Use of the passive, Doctor Matet – 'is being reckoned'." . . .
> "And there are people who claim that French is the subtle language," he said.
> (Wyndham, 1970, pp. 101-102)

Seeing how the status of and beliefs about the validity of knowledge claims are crucial in mathematics, again it seems curious to me that more is not known about how these pragmatic utterances are made. Though it must be said this forms a subtle part of communicative competence. Recently, a similar shift of focus and concern has occurred in mathematics education to that from syntactic to semantic and then to the burgeoning area of pragmatic issues present in linguistics itself. I predict the extremely subtle pragmatic interpretative judgements regularly made by both teachers and students in the course of mathematics teaching and learning in classrooms will move steadily to the fore as a research topic.

4.3 Force

My current thesis is quite simple. All that hearers have direct access to in the classroom is the *form* of any utterance. But that form is influenced and shaped by the intended *function* of the utterance (some particular examples of general teacher functions include: keeping in touch, to attract or hold student attention, to get them to speak or be quiet, to be more precise in what they say). And form is also shaped by personal *force*, the inner purposes and intentions of the speaker, usually in this case what the teacher is about both as a teacher and a human being.

I am currently exploring some aspects of mathematics classroom discourse with regard to:

1. *Linguistic form* (all that is actually readily available to the external ear and eye): for instance, pronominal usage and deixis (Pimm, 1987, on "we"; Rowland, 1992, on "it"). Mathematics has a problem with its referents, so the ways in which language is made to point is of particular interest.

2. *Some of the apparent or hoped-for functions* (quite common and general ones, such as, for the teacher, having students say more or less, deflecting questions; or for the student, avoiding exposure, engaging with the content, finding out what is going on).

3. *Force*. The personal, individual intents (conscious and unconscious) that give rise to the desire to speak. I start from the premise (that of Anna Lee, founder of the Shakers) that "Every force evolves a form."

I believe that force and function combine to shape form, but also that the existence of conventional forms of speaking, the pressure of certain classroom discourse patterns, can actually interfere with expression. I am also becoming increasingly interested in how the notion of *force*, of necessity must include "unconscious force." (See Blanchard-Laville, 1991, 1992, for a lucid account of her work on some of the psychological elements at work in mathematics teachers. There is also a special issue – Vol. 13(1), 1993 – of the journal *For the learning of mathematics* devoted to aspects of unconscious elements in mathematics education.)

Some "anomalous" examples of student discourse can be used to argue for "unconscious" forces also being present: for instance, a student offering the word "fidelity" rather than "infinity," or (from Tom Kieren) another explaining that "Four-fifths is my favourite fraction, because it gives me a lot to think about. There are five in our family but only four are here. It is my mother who is gone." A third possible instance comes from an interview carried out by Lesley Lee in which "odd" numbers are referred to consistently as "bastards" (see Pimm, 1991, 1993; Tahta, 1991). Despite such instances being difficult to discuss, let alone submit to systematic exploration, I predict an increasing attention to unconscious elements will also emerge.

Learning to speak mathematically involves stressing and ignoring and is achieved only at a cost. How aware are students of teachers' intentions and that the nature of the latter's classroom talk is closely related to them? What are some of the relations among teacher focus and student learning in mathematics? While the external form of the discourse is all that is observable, it is how that form relates to, and is successfully generated by, inner phenomena that should be one of the prime considerations of mathematics education. Learning to use mathematical language successfully is not solely the learning of forms in themselves, yet control over the forms is one product of that learning. How can we assist our students in discerning our intents through the forms – the forms that can necessarily be their sole *external* experience?

REFERENCES

Aiken, L. (1972). Language factors in learning mathematics. *Review of Educational Research, 42*, 359-385.

Ainley, J. (1987). Telling questions. *Mathematics Teaching, 118*, 24-26.

Ainley, J. (1988). Perceptions of teachers' questioning styles. In E. Borbás (Ed.), *Proceedings of PME XII Conference (pp. 92-99).* Veszprem: OOK Printing House.

Barham, J., & Bishop, A. (1991). Mathematics and the deaf child. In K. Durkin & B. Shire (Eds.), *Language in mathematical education* (pp. 179-87). Milton Keynes: Open University Press.

Barnes, D. (1976). *From communication to curriculum.* Harmondsworth: Penguin.

Blanchard-Laville, C. (1991). La dimension du travail psychique dans la formation continue des enseignant(e)s des mathématiques. In F. Furinghetti (Ed.), *Proceedings of PME XV* (pp. 152-159). Assisi: Programme Committee of the 15th PME-Conference.

Blanchard-Laville, C. (1992). The dimension of psychic work in the in-service training of teachers. *For the learning of mathematics, 12*(3), 45-51.

Borasi, R., & Rose, B. (1989). Journal writing and mathematics instruction. *Educational Studies in Mathematics, 20*(4), 347-365.

Borasi, R., & Siegel, M. (1990). Reading to learn mathematics: New connections, new questions, new challenges. *For the learning of mathematics, 10*(3), 9-16.

Brown, G. (1982). The spoken language. In R. Carter (Ed.), *Linguistics and the teacher.* London: Routledge & Kegan Paul.

Chapman, A. (1993). *Language practices in school mathematics: A social semiotic perspective*. Unpublished doctoral dissertation. Murdoch University, Perth, Australia.
Cocking, R., & Mestre, J. (Eds.). (1988). *Linguistic and cultural influences on learning mathematics*. Hillsdale, NJ: Erlbaum.
DES (1982). *Mathematics counts*. London: HMSO.
Durkin, K., & Shire, B. (Eds.). (1991). *Language in mathematical education*. Milton Keynes: Open University Press.
Edwards, D., & Mercer, N. (1988). *Common knowledge*. London: Methuen.
Ellerton, N., & Clements, M. (1991). *Mathematics in language: A review of language factors in mathematics learning*. Geelong, Australia: Deakin University Press.
Frye, N. (1963). *The educated imagination*. Toronto: CBC Enterprises.
Garfinkel, H., & Sacks, H. (1970). On formal structures of practical actions. In J. McKinney & E. Tiryakian (Eds.), *Theoretical sociology: Perspectives and developments* (pp. 337-366). New York: Appleton-Century-Crofts.
Grice, P. (1989). *Studies in the way of words*. Harvard, MA: Harvard University Press.
Hodge, R., & Kress, G. (1988). *Social semiotics*. Cambridge: Polity Press.
Laborde, C. (1990). Language and mathematics. In P. Nesher & J. Kilpatrick (Eds.), *Mathematics and cognition* (pp. 53-69). Cambridge: Cambridge University Press.
Laborde, C. (1991). Lecture de textes mathématiques par des élèves (14-15 ans): Une experimentation. *Petit x, 28*, 57-90.
Lakoff, G. (1972). Hedges: A study in meaning criteria and the logic of fuzzy concepts. *Chicago Linguistic Society Papers*. Chicago, IL: The Society.
Love, E., & Mason, J. (1991). *Teaching mathematics: Action and awareness*. Milton Keynes: Open University.
NCTM (1989). *Curriculum and evaluation standards*. Reston, VA: NCTM.
Pimm, D. (1987). *Speaking mathematically*. London: Routledge & Kegan Paul.
Pimm, D. (1991). Signs of the times. *Educational Studies in Mathematics, 22*(4), 391-405.
Pimm, D. (1992). "Why are we doing this?" Reporting back on mathematical investigations. In D. Sawada (Ed.), *Communication in learning mathematics* (pp. 43-56). Edmonton, Alberta: MCATA.
Pimm, D. (1993). The silence of the body. *For the learning of mathematics, 13*(1), 35-38.
Pirie, S., & Schwarzenberger, R. (1988). Mathematical discussion and mathematical understanding. *Educational Studies in Mathematics, 19*(4), 459-70.
Prince, E. F., Frader, T., & Bosk, T. (1982). On hedging in physician-physician discourse. In R. J. di Pietro (Ed.), *Linguistics and the professions* (pp. 83-98). Norwood, NJ: Ablex.
Rowland, T. (1992). Pointing with pronouns. *For the learning of mathematics, 12*(2), 44-8.
Sinclair, J., & Coulthard, M. (1975). *Towards an analysis of discourse*. London: Oxford University Press.
Stubbs, M. (1975). *Organizing classroom talk*, Occasional paper 19, Centre for Research in the Educational Sciences, University of Edinburgh, Scotland.
Stubbs, M. (1980). *Language and literacy*. London: Routledge & Kegan Paul.
Stubbs, M. (1983). *Discourse analysis*. Oxford: Basil Blackwell.
Stubbs, M (1986). A matter of prolonged fieldwork: Notes towards a modal grammar of English. *Applied Linguistics, 7*(1), 1-25.
Tahta, D. (1991). Understanding and desire. In D. Pimm & E. Love (Eds.), *Teaching and learning school mathematics* (pp. 220-246). London: Hodder & Stoughton.
Waywood, A. (1990). Mathematics and language: Reflections on students using mathematics journals. In G. Davis & R. Hunting (Eds.), *Language issues in learning and teaching mathematics*. Bundoora, Australia: La Trobe University.
Waywood, A. (1992). Journal writing and learning mathematics. *For the learning of mathematics, 12*(2), 35-43.
Wyndham, J. (1970). *The Kraken wakes*. Harmondsworth: Penguin.
Yates, J. (1978). *Four mathematical classrooms*. Technical report, available from Faculty of Mathematical Studies, University of Southampton, Southampton, England.

CHAPTER 4

TECHNOLOGY AND MATHEMATICS TEACHING

edited and introduced
by
Bernard Winkelmann
Bielefeld

Technology always has had great influence on teaching in general and on mathematics teaching in particular. On a more general level, we may think of printed textbooks, paper and pencil, blackboards, ready-made or teacher-prepared overhead transparencies, or videotape sequences illustrating mathematical concepts and relationships, as well as the use of standard software by the teacher to produce worksheets, store students' data, correct examination tasks, search for mathematics-related information from encyclopedias on CD-ROM, or get real data for statistical analysis in wide area networks. On a more mathematical level, there are various mathematical instruments and tools such as drawing instruments for geometry, logarithm tables, slide rules, pocket calculators, and simple or sophisticated mathematical software on desktop or portable computers. Even the mental techniques of writing decimal numbers or performing calculating algorithms, using the notations of algebra and calculus, may be regarded as belonging to this realm.

This chapter concentrates on the impact of *computers* on mathematics teaching, and especially on the use of software in the process of teaching

R. Biehler, R. W. Scholz, R. Sträßer, B. Winkelmann (Eds.),
Didactics of Mathematics as a Scientific Discipline, 171-175.

and learning mathematics, since this has had the most dramatic effect on discussions on the goals and methods of mathematics education at all levels in the last decade and will continue to be one focus of didactical research and development. The short history of the struggle of didactics with software relevant for mathematics education may be sketched as follows:

Ideas, considerations, reflections, and concrete suggestions for the use of computers in teaching mathematics depend on the knowledge about and experience with such instruments shared by mathematical educators and teachers. Fifteen years ago, these people had access to computers mostly as *programmers in numerically oriented languages*. Thus computing power was mainly used for numerical algorithms, for instance, in the form of short BASIC programs. Ten years ago, another step – but again in the algorithmic spririt – was taken with the availability of *Logo* on various personal computers. Logo introduced its underlying philosophy of exploring mathematics in specially designed microworlds and of learning mathematics by teaching it to the computer; it also included the use of geometry and symbolic manipulations. The proliferation of so-called *standard software* on personal computers in the last decade led to new considerations and experiments, especially with spreadsheets, programs for data representation, statistical and numerical packages, databases, CAD (Computer Aided Design)-software, and computer algebra systems. But such software was at first not very user-friendly, and became too complex afterwards. The need for *special school adaptations* soon became obvious; these ideally allowed easy specializations, employed mathematical notations similar to those used at school, and used powerful and helpful metaphors, so that even users with little training and only occasional practice (as is typical of school users) could handle them successfully. This led to the creation of general and *didactical software tools* that sometimes also had a tutorial component, thereby integrating some traditions of computer-aided instruction (CAI). All these forms of using the computer came into being in sequence, but can now be found simultaneously in discussions about teaching mathematics (cf. Graf, Fraser, Klingen, Stewart, & Winkelmann, 1992, pp. 57-58).

Those developments impact on the different actions in curriculum development, such as discussions on content/process goals, on teaching/learning styles, and on means of assessing not only specific mathematical/computational activities such as numerical, graphical, and symbolic computations but also multiple representations of information (cf. Fey, 1989).

In accordance with the postulated changing demands of a computerized society (cf. Niss, this volume), increasingly less attention is being given to those aspects of mathematical work that are readily done by machines, while increasing emphasis is being placed on the conceptual thinking and planning required in any tool environment. In addition, students should know not only which mathematical activities could be given to machines to solve and which not but also, for example, which kind of preparations and answers could be expected by using numerical or symbolic computations (cf. Graf, Fraser, Klingen, Stewart, & Winkelmann, 1992, p. 58).

There is also a certain shift toward mathematical ideas and applications of greater complexity than those accessible to most students via traditional methods, such as system dynamics, data analysis, simulations, and a general trend toward more experimental mathematics (cf. Cornu & Ralston, 1992).

While these considerations belong to the domain of context/process goals, the papers in this chapter are generally more concerned with the new possibilities to enhance the teaching-learning process in mathematics opened up by computers with modern software. The first three papers throw a specific light on the issue of preparing mathematics for students (cf. chapter 1): They describe impacts not only on possibilities and implementations of mathematical teaching methods but also on the problems of justification of certain contents. This is most explicit in the paper by Dreyfus.

The activity of *programming* is not just present in the use of special programming languages but also in the use of most other mathematical software. Most Computer Algebra Systems include a programming possibility – normally on a higher level than general-purpose programming languages. Other mathematical software environments allow for the relatively unconstrained creation or definition of certain objects such as functions, geometric figures, geometric constructions, simulations of data sets, calculation and drawing procedures, and sometimes also of transformations regarding these objects. Such activities are normally subsumed under the general concept of programming, especially if the algorithmic character of the activity is evident.

In her paper on *the role of programming in mathematical education*, *Rosalind Sutherland* concentrates on the effect of programming environments such as Logo, BASIC, or spreadsheets on learning fundamental mathematical concepts such as variables. She clearly points out the different needs and habits of programming in mathematics education versus the mainframe habit (considered outdated) of most programming teachers who favor top-down programming and thinking in advance in contrast to the interactive style of work in mathematical programming that has proven so successful.

By presenting examples of students' work with Logo and spreadsheets, the author shows that it may be mistaken to assume that students can first express a general relationship in natural language and then somehow translate this into computer language. When working on a new and challenging problem, students tend to formulate general relationships by interacting with the computer language. The computer-based language becomes incorporated into their thinking and communication and helps to structure the generalizing process. In the spreadsheet environment, the use of pointing (to different cells on the screen) is also an important mediator in the generalizing process. By directly interacting with the language whilst working at the computer, students develop a way of using the language to express their mathematical ideas.

David Tall, in his paper on *computer environments for the learning of mathematics,* describes the growth of mathematical knowledge in students as vertical growth – encapsulation of processes into concepts – and horizontal growth – combining and understanding the linking of different representations of the same concept. Carefully designed computer environments may take a specific role between the inanimate natural environment and interpersonal communications: In a cybernetic mode, they may react according to preordained rules. Examples in the paper range from *simulative explorations* in Newtonian mechanics over geometric environments, which allow *enactive and visual manipulations*, arithmetic understanding through *multiple-linked representations,* to *generic organizers* in calculus, which help the student to build the first steps in more subtle understandings of the concept of differentiability. The author shows the possibilities and specific design criteria such as *selective construction*: To help the learner cope with the cognitive load of information processing, the computer can be used to carry out specific operations internally so that the student can focus on the others and on the conceptual outcome of those operations; at different times in the learning process, the student can focus on different aspects of the knowledge structure. Some dangers are also pointed out that often result from the differences between the concepts in the mathematical mind and the only approximating and finite representations by the computer.

The role of cognitive tools in mathematics teaching is dealt with in the paper by *Tommy Dreyfus.* He explicitly discusses the possibilities and issues raised by the growing number of mathematically based and didactically based tools available in mathematics teaching such as Computer Algebra Systems or David Tall's Graphics Calculus. He starts with the discussion of an introductory example: the use of a general purpose spreadsheet for learning about some aspects of discrete dynamical processes in one dimension. On the basis of the example, the author points out that computer tools should act not only as amplifiers (saving time on computations and making graphing easy in the above example) but also, and more importantly, as reorganizers. Thereby mathematics itself becomes different for the learner: New tools change cognition. This introduces new opportunities, but also new problems and new tasks (for curriculum developers, teachers, and students). As problems, the issue of why and how to learn mathematical techniques that are routinely solved by computers, the proper design of unified or diversified, mathematically or didactically based tools, and the black box problem are discussed: How much of the inner working of a tool should the student know in order to understand the mathematics and efficiently use the tool? All three problems have no strict solutions; they need to be studied in concrete settings of concrete curricula, and, on the other hand, they pose deep questions to the process of constructing curricula itself.

In contrast to the first three papers in this chapter, which describe the actual use of computers in the mathematical classroom and the problems and controversies involved, the closing paper by *Gerhard Holland* on *intelligent tutorial systems* is more concerned with potential uses and developments for the future. The author names the reasons why tutorial systems still have little impact on everyday mathematics teaching and learning: the demands they exert on hard- and software, and the reluctance of teachers and didacticians toward tutorial systems caused by negative experiences with (unintelligent) programmed instruction. The paper aims at initiating a qualified debate about the significance of tutorial systems for mathematics instruction and for research into mathematics education. It describes the classical architecture of an intelligent tutorial system as an integrated information-processing system having an expert module, an environmental module, a module for student modeling, and a tutor module. This is exemplified by the system HERON for solving word problems; and the paradigm of an intelligent tutorial system as a private teacher is opposed to the concept of a mathematical microworld with tutorial support. Then, to some extent, the author's own approach to solve the implementation problem of such tutorial systems is presented as a somewhat simplified architecture of a task-oriented intelligent tutorial system that reduces development costs and demand on system resources by concentrating on more narrowly defined goals in the realm of exercising the use of concepts that are already understood in principle. So not only didactical and technical problems of tutorial systems are discussed but also possible solutions that might have greater impact on didactical research and development in the near future.

Because technology, and especially computers, are nowadays a main force of innovation and a challenging field of research, the topic is also dealt with in papers in other chapters of this book. I shall just name the paper by *James T. Fey,* who discusses specific influences of computers, and that of *James J. Kaput,* whose discussion on representations is deeply concerned with computerized environments.

REFERENCES:

Cornu, B., & Ralston, A. (Eds.). (1992). *The influence of computers and informatics on mathematics and its teaching*. Paris: UNESCO.

Fey, J. (1989). Technology and mathematics education: A survey of recent developments and important problems. *Educational Studies in Mathematics, 20,* 237-272.

Graf, K. D., Fraser, R., Klingen, L., Stewart, J, & Winkelmann, B. (1992). The effect of computers on the school mathematics curriculum. In B. Cornu & A. Ralston (Eds.), *The influence of computers and informatics on mathematics and its teaching* (pp. 57-79). Paris: UNESCO.

THE ROLE OF PROGRAMMING: TOWARDS EXPERIMENTAL MATHEMATICS

Rosamund Sutherland
London

1. INTRODUCTION AND BACKGROUND

Within this chapter, I shall discuss the developing use of computer programming within mathematics education, describing what are, in my view, the important aspects of programming from the point of view of teaching and learning mathematics. By programming, I mean a means of communicating between the user and the binary code of the computer. From this perspective, a programming language must have some notation that is related to the set of problems to be solved. Programming is essentially problem-solving that involves defining and refining a problem and trying out a range of solutions. It also involves identifying the relevant variables in a problem and expressing relationships between these variables. Dividing a problem into smaller and more manageable parts is a valuable problem-solving and programming activity. Logo, for example, is a language in which the user can write procedures (sequences of code) to solve separate parts of the problem to be solved. These procedures are given names that can then be used within other procedures. In some programming environments, the word macro is used to describe a sequence of instructions that can be named so that the programmer can use the macro without having to think about the details of its definition. In the computer programming world, there are often standard ways of solving particular problems, for example, the problem of sorting a set of numbers. The word algorithm is often used to describe a series of instructions to solve a specific problem. From a programming point of view, some algorithms are more efficient than others (e.g., in terms of time and memory). An emphasis on standard algorithms, pre-written macros and efficiency is clearly important for effective computer programming, but is not, I suggest, where the emphasis should be placed when programming in the mathematics classroom.

My own personal experience of computer programming illustrates the dramatic way in which it has changed over the last 25 years. In 1966, as a university student, I attended a one-week Algol programming workshop, which consisted of lectures and hands-on experience. This hands-on experience involved spending hours typing a program on a set of punched cards (a

R. Biehler, R. W. Scholz, R. Sträßer, B. Winkelmann (Eds.),
Didactics of Mathematics as a Scientific Discipline, 177-187.

piece of card containing data in the form of punched holes) and waiting at least overnight for the program to run only to discover that typing errors had been made, errors that were difficult to identify because the punched code had to be translated into the computer language before it could be read. So, at this time, it was very important to plan a program in advance, and it was very important not to make syntax errors because these cost time. In no way was it possible to interact with the computer code as it was interpreted and evaluated by the machine. Things began to change with teletype terminals, which were attached to mainframe computers, but these were very unfriendly, feedback could be slow, and the link to the mainframe computer was often fragile. Nowadays, we can write sophisticated programs on a portable computer, interacting with the language in a negotiating way. Professional programmers have responded to these technological changes, but in the educational world (i.e., the world of teaching and learning programming), a "mainframe mentality" often prevails. This can result in an over-emphasis on planning away from the computer and an over-emphasis on a directed form of teaching. Nowadays, there are many possible ways of interacting with a computer program, and so it is interesting to question why so many university computer programming courses are still taught in ways that are similar to those used 25 years ago. Lack of computer provision, or student numbers, is often given as a reason, but, in my opinion, the reason is more related to the need of the teacher to hold onto knowledge as a means of power and control. Also, if, as a teacher, you have a strong model of learning as being related to both the ability and developmental stage of a student (possibly influenced by Piaget's theories), then you have more or less rid yourself of the responsibility of changing your teaching method. We now know that elementary school children can program in Logo (Noss, 1985). This knowledge has not revolutionized the teaching of programming, it has merely resulted in the marginalization of Logo as a programming language.

2. PROGRAMMING IN THE UK MATHEMATICS CLASSROOM

In the UK, programming in school was firstly the province of school computer science courses, a new subject taught and examined to 14- to 16-year-olds. This subject was often taught by the mathematics teacher, and the programming language used was almost always BASIC. So there developed, in the UK, a body of secondary school mathematics/computer science teachers with an expertise in BASIC programming. Most of these teachers were men and most of the students studying computer science were boys. BASIC programming began to be used by mathematics teachers in the mathematics classroom, and it was this activity that was greeted with such enthusiasm by the mathematics inspectorate as expressed by Fletcher:

> Some years ago I saw the heartening, indeed amazing, response when microcomputers were first introduced into schools . . . excellent work was done when stu-

> dents were encouraged to explore, to investigate things which interested them and to find their own way forward. (Fletcher, 1992, p. 1)

When Logo became available on small computers (in about 1982) and started to be used in schools, it challenged the BASIC programming community for a number of reasons: Firstly, young children began to learn computer programming, and, secondly, Logo was difficult to learn for those who had previously programmed only in BASIC. This relates to the recursive control structure of Logo, which cannot easily be followed in a step-by-step way. Thirdly, Logo came with a whole set of ideas about the philosophy of teaching, ideas that have become polarized as learning by discovery. Many of us who have carried out research and development with Logo no longer accept this polarized view of learning and have extensively written about the issues surrounding the teaching and learning of Logo (Noss & Hoyles, 1992; Sutherland, 1993).

The tensions and debates about the relative value of Logo and BASIC in the UK mathematics curriculum, which now seem very outdated, have nevertheless resulted in an equal share being given to both programming languages in the new National Curriculum for Mathematics. For example, in the strand related to algebra, it states that students are expected to *follow instructions to generate sequences* as illustrated by the following example:

Follow the instructions to find all the square numbers between 0 and 100
10 FOR NUMBER = 1 TO 10
*20 PRINT NUMBER * NUMBER*
30 NEXT NUMBER
40 END

In the strand related to problem-solving, it states that students are expected to identify and obtain information necessary to solve problems. This is elaborated as: When trying to draw repeating patterns of different sizes using Logo, realize the need for a procedure to incorporate a variable, and request and interpret instructions for doing it.

The whole nature of this UK National Curriculum is such that it fragments mathematics, and, as can be seen from the above example, ideas from computer programming have become so fragmented as to be almost pointless. But computer programming in schools predates the National Curriculum, and I am optimistic enough to believe that some of the absurdities in this new curriculum will change with time. Over the last 10 years, computer provision in schools has changed dramatically. Ten years ago, we had to provide the computers in order to carry out our research in the classroom. Nowadays, we can easily find schools with adequate computer provision. The school in which I recently completed a project has three computer rooms full of networked computers and a computer in each mathematics classroom. Many secondary schools in the UK now have good computer facilities, but the mathematics teachers still need considerable support to make use of these facilities for teaching and learning mathematics (Sutherland, Hoyles, & Noss, 1990).

3. COMPUTER PROGRAMMING AND LEARNING ALGEBRA

For a number of years I have been working on the ways in which programming influences students' developing use and understanding of algebraic ideas. This work was initially influenced by the considerable research on students' learning of algebra (e.g., Küchemann, 1981), which reported that students find it difficult to understand that a letter in algebra can represent a range of numbers and to accept "unclosed" expressions in algebra (e.g., $x + 4$). Most of this work on childrens' understanding of algebra was influenced by a Piagetian perspective. The implicit assumption often made was that if students cannot perform satisfactorily on certain algebraic tasks, then they have not reached the stage of formal operations. Results from work in computer programming environments conflict with many of the established results on the learning of traditional algebra (Sutherland, 1992; Tall, 1989)

4. LOGO PROGRAMMING

Our first study carried out with the programming language Logo (Sutherland, 1989) as part of the Logo Maths Project (Hoyles & Sutherland, 1989) showed that, with Logo programming experience, students develop a different view of literal symbols from those developed within school algebra. Tall also found similar results working with the BASIC programming language (Tall, 1989). In the programming environment, students know that any name can be used for a variable, that a variable name (either a word or a literal symbol) represents a range of numbers, and readily accept the idea of working with unclosed, variable-dependent expressions. Moreover, many students can use these programming experiences and more traditional algebra situations (Sutherland, in press). But the most important result from this work, which influenced the direction of our ongoing research, was that the algebra understandings that students develop depend very much on the nature of their Logo programming experiences, and this is influenced by the way the teacher structures the classroom situation. In retrospect, this seems like common sense, but, at the time, the prevalent theoretical view, influenced by the theories of Piaget, was that algebraic understandings depend more on the developmental stage of the child. Initially in the Logo Maths Project, we had been cautious about introducing the idea of variable to students because of an awareness of the negative attitudes many students have about algebra. So, in the first instance, we waited for students to choose goals that needed the idea of variable, and only changed this strategy when it became clear that most of them would not do this spontaneously. The development in our teaching approach and how it changed within two subsequent projects has been described in Sutherland (1993).

When a whole class of students are working on computer programming activities, they can be actively engaged in their own process of problem-solving. The teacher's role ought to be one of providing problems to be solved, or letting students choose their own problem, giving support with syntax, discussing a problem solution, but essentially devolving much of the responsibility to the students themselves. It seems that the crucial factor here, from the point of view of mathematics education, is that the students construct a problem solution themselves. This contrasts with the idea of giving students a preprogrammed algorithm, which is more prevalent in the teaching of BASIC than in the teaching of Logo. Presenting students with standard solutions is also part of school mathematics practice, and Mason (1993) has criticized the fact that, in much of school algebra, students are presented with someone else's solution to a problem and are not given the opportunity to construct their own solutions. Interactive programming languages provide an ideal setting for students to construct their own programs, so it is interesting to question why teachers so often provide programming solutions for their students, either in the form of pre-written macros or standard algorithms. It may result from a lack of confidence, on the part of the teacher, that students will be able to construct their own programs – often a projection of the teacher's own lack of confidence and expertise onto the students. Another reason relates to the "mainframe mentality" and the idea that a program solution must be planned away from the computer.

5. A SPREADSHEET ENVIRONMENT – EXCEL

More recently, I have been working with the spreadsheet Excel with groups of 10-year-olds, 11- to 13-year-olds and 14- to 15-year-olds. Here I will discuss the work with the older group of students who were chosen because they had all experienced considerable difficulty with school mathematics – many of them were disaffected with mathematics and disaffected with school, and all of them had very little previous experience of algebra. All students were interviewed at the beginning and end of the study in order to trace their developing use of algebraic ideas. The majority of the 14- to 15-year-olds could not answer any of the pre-interview questions that focused on the algebraic ideas of: expressing generality; symbolizing a general relationship; interpreting symbolic expressions; expressing and manipulating the unknown; function and inverse function. All of the students had great difficulty in expressing very simple general rules in natural language (e.g., "add 3"), and none of them were able to answer questions on inverse functions. The majority were unfamiliar with literal symbols exhibiting the classic "misconceptions" reported in a number of algebra studies (e.g., Küchemann, 1981). For example, Jo thought that the higher the position in the alphabet the larger the number represented. This clearly related to experiences from primary school: *"A starts off as one or something . . . when we were little we used to do a code like that . . . A would equal 1 . . . B equals 2 . . . C equals 3."*

The spreadsheet activities centred around the following mathematical ideas:

Function, inverse function and equivalent expressions. Students were introduced to the ideas of: entering a rule; replicating a rule; function and inverse function; symbolizing a general rule; decimal and negative numbers; equivalent algebraic expressions (e.g., $5n$ and $2n + 3n$). They worked on a range of problems, most of which were taken from the book *Exploring Mathematics with Spreadsheets* (Healy & Sutherland, 1990).

Algebra story problems. Students used a spreadsheet to solve algebra story problems by: representing the unknown with a spreadsheet cell; expressing the relationships within the problem in terms of this unknown; varying the unknown to find a solution by changing the value in the spreadsheet cell (see, e.g., Figure 1).

Rectangular Field

The perimeter of a rectangular field measures 102 metres.
The length of the field is twice as much as the width of the field.
Use a spreadsheet to work out the width and the length of the field.

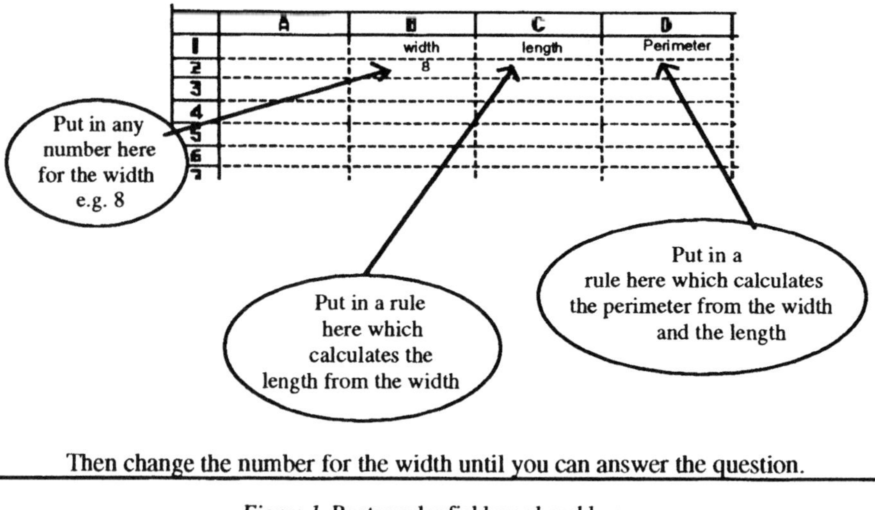

Then change the number for the width until you can answer the question.

Figure 1. Rectangular field word problem.

It is important to stress that students were initially taught to enter a spreadsheet rule by pointing with the mouse to the cell that was being referenced. They were never explicitly taught to type in the spreadsheet-algebraic code (e.g., *A5*), although they had been explicitly shown how to display the "formulae" produced by the spreadsheet. Analysis of transcripts of the conversation between pairs of students indicated that they used this code in their talk (*"so what will it be . . . B2 take 4"*), and further questioning of the students in the final interviews revealed that they all knew the code for the spreadsheet formulae that they had entered with the mouse. They also knew how this code changed when being copied using relative referencing (e.g., from $A3 + 1$ to $A4 + 1$). The fact that they noticed and knew this code is, I suggest, related to the nature of the Excel spreadsheet environment in which the spreadsheet code is transparently displayed in the formula bar. Students learned that this was the language to communicate with the computer and began to use it as a language to communicate with their peers.

Analysis of the results from the final interview revealed that the spreadsheet-algebraic code played a mediating role in students' developing ability to solve the algebra problems that were the focus of this study. In the post-test, the majority could express a general rule for a function and its inverse and often expressed these rules in spreadsheet-algebraic code. This contrasts with their performance on the pre-test. When asked how she could answer so many questions successfully in the post-test, when she had not been able to answer any in the pre-test, Jo said *"because you have to think before you type it into the computer anyway . . . so it's just like thinking with your brain."* Students said that they thought of a spreadsheet cell as representing any number, and many of them were able to answer traditional algebra questions in the post-test. The following problem was given to the students in the post-test and is similar to the Block 2 algebra story problems:

> 100 chocolates were distributed between three groups of children. The second group received 4 times the chocolates given to the first group. The third group received 10 chocolates more than the second group. How many chocolates did the first, the second and the third group receive?

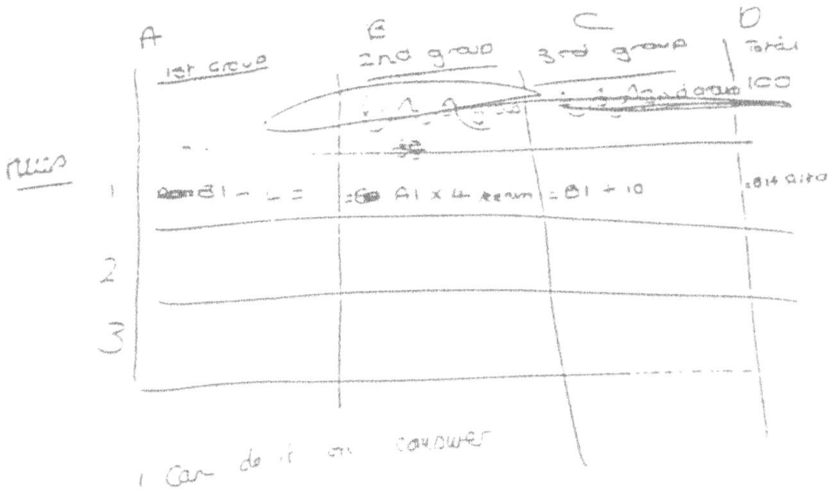

Figure 2. Ellie's solution to the measurement of a field problem (away from computer).

Ellie's solution (with no computer present) illustrates the way in which the spreadsheet code played a mediating role in her solution process.

In the post-interview, Ellie was asked "If we call this cell *X*, what could you write down for the number of chocolates in the other groups," and she wrote down:

$$=X \qquad\qquad =X \cdot 4 \qquad\qquad =X \cdot 4 + 10$$

Many of these students were able to represent the relationships in the word problems in traditional algebra language. Collaborative and parallel studies (with similar results) have been carried out by Teresa Rojano in Mexico (Rojano & Sutherland, 1993; Sutherland & Rojano, in press).

6. PROGRAMMING AS A MEANS OF EXPRESSING AND EXPERIMENTING WITH MATHEMATICAL IDEAS

Within all of the studies discussed in this chapter, we have made video- or audiotape recordings of groups of students as they work in pairs on the programming activities. The programming language itself and the ways in which students interact with the language and use it in their talk to communicate with their peers play an important role in the student constructions. Most of the problems presented to the students are challenging in that they do not know how to solve the problem before working at the computer. So, for example, in Logo, students might be constructing a general Logo procedure to produce geometrical images in proportion without knowing rules for ratio and proportion. These rules are constructed by the students as they work at the computer. They learn from the visual image on the screen that *"take does not always work . . . times is better."* In this sense, they are aware of the global geometric constraints of the problem: *"well the two sides there stay the same . . . it would still be the same distance between here and here."* When constructing the function and inverse function shown in Figure 3, students used the spreadsheet to help them find the rule. The majority of the 10-year-old group of students and the 14- to 15-year-olds (with low mathematical attainment) did not immediately program the correct rule for this problem. Many of them entered a rule of the form "*A*3 - 0.5" to produce the *Y* values, and then, when they copied this rule (in the column labelled *Y*) realized that this was not correct. They usually tried out a number of other rules before finding the correct one (of the form "*A*3/2"). But after this experimental work at the computer, both groups improved on these types of problem in a post-test carried out away from the computer. This may seem a trivial problem, but it illustrates the important idea that students can negotiate a general rule whilst working at the computer.

The idea of experimenting in mathematics is new and contentious. As Epstein points out:

> Originally, experimenting would have been doing calculations with a pen and trying out various special cases of a theorem you think might be true. Then when

you've found enough cases to convince you that it is true you try to prove it. This is the method Gauss used a lot. His private notebooks are just covered by huge numbers of calculations. (quoted in Bown, 1991, p. 35)

Use the spreadsheet to help you construct the following tables of numbers

X	Y	X
1	0.5	1
2	1	2
3	1.5	3
4	2	4
5	2.5	5
6	3	6
7	3.5	7

Write down the rule that tells you how to get the *Y* number from the *X* number

Y = ...

Write down the rule that tells how to get the *X* number from the *Y* number

X = ...

Figure 3. Function and inverse function problem.

Epstein goes on to discuss how mathematicians have traditionally hidden this experimental work:

> A typical example is a 140-page paper I wrote and won a prize for. The whole thing is based on computer work but the paper just goes on and on with theory . . . the whole direction of the research, how I decided which thing to try and do next was determined experimentally. (quoted in Bown, 1991, p. 35)

Programming is an ideal environment for developing an experimental mathematics. Different languages and problems allow the student to experiment with different types of object. In a spreadsheet, the focus of experimentation can be with the algebraic code, or with the graphical representation, depending on the type of problem. The language used will depend on the problem and will include such environments as Cabri Géomètre (Laborde & Strässer, 1990) and computer algebra systems like Maple. In the past, we have not paid enough attention to how students justify the results of their experimentation (actually, in the traditional mathematics classroom, it has often been the teacher or the answers in the book that provide the justification). Students are much more likely to invest time in a proof if they are convinced (by means of experimentation) that their conjectures are correct. Programming involves the use of a formal language, and this language can be the basis for justification and proof, but students will not do this spontaneously. Here again, the teacher will have a critical role.

7. A CONCLUDING REMARK

In the future, students are likely to have their own portable computer, which will be powerful enough to support a range of programming environments. The majority of students will not spontaneously use their computers for mathematical experimentation unless this is supported by the culture of the school mathematics classroom. With this support, there will be more students like Sam who learned to program at home and at the age of 10 said:

> there's quite a lot of maths involved in it. I did a program that calculates your age . . . it's still a bit faulty at the moment . . . but what it does you enter in your age in years and the date . . . well just the date and the month that you were born and it calculates the year you were born and how many years and days old you are.

Of course there are standard and efficient algorithms to calculate age from date of birth, but, for Sam, it was important to construct the program for himself. Interactive programming offers the potential for trying out and refining problem solutions, and all the evidence from classroom work suggests that students are remarkably successful at this activity. I suggest that most of the potential of programming within mathematics education will be lost if teachers over-direct students' problem solutions by an overemphasis on pre-written macros, standard algorithms and work away from the computer. In my work in schools, I have focused on relatively unsophisticated uses of computer programming, because I believed that these needed attention. This work has shown that students can construct programs and experiment mathematically, but rather more work still needs to be done to flexibly integrate these activities into the mathematics curriculum.

REFERENCES

Bown, W. (1991). New-wave mathematics, *New Scientist, 131*(1780)

Fletcher D. (1992). Foreword. In W. Mann (Ed.), *Computers in the mathematics curriculum. A report of the mathematical association.* Leicester: Mathematical Association.

Healy, L., & Sutherland, R. (1990). *Exploring mathematics with spreadsheets.* Hemel Hempstead: Simon & Schuster.

Hoyles, C., & Sutherland, R. (1989). *Logo mathematics in the classroom.* London: Routledge.

Küchemann, D. E. (1981). Algebra. In K. Hart (Ed.), *Children's understanding of Mathematics* (pp. 11-16). London: Murray.

Laborde, J., & Strässer, R. (1990). Cabri-Géomètre: A microworld of geometry for guided discovery learning. *Zentralblatt für Didaktik der Mathematik, 90*(5), 171-177.

Mason, J. (1993, May). *Expressing generality and roots of algebra.* Paper presented at the conference on Research Perspectives on the Development and Emergence of Algebraic Thought, Montreal.

Noss, R. (1985). *Creating a mathematical environment through programming: A study of young children learning Logo.* Umpublished Master's thesis, Institute of Education, University of London.

Noss, R., & Hoyles, C. (1992). Looking back and looking forward. In C. Hoyles & R. Noss (Eds.), *Learning mathematics and Logo.* Cambridge; MA: MIT Press.

Rojano, T., & Sutherland, R. (1993). *Towards an algebraic approach: The role of spreadsheets.* Proceedings of the 17th International Conference for the Psychology of Mathematics Education, Japan.

Sutherland, R. (1989). Providing a computer-based framework for algebraic thinking. *Educational Studies in Maths, 20*(3), 317-344.
Sutherland, R. (1992). Some unanswered research questions on the teaching and learning of algebra. *For the Learning of Mathematics, 11*(3), 40-46.
Sutherland, R. (1993). Connecting theory and practice: Results from the teaching of Logo. *Educational Studies for Mathematics, 24*, 1-19.
Sutherland, R., Hoyles, C., & Noss, R. (1991). *The microworlds course: Description and evaluation.* Final Report of the Microworlds Project, Volume 1. Institute of Education, University of London.
Sutherland, R., & Rojano, T. (in press). A spreadsheet approach to solving algebra problems. *Journal of Mathematical Behaviour.*
Tall, D. (1989). Different cognitive obstacles in a technological paradigm. In S. Wagner & C. Kieran (Eds.), *Research issues in the learning and teaching of algebra.* Hillsdale, NJ: LEA.

COMPUTER ENVIRONMENTS FOR THE LEARNING OF MATHEMATICS

David Tall
Warwick

1. INTRODUCTION

Computer software for the *learning* of mathematics, as distinct from software for *doing* mathematics, needs to be designed to take account of the cognitive growth of the learner, which may differ significantly from the logical structure of the formal subject. It is therefore of value to begin by considering cognitive aspects relevant to the use of computer technology before the main task of focusing on computer environments and their role in the learning of mathematics.

2. THE GROWTH OF (MATHEMATICAL) KNOWLEDGE

The human brain is remarkable in its ability to store and retrieve complex information, but it is correspondingly limited in the quantity of independent pieces of data that may be manipulated in conscious short-term memory. To minimize the effects of these limitations, one method is to "chunk" the data by using an appropriate representation that is easier to manipulate. For instance, standard decimal notation is a compact method of representing numerical quantities of any size with corresponding routines for manipulation; algebraic notation can be used to formulate and manipulate certain types of data for problem-solving; graphical representations are appropriate for other tasks such as representation of complex data in a single gestalt.

Traditional mathematics often consists in performing algorithms using these representations, minimizing the cognitive strain by routinizing the procedures so that they become automatic and require less conscious memory to perform. A more subtle transformation also occurs in which the symbols used to evoke a mathematical process begin to take on a life of their own as mental objects, so that processes become encapsulated as objects. Thus, counting using the number words gives the numeric symbols a related meaning as numbers, the process of addition becomes the concept of sum, repeated addition becomes product, and so on. This long-term cognitive process in which procedures are routinized to become more compressed and then encapsulated as mathematical objects in their own

R. Biehler, R. W. Scholz, R. Sträßer, B. Winkelmann (Eds.),
Didactics of Mathematics as a Scientific Discipline, 189-199.

right is referred to by Piaget and subsequent authors as *vertical* growth, in contrast to the *horizontal* growth of relationships between different representations.

Both vertical and horizontal growth impose difficulties on the individual. Vertical growth requires ample time for familiarization with the given process to enable it to be interiorized and also for the cognitive re-organization necessary during encapsulation of process as object. Horizontal growth requires the simultaneous grasping of two or more different representations and the links between them, which is likely to place cognitive strain on short-term memory resources.

These difficulties may be alleviated in various ways by using a computer environment to provide support. Software may be designed to carry out some of the processes, leaving the learner to concentrate on others chosen to be the selected focus of attention. The sequence of learning in vertical growth may be modified by providing environments that allow the study of higher-level concepts in an intuitive form before or at the same time as they are constructed through encapsulation. Horizontal linkages between different representations may be programmed so that the individual operates on one representation and can see the consequences of this act in other linked representations. Moreover, because the computer can be programmed to respond in a pre-ordained manner, it can provide an environment in which the learner can explore the consequences of selected actions to predict and test theories under construction.

3. THE COMPUTER AS A PREDICTABLE ENVIRONMENT FOR LEARNING

Skemp (1979, p. 163) makes a valuable distinction between different modes of building and testing conceptual structures (Table 1).

The introduction of computer technology brings a new refinement to this theory. Whereas Mode 1 is seen as the individual acting on and experimenting with materials that are largely passive, a computer environment can be designed to *re*-act to the actions of the individual in a predictable way. This new form of interaction extends Skemp's theory to four modes (Tall, 1989) in which building and testing environments are:

1. *Inanimate*: The stimuli come from objects in actuality that the individual may also be able to manipulate.
2. *Cybernetic:* The stimuli come from systems that are set up to react according to pre-ordained rules.
3. *Interpersonal*: The stimuli come from other people.
4. *Personal*: The stimuli are from the individual's own cognitive structure.

The new cybernetic mode of building and testing concepts affords rich possibilities for the learning of mathematics.

Table 1. *Modes of building and testing personal constructions*

	Reality construction	
Mode	Reality building	Reality testing
1	From our encounters with actuality: *experience*	Against experience of events in actuality: *experiment*
2	From the reality of others: *communication*	Comparison with the reality of others: *discussion*
3	From within, by formation of higher-order concepts, by extrapolation, imagination, intuition: *creativity*	Comparison with one's own existing knowledge and beliefs: *internal consistency*

4. MICROWORLDS

The term *microworld* was originally used by Papert to describe "a computer-based interactive learning environment where the pre-requisites are built into the system and where learners can become active, constructing architects of their own learning" (Papert, 1980, p. 117). Initially the term microworld was used specifically for programming environments (often in the computer language Logo).

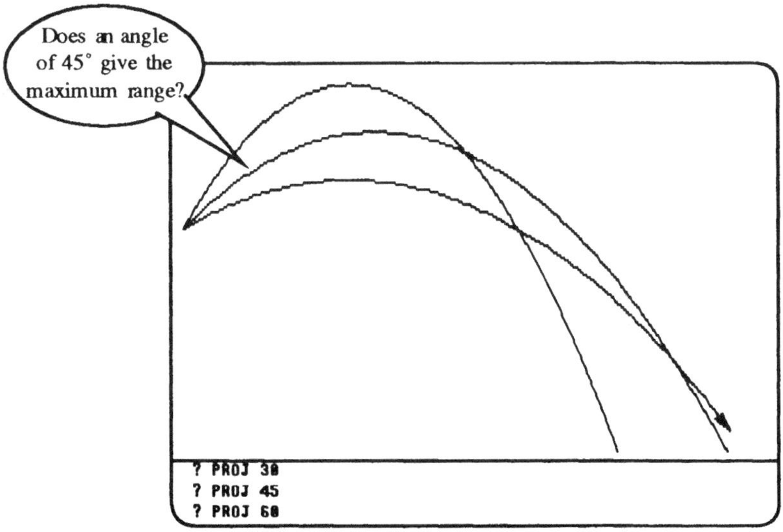

Figure 1. Investigating the range of a projectile for different angles of projection.

For instance, the program *Newton* (Pratt, 1988) is a microworld designed so that turtles move according to Newton's laws, allowing investigations of a variety of topics including motion under a central force. In Figure 1, the student has designed an experiment to model an object being projected from a point above a plane to investigate the angle that gives the maximum range; it turns out differently from the expected 45°.

Such an environment provides facilities to construct ways of formulating and testing conjectures. In the early stages, Papert considered such environments to encourage what he termed "Piagetian learning," or "learning without a curriculum," or "learning without being taught" (Papert, 1980, p. 7). Children are often highly creative within such environments, but "powerful ideas," particularly vertical growth of concepts, do not readily occur spontaneously, and long-term curriculum objectives require external guidance and support.

5. ENVIRONMENTS FOR ENACTIVE AND VISUAL MANIPULATION

More sophisticated computer environments have been designed in recent years that take advantage of flexible computer interfaces. Geometric software such as *Cabri Géomètre* (1987) or *The Geometer's Sketchpad* (1992) allows figures to be drawn with specific relationships defined, such as a given point must always lie at the midpoint of a given line-segment or be constrained to lie on a given circle. Then the figure may be pulled around enactively retaining all the defined constrains to investigate possible consequent relationships.

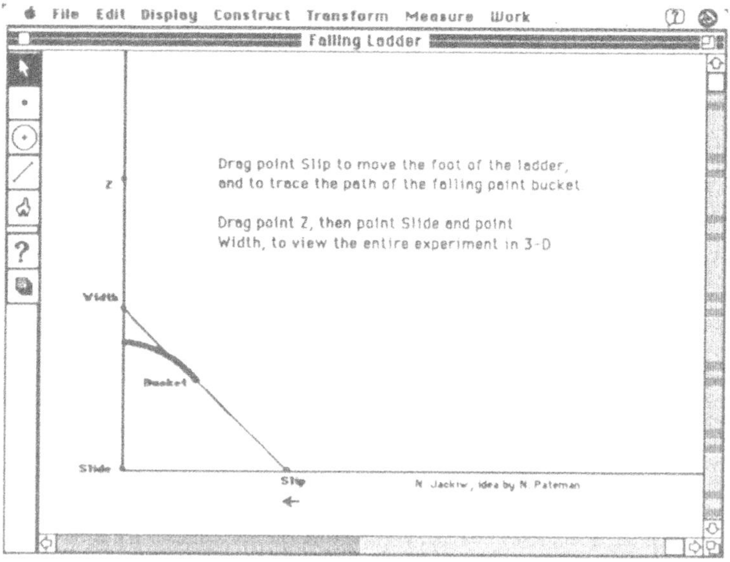

Figure 2. The locus of a point on a line of fixed length, constrained with its endpoints on two perpendicular lines.

Figure 2 shows a model of a bucket on a ladder set against a vertical wall and sketches the path taken by the bucket as the ladder slides. Such software may be used to gain enactive visual support in conjecturing and testing geometric theorems, enabling students to take an active part in the construction of their own knowledge, though, once again, the formal proof structure of geometry will need separate consideration.

6. MULTIPLE LINKED REPRESENTATIONS

Computer environments can be set up to link different representations of the same concept. The *Blocks Microworld* of Thompson (1992) is designed to link screen representations of Dienes' multibase blocks to numerical representations (Figure 3). In the top right of the window are representations of different units in base 10, comprising a single, long (10 singles in a line), flat (10 longs in a square), and block (10 flats to make a larger cube). As the user selects one of these and pulls a copy to the lower part of the screen to build up collections of blocks, the corresponding numerical display is simultaneously updated. If the blocks in the figure representing 78 and 45 are combined by removing the vertical separator between them, the resulting collection of 11 longs and 13 singles can be re-organized by the learner to give 1 flat, 2 longs, and 3 singles (123).

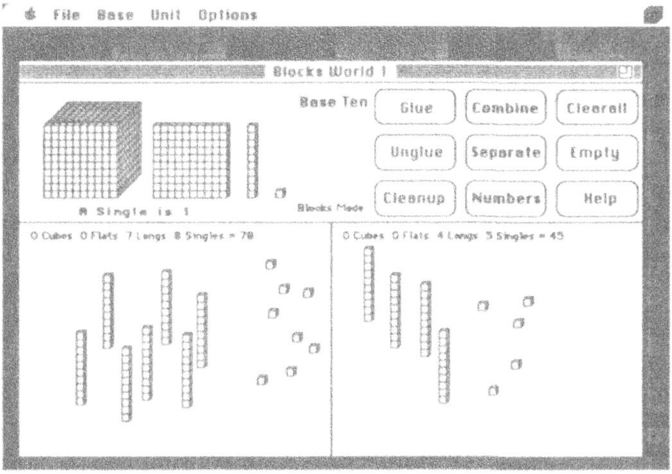

Figure 3. Manipulating Dienes' blocks on-screen, with the corresponding linked numerical representation.

This environment may be used to give a direct link between physical experience and the formal symbolic notation, allowing children to explore their own algorithms for, as well as giving meaning to, the formal routines for addition and subtraction.

7. THE PRINCIPLE OF SELECTIVE CONSTRUCTION

What has been exemplified in all the environments described so far is the way in which the software can be programmed to carry out internal algorithms, leaving the learner free to explore other aspects. This can occur in horizontal growth of knowledge, in which the learner builds links between different representations, but it is even more powerful in vertical growth. Whereas a traditional development would almost always require the learner to become familiar with a given process and routinize it before beginning to consider the consequences, computer environments may carry out the processes and allow the user to explore the resultant concepts either before, after, or at the same time as the processes. This ability to reorganize the curriculum to allow the learner to focus on one aspect of cognitive growth whilst the computer carries out others, I term the *principle of selective construction*.

In carrying out such a principle, it is important to consider the concept imagery that it may generate in the learner and the type of insight that such interaction may bring. Tall and Winkelmann (1988) described three different kinds of insight: external, analogue, specific.

External insight occurs when the user has no idea what is going on inside the software, but has knowledge that allows him or her to check that the results are sensible; *analogue* insight occurs when the user has an idea of the type of algorithm in use; and *specific* insight is when the user is fully aware of how the software is programmed.

Specific insight into computer software is rarely possible or even desirable for the majority of computer users, but it is helpful for the student to have at least external insight or, preferably, analogue insight. The concept image of a cybernetic system constructed in the mind of the user is likely to be idiosyncratic, and a teacher has a fundamental role to play through guidance and discussion. Tall (1989) describes the combination of a human teacher as guide and mentor using a computer environment for teaching, student exploration, and discussion as the *Enhanced Socratic Mode* of teaching and learning. It combines the *interpersonal* interactions between student and teacher, the *cybernetic* interactions with the computer environment to give an independent source of consistent evidence, and the *personal* constructions of the learner in building and relating together the different parts of the knowledge structure.

8. GENERIC ORGANIZERS

Ausubel, Novak, and Hanesian (1978) defined an *advance organizer* as

> Introductory material presented in advance of, and at a higher level of generality, inclusiveness, and abstraction than the learning task itself, and explicitly related both to existing ideas in cognitive structure and to the learning task itself . . . i.e. bridging the gap between what the learner already knows and what he need to know to learn the material more expeditiously. (p. 171)

Such a principle requires that the learner *already has the appropriate higher-level cognitive structure* available to him or her. In situations in which this may be missing, in particular, when moving on to more abstract ideas in a topic for the first time, a different kind of organizing principle will be necessary. To complement the notion of an advance organizer, a *generic organizer* is defined to be an environment (or microworld) that enables the learner to manipulate *examples* and (if possible) *non-examples* of a specific mathematical concept or a related system of concepts (Tall, 1989). The intention is to help the learner gain experiences that will provide a cognitive structure on which the learner may reflect to build the more abstract concepts. I believe the availability of non-examples to be of great importance, particularly with higher-order concepts such as convergence, continuity or differentiability in which the concept definition is so intricate that students often have difficulty in dealing with it when it fails to hold.

A simple instance of a generic organizer embodying both examples and non-examples is the *Magnify* program from *Graphic Calculus* (Tall, Blokland, & Kok, 1990) designed to allow the user to magnify any part of the graph of a specified function (Figure 4).

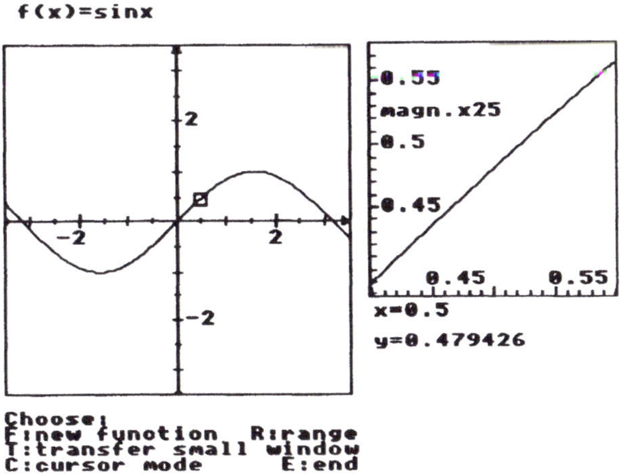

Figure 4. A "locally straight" graph.

Tiny parts of certain graphs under high magnification eventually look virtually straight, and this provides an anchoring concept for the notion of differentiability. Non-examples in the program are furnished by graphs that have corners or are very wrinkled so that they never look straight, providing anchoring concepts for non-differentiability (Figure 5).

The gradient of a "locally straight" graph may now be seen graphically by following the eye along the curve, or a piece of software may be designed that traces the gradient as a line through two close points on the graph that moves along in steps (Figure 6).

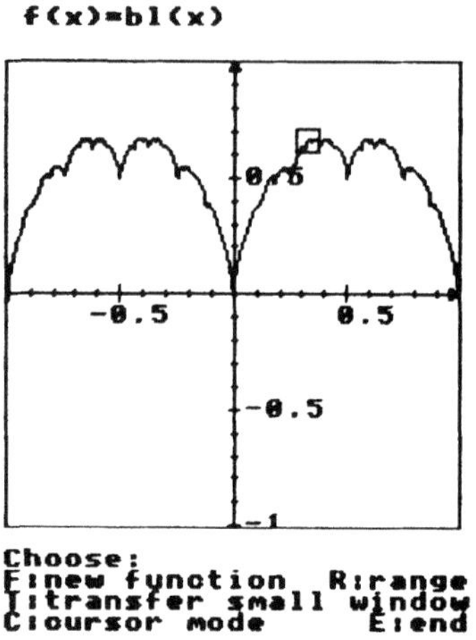

Figure 5 : A highly wrinkled graph.

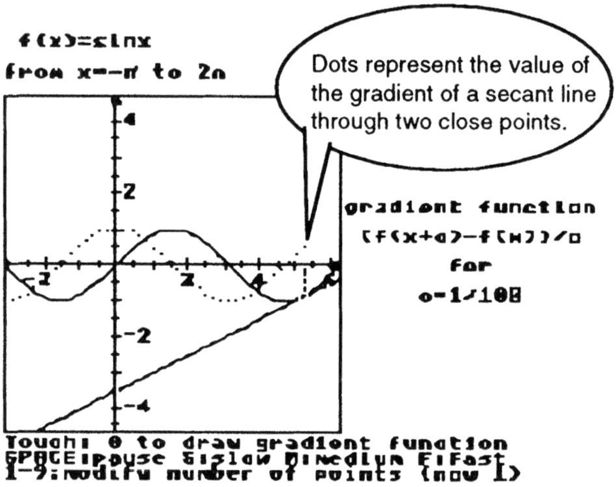

Figure 6. Drawing the gradient (dotted) of sin*x*.

In this way, a student with some experience of the shape of trigonometric curves will be able to conjecture that the derivative (gradient) of sin*x* is cos*x* from the shape of the dotted gradient, even though the manipulation of trigonometric formulae and the formal notion of limit is at present beyond his or her capacity.

9. GENERIC DIFFICULTIES

Given the human capacity for patterning, and the fact that the computer model of a mathematical concept is bound to differ from the concept in some respects, we should be on the lookout for abstraction of inappropriate parts of the model. Visual illusions in interpreting graphs have been documented by Goldenberg (1988) and by Linn and Nachmias (1987). In the latter case, one third of the students observing a cooling curve of a liquid on a computer VDU interpreted the pixellated image of the graph as truly representing what happened to the liquid – constant for a time, then suddenly dropping a little (to the next pixel level down).

Working with older students, the inadequacy of the representation may prove to be an advantage. It can be source of discussion that the jagged pixellated imagery does not represent the true conceptualization in the mind, encouraging the student to make personal mental constructs of a more platonic form of the theory. For instance, free play with a gradient-drawing program may lead the student to think that all reasonable looking graphs are differentiable, but this view may be challenged by being confronted with Figure 7.

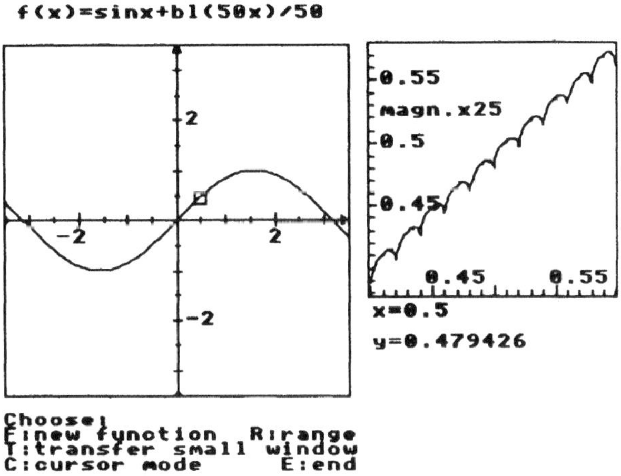

Figure 7. A graph with a difference.

This graph looks very similar to that in Figure 4, but under high magnification, the wrinkles produced by the tiny added blancmange become apparent. Simple visualization at a fixed scale is therefore inadequate: two graphs may *seem* to be similar at one level, yet, at a deeper level, one is differentiable everywhere and the other nowhere. In this way the generic organizer reveals itself as only a step along the path of cognitive growth. The student progressing to more formal study has the opportunity to develop flexible concept imagery showing the necessity for more subtle symbolic representation of the mathematics, whilst the student who is only using the calculus in its applications has at least an intuitive appreciation of the possible theoretical difficulties.

10. REFLECTIONS

In considering the way in which computer environments can be used in the learning of mathematics, we see the possibility of providing cybernetic environments that react in a predictable manner to help the learner build and test his or her own mental constructions. The computer can carry out internal procedures, allowing the learner to focus on other facets of importance in the cognitive growth of mathematical knowledge. This can help develop a concept image of higher-order concepts in a different sequence from the traditional method of routinization and encapsulation. It must be noted that the mental objects may not have the same structure as is given by traditional learning sequence, and that such exploration may give gestalts that do not link directly to the sequence of definitions and logical deductions in the formal theory. However, insights are possible for students who might not attain such a level in a traditional approach, while those who are able to move to higher levels may have more appropriate concept imagery available to give a more rounded mental picture of the theory. The software described in this chapter invariably needs to be embedded in a wider conceptual context in which the powerful ideas are made the explicit focus of attention. This is usually provided by prepared materials or by the teacher as mentor, although a solution has long been sought in which the computer itself can play the guiding role in a more intelligent manner (see section 4). Meanwhile, interactive video is beginning to provide flexible environments in which the study guide offers the student deeper levels of information as required with interactive animated graphics and flexible computer environments of the type described in this chapter. As technology grows more sophisticated, such developments are likely to play an increasing role in the learning of mathematics.

REFERENCES

Ausubel, D. P., Novak, J. D., & Hanesian, H. (1978). *Educational psychology: A cognitive view* (2nd ed.). New York: Holt, Rinehart & Winston.

Cabri Géomètre (1987). [Computer program]. Université de Grenoble, France (IMAG, BP 53X).

Goldenberg, P. (1988). Mathematics, metaphors and human factors: Mathematical, technical and pedagogical challenges in the educational use of graphical representations of functions. *Journal of Mathematical Behaviour, 7*(2), 135-173.

Linn, M. C., & Nachmias, R. (1987). Evaluations of science laboratory data: The role of computer-presented information. *Journal of Research in Science Teaching, 24*(5), 491-506.

Papert, S. (1980). *Mindstorms*. Brighton, Sussex: Harvester Press.

Pratt, D. (1988). Taking a dive with Newton. *Micromath, 4*(1), 33–35.

Skemp, R. R. (1979). *Intelligence. Learning and action.* Chichester, Sussex: Wiley.

Tall, D. O. (1989). Concept images, generic organizers, computers and curriculum change. *For the Learning of Mathematics, 9*(3), 37–42.

Tall, D. O., & Winkelmann, B., (988). Hidden algorithms in the drawing of discontinuous functions. *Bulletin of the I.M.A., 24,* 111-115.

Tall, D. O., Blokland, P., & Kok, D. (1990). *A graphic approach to the calculus.* Pleasantville, NY: Sunburst. [also published in German as *Graphix* by CoMet Verlag, Duisburg, and in French as *Graphe*, by Nathan, Paris]
The Geometer's Sketchpad. (1992). [Computer program]. Visual Geometry Project. Berkeley, CA: Key Curriculum Press.
Thompson, P. (1992). *Blocks microworld.* [Computer program]. University of California, San Diego, CA.

THE ROLE OF COGNITIVE TOOLS IN MATHEMATICS EDUCATION

Tommy Dreyfus
Holon

1. INTRODUCTION

Imagine a group of junior high school teachers or students; suppose you are asked to teach them something relevant and interesting and you decide to introduce them to some elementary notions about chaotic dynamical systems. One possible way to do this would be to roughly follow the approach taken by Devaney (1990); this approach starts by letting students explore what can happen when a function such as $f(x) = cx(1-x)$ is repeatedly applied to an initial value x_0; among the observed phenomena are attractive and repulsive fixpoints and periodic cycles as well as chaotic behavior.

A typical activity in investigating the behavior of iterated applications of a function might include, as a first stage, the computation of long sequences of numbers for various values of x_0. Because the structure of such a number sequence is grasped more easily in a holistic representation, it would be advantageous, in a second stage, to graph the sequence as a function of the number of iterations. Moreover, in a third stage, the parameter c needs to be varied, and the effects of this variation investigated. One might want to do this dynamically by looking at the effect of continuously changing the parameter c on the global shape of the graph of the sequence. Finally, in a fourth stage, one might want to show that fixpoints, cycles, attraction, and repulsion can be explained by using a completely different graphical representation of the process, namely spiderweb diagrams; these are diagrams obtained by finding and connecting the sequence of points $(x_0, f(x_0))$, $(f(x_0), f(x_0))$, $(f(x_0), f(f(x_0)))$, $(f(f(x_0)), (f(f(x_0)))$, $(f(f(x_0)), f(f(f(x_0))))$ in a Cartesian coordinate system in which the graphs of $y = f(x)$ and $y = x$ have been drawn.

Let us now look at the support provided by a computer tool in each of the four stages. The first two stages – computing the sequences and graphing them – are so time-consuming as to make them virtually impossible without the computational power of a computer. But computer use in these stages is trivial, in the sense that the computational power only helps one to carry out many more explorations much more quickly than would otherwise be possible. The computer acts as an amplifier. In the third and fourth stages, how-

R. Biehler, R. W. Scholz, R. Sträßer, B. Winkelmann (Eds.),
Didactics of Mathematics as a Scientific Discipline, 201-211.

ever, the computer's function is one of reorganizing the knowledge (Pea, 1985); it allows one to act on the sequence as a whole and to transform it by changing the parameter; it allows one, furthermore, to switch and establish links between a numerical and two graphical representations; it finally allows one to reason about the phenomena in a qualitative manner based on the spiderweb diagrams.

Computer tools thus enable us to approach mathematics from different angles than is traditionally done. In the present case, learners may develop a view of a dynamical process that incorporates, in the process of repeated application of a function, numerical sequences, various graphical representations, dependence on parameters such as *c* in the above example, and so forth. Many of these aspects of dynamical processes can be described in qualitative rather than quantitative terms. Computer tools may thus change the quality of the mathematical objects and processes the learner experiences (Dörfler, in press): Computer tools may become cognitive tools. In this chapter, several general issues about the use of cognitive tools for learning mathematics will be raised and discussed.

2. AN EXAMPLE

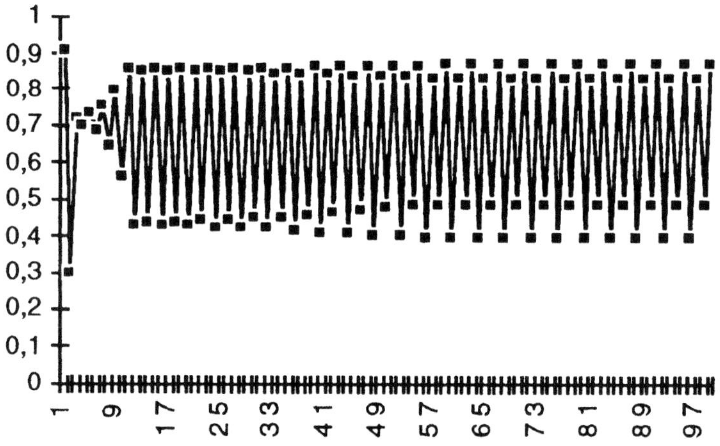

Figure 1. EXCEL output of a discrete algebraical system.

For the teacher who intends to teach about dynamical systems, the question naturally arises which computer software to use as a tool. One choice is to use only a programming language and let the students program. For teaching dynamical processes, this would be a rather confining choice – both in terms of the student population and of the screen representations that could realistically be expected. The use of a spreadsheet is one viable alternative. Spreadsheets provide both the power to quickly compute the necessary se-

quences of numbers and the possibility to graph what has been computed. Therefore, a spreadsheet appears to be a natural choice.
In fact, the spreadsheet EXCEL has been used with groups of teachers and allowed them to quickly make some progress in understanding iterated applications of functions – as far as the first two stages mentioned above. For example, cycles of length two, four, and eight are easily identified. Figure 1 shows the graph of a sequence with a cycle of length four (it is the graph of the first 100 iterations of the function $f(x) = cx(1 - x)$ for $c = 3.48$ and $x_0 = 0.907$). The teachers also had to contend with quite a few idiosyncrasies of the software in handling such simple operations as entering a fraction like 7/3 (which EXCEL insisted on interpreting as July 3) and even with mistakes, such as the graph presented in Figure 2 (which was obtained for $c = 1.25$ and $x_0 = 1.000001$ and is supposed to represent a function exponentially decreasing to minus infinity).

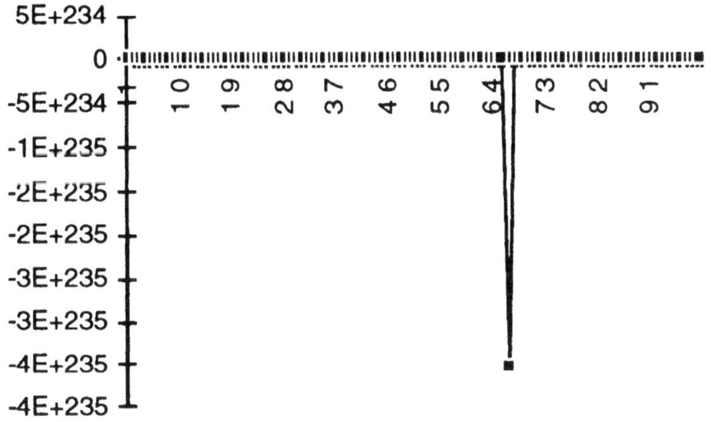

Figure 2. EXCEL output of negative exponential sequence.

But there are matters that are, from a didactic point of view, far more important than these technical details. A curriculum designer may want the power to decide on any of the following: the kind and presentation of the graphs to be used; simultaneous display of the numerical and graphical information; introduction of sophisticated, didactically motivated representations such as a spiderweb diagram; links between any two representations, for example, by highlighting the corresponding part of the graph when a portion of the numerical table is selected; coupling and decoupling of representations, and so forth. Some of these options happen to be available in EXCEL, others are not. Even those that are available may only be accessible to the user who has an intimate knowledge of the spreadsheet, or to the user who is given a spreadsheet that has been suitably prepared.

The example, using a spreadsheet for dynamical systems, is indicative for a general situation: Readily available software may have a lot of the power required from a didactical point of view, but it may also have drawbacks due to the fact that it has been prepared for other purposes and, more importantly, it may lack some features that are essential from a didactic point of view. Thus, computer tools introduce into mathematics education new opportunities, but also new issues to be resolved. In the next section, I briefly mention some of the opportunities. In section 4, some of the issues will be discussed.

3. OPPORTUNITIES

One of the most frequently mentioned opportunities offered by computer tools is their potential for using multiple-linked representations; for example, a numerical and two graphical representations were described in section 1. Kaput (this volume) gives some of the arguments that have been made in support of the expectation of a significant effect of multiple-linked representations on students' understanding of mathematical concepts such as ratio and function. The idea is to use several representations of the same concept in such a way that different aspects of the concept are stressed in different representations, and that students are helped to conceptually link corresponding aspects in different representations. At least in a number of specific cases that have been systematically investigated, many students succeeded in integrating information from several representations in a meaningful way (e.g., Schwarz & Dreyfus, in press).

One of the reasons computers have increased the potential of multiple-linked representations is computer graphics, which make powerful diagrammatic representations possible. Even without necessarily being linked to other representations, reasoning with diagrammatic representations has recently received much attention from researchers. Koedinger (1992), for example, has identified several properties of diagrams that make them superior to a sentential (linear) representation of information for many reasoning and learning activities. These properties are of two types: structural and emergent. Structural refers to the spatial arrangement of information in a diagram, for example, distance between related elements and whole-part relationships. Emergent refers to the potential of perceptually realizing relationships that might otherwise (in a nondiagrammatic representation) escape attention.

Computers make it possible to represent mathematics visually, by means of diagrams, with an amount of structure not offered by any other medium. Graphic computer-screen representations of mathematical objects and relationships allow for direct action on these objects (rather, their representatives) and observation of the ensuing changes in the diagrammatically represented relationships; this, in turn, may help the student to realize the existence and understand the nature of relationships. It may be didactically more effective to invert the task, that is, to let students investigate the question which actions will lead to a given change in the relationships. The result of such action can often be implemented dynamically; actions can be repeated at liberty, with or without changing parameters of the action, and conclusions can be drawn on the basis of the feedback given by the computer program. The power of the computer for supporting diagrammatic reasoning in mathematics derives from these possibilities.

Tall (this volume) provides a case in point. As an example, in Graphic Calculus, local straightness rather than a limiting process is suggested as a basis for developing the notion of derivative; Tall stresses that the goal is not only to provide solid visual intuitive support but also to sow the seeds for understanding the formal subtleties that occur later. This implies that the students learn to reason on the details of screen representations of concepts such as function, secant, tangent, gradient, gradient function, and so forth. Other projects that induce students to analyze the details of the relationships contained in screen diagrams and to reason based on such analysis have been reported by Kaput (1989), Yerushalmi and Chazan (1990), Shama and Dreyfus (in press), and others.

A further tool-based opportunity for mathematics education is due to the possibility to let computers do the "trivial computations" such as the repeated application of the function in the dynamical processes example. The idea is for students to operate at a high conceptual level; in other words, they can concentrate on the operations that are intended to be the focus of attention and leave the lower-level operations to the computer. For example, when learning algebraic manipulation, they can leave numerical computations to the computer. Thus, they are enabled to operate on a high level in spite of a lack of lower-level skills. This gives a chance to remedial students to reenter the mathematics curriculum without necessarily first closing all gaps (Hillel, Lee, Laborde, & Linchevski, 1992).

4. ISSUES

The very same possibility, which was presented in the previous paragraph as an opportunity, may also be seen as causing a problem. Leaving numerical computations to the computer during activities that aim at learning about algebraic manipulation can be considered as one step on a hierarchically ordered sequence of levels:

1. learn about numbers;
2. automatize number computations for use when learning algebra;
3. automatize algebraic manipulations for use when learning calculus;
4. automatize integration for use when learning differential equations;
5. automatize the solution of differential equations for use when learning dynamics.

This hierarchy could be made finer and far more extensive; it is, in fact, a subset of a partially ordered hierarchy; algebraic manipulations, for example, are needed not only in calculus but also in linear algebra, statistics, and so forth. But the point here is not to present a complete hierarchy; it is rather to focus attention on a problem that may arise when students are using computer tools with such hierarchies of capabilities: How do we prevent students from also using the computer for doing the algebra while they are supposed to be learning algebraic manipulations? More fundamentally: *Should* we prevent them? Later in life, they will hopefully have a computer algebra system at their disposal whenever they need one – so why not in school? But this raises the question whether and how it is possible to learn about algebra with an algebraic manipulator at one's fingertips (and analogous questions about number operations, calculus, etc.). Trying to answer this, one is led to the old issue about the relationship between skills and understanding: whether and to what extent are manipulations necessary for conceptual understanding (see, e.g., Nesher, 1986).

No generally accepted answer to this complex issue has been given yet, and none is to be expected in the near future. On the other hand, curriculum developers and teachers continue teaching and thus have to take decisions. At least two options are available: One is to attempt to develop curricular materials appropriate for use with a general computer algebra system and to investigate the effects. This approach has been taken mainly at the college level (Hillel, Lee, Laborde, & Linchevski, 1992; Karian, 1992). The other option is to design specific computer tools for use in educational settings. This approach seems to be predominant at the K-12 level; examples abound (e.g., Dreyfus, in press; Thompson, 1985; Yerushalmi & Schwartz, 1989).

4.1 Mathematically Versus Didactically Based Tools

A dichotomy between mathematically based tools and didactically based tools thus becomes apparent. Mathematically based tools such as computer algebra systems and spreadsheets are constructed to conform to the inner logic and structure of the content area. They respect the logical (but not necessarily the psychological) order and structure inherent in the mathematical content area. They are applicable in a wide range of situations, which is not limited to educational ones. If, for example, students learn about derivatives or integrals with a computer algebra system like Maple, they are likely to acquire the ability to use that tool for finding and using derivatives and integrals beyond the specific calculus course within which the tool was used. More than that, they also acquire some familiarity with a mathematical software tool that has capabilities far beyond the ones under direct consideration, and they can potentially exploit these capabilities.

On the other hand, students may become very apt at using derivatives or integrals in the particular given mathematically based tool within which they have learned about them, but not even recognize these concepts outside of the tool – conceptual transfer is notoriously weak. The notion of, say, derivative may be linked for these students to the tool within which they have learned about the notion. Moreover, this tool may not be didactically appropriate in the sense that it supports the execution of procedures while neglecting the underlying conceptual structure. Specifically, a mathematically based tool will presumably be able to carry out computations and draw graphs very efficiently, but it will not usually take into account any of the conceptual difficulties arising for the student who grapples with the construction of an appropriate mental image for, say, the notion of limit or derivative. And it is exactly with these specific, in some cases, well-known difficulties in mind that didactically based tools like Graphic Calculus have been designed. Such tools aim at the creation of learning experiences that promote the progressive construction by the student of flexible and widely applicable concept images of such notions as ratio, function, derivative, and so forth. One aim of the construction of such concept images is flexibility in problem-solving. Another, related aim is to establish connections: The concept will probably come up in a different framework some time later, and we may hope the student will recognize it as the same concept, exactly because of the flexibility of thought that was inherent in the learning experience. If local concept acquisition is the main goal of a curriculum, a didactically based tool may thus be the correct choice.

But precisely this same feature is a main problem of didactically based tools: They may be too local, too specifically designed, and adapted to a particular concept or cluster of concepts or to a particular curriculum. As curriculum designers, can we afford a different tool for every concept? Clearly, questions about goals are involved here: What is the curriculum driving at? A didactically based tool can be designed to be adapted to a particular curriculum with its specific learning goals (Dreyfus, in press). It becomes an organic component of that curriculum. A mathematically based tool, on the other hand, has to be used by the curriculum as it has been produced and brought to the market. In didactically based tools, we can deal with didactical design (Dugdale, 1992). Are we looking for cognitive tools for learning mathematics, or is the aim for the students to learn to use (computerized) mathematical tools? Should the mathematics that students learn depend on the tool, or should the tool depend on the mathematics to be learned? While, today, the answer, at least from the point of view of a mathematics educator, might still seem quite clear – the mathematical concepts should be the primary objective and should determine the tools – the distinction between these two poles has decreased progressively over the past few years and might disappear almost completely in the (not too far) future. Biehler (in press) has suggested, for the domain of statistics, to build didactically based elements onto a mathematically based tool. Mathematics, at least the mathematics to be taught in school, might become more tool-oriented, and, at the same time, the general-purpose tools might become more didactically appropriate.

In the next subsection, one specific design issue will be discussed in more detail in order to illustrate the dichotomy between general-purpose, mathematically based software tools and didactically based learning environments.

4.2 The Black Box Issue

Any computer program, whether or not intended for didactic use, is a black box to the user at some level of depth. Two extreme examples are a simple drill-and-practice program at one end of the spectrum and a Logo microworld at the other end. The drill-and-practice program is "black," that is, inaccessible and opaque, to students at a very high level; they only know whether their answers were right or wrong, but do not get any access or insight to the mathematical content behind; not to speak of the way the content is structured, the reasons for this structure, or how it is implemented in the computer program. Some Logo microworlds, on the other hand, can be thought of as learning environments left completely open to the students; namely, they may not only enter and analyze the Logo code constituting the microworld but may even reprogram it, thus changing the microworld itself. (Obviously, this environment is also "black" at some level: Most students do not know how the Logo interpreter works.)

Mascarello and Winkelmann (1992) have posed the question at what level of depth the black box should be. How much of the inner workings of a computer tool do students need to know? How much of it *should* they know in order for the learning experience to be maximally effective? In other terms, what types of actions should be available to the student who interacts with a tool, and what types should not be available? This complex of questions is the "black box issue."

Various possible levels that one could imagine being or not being influenceable by the student are: the tasks given to the student, the mathematical objects and operations available in the tool, the representations being used, and the mathematical topic being considered. If the designer wants a tool to offer students the possibility to investigate questions that they ask themselves, the choice of task must not be "black," it should be accessible. (In many drill-and-practice programs, this is not the case.) On the other hand, if the designer wants a curriculum to be reflected in the tool, it must be the curriculum that determines at least the mathematical topic to be dealt with, and, in fact, much more than that, namely, an approach to the topic that is consistent with the general philosophy of the curriculum. In this case, it is insufficient to simply give the student a programming language or a spreadsheet as tool. That does not mean that there are no good educational uses of programming languages or spreadsheets in mathematics classes; but it does mean that if a programming language or spreadsheet is to be used within a given curriculum, it needs, in some way or other, to be invested with some specific mathematics and some specific didactical approach. From here, the black box issue leads to the question whether the specific mathematics and the didactical approach should be internal or external to the software. And this possibly depends not only on mathematical and didactical considerations but also on organizational and economic ones.

Thus the black box issue appears to have no generally valid answer; it must be dealt with after goals of instruction are set, that is, within the framework of a curriculum. What, then, are the didactic considerations that determine at what level the black box should be for any specific tool?

One may try to answer this question in terms of possible student activities with the tool. Many didactically based learning environments are closed, fixed, whereas the student activity is, at least potentially, open. Mathematically based tools such as spreadsheets, computer algebra systems, even programming languages are also fixed; in this sense, the situation is in fact quite parallel. Furthermore, a mathematically based tool allows one to create within it. Similarly, within most computerized learning environments, the student can create, namely, new problems and, in many cases, new mathematical objects, such as functions, transformations, and so forth. A certain number of these will naturally be available in any environment. In order to give students the possibility to find out about the behavior of mathematical objects in the domain they are investigating, most tools allow the creation of additional objects and transformations (Thompson, 1985). The question is thus not one of choosing between extendable and fixed tools. Rather it is: What tools for creation are at the students' disposal? Are these tools sufficiently flexible to allow for mathematical creativity on the part of the students? Are they sufficiently specific to be useful to them? And how well-designed are these tools from the didactic point of view?

Here the discussion of the black box issue returns to the dichotomy between mathematically and didactically based tools. For example, in a very transparent tool such as Logo, distraction and lack of focus are likely to occur: The tools at the students' disposal are the Logo commands; these are not very specific in terms of any mathematical concept. Therefore, students might easily go off on a tangent when programming; they are likely to deal with syntax questions ("where is the colon missing?") rather than with conceptual ones. In an environment such as Graphic Calculus, on the other hand, students may well be limited by the fact the the designer's choices do not do justice to their ideas and ways of thinking. The environment may force a certain way of thinking onto the students, thus limiting their creativity.

In summary, it might seem that, in terms of didactic efficacy, there are advantages to custom-designing tools and making them didactically based: They can be custom-made to give exactly the didactically "ideal" amount of transparency. But the term didactically "ideal" is not a constant; it certainly depends on the curriculum if not on the teacher and even the student. Therefore, at present, this discussion remains inconclusive.

5. CONCLUSION

It is generally agreed that learning mathematics is not a spectator sport, but requires active involvement on the part of the learner; for learning abstract mathematical concepts, such activity is usefully described in terms of student actions on mathematical objects and relationships; these objects and relationships are necessarily given in some representation, which incorporates, or omits, links between them. The point has been made above that computer tools have the potential to contribute to the learning process not only as amplifiers (saving time on computations and making graphing easy in the above example) but also, and more importantly, as reorganizers: Mathematics itself becomes different for the learner; new tools change cognition. Representations can be linked. Diagrammatic and qualitative approaches can be taken.

One of the central questions to be answered by any cognitive tool concerns the cognitive appropriateness of these representations (Dörfler, in press): What are the advantages and disadvantages of various representations for implementing a certain concept, certain aspects of a concept, or certain relationships between concepts? For example, which representations are appropriate to help a student learn about the notion of increase of a function; and what needs to be the nature of linkage between the different representations in the same tool in order to help the student to establish connections between them with respect to the notion of increase? And how does the nature of the concept generated in the student's mind, the concept image, depend on these representations? These questions have both epistemological and cognitive components; they are deep questions, requiring both theoretical and empirical investigation. Moreover, they are very complex questions: Answers depend quite strongly on the intended student population, their age, experience, mathematical maturity, and so forth.

While these questions are of central importance for judging the appropriateness of a cognitive tool, they obviously cannot be investigated empirically without existing cognitive tools. Design and implementation of such tools, didactically and mathematically based ones, is therefore a largely empirical undertaking that continuously informs and is informed by progress on the theoretical, epistemological, and cognitive research questions. Only in the framework of a teaching-learning experiment can the didactic effectiveness of a given tool be investigated. Only within a curriculum with its specifically defined goals can one undertake the epistemological analysis mentioned above. And only when the tool is actually used at least in a laboratory situation with students can the corresponding cognitive analysis be started. Given enough thought, effort, and time, such analyses can be expected to contribute to the resolution of the issues raised above such as the black box issue and, more generally, the dichotomy between mathematically and didactically based tools.

REFERENCES

Biehler, R. (in press). Software tools and mathematics education: The case of statistics. In C. Keitel, & K. Ruthven (Eds.), *Learning from computers: Mathematics education and technology.* Berlin: Springer.

Devaney, R. (1990). *Chaos, fractals, and dynamics: Computer experiments in mathematics.* Menlo Park, CA: Addison-Wesley.

Dörfler, W. (in press). Computer use and views of the mind. In C. Keitel, & K. Ruthven (Eds.), *Learning from computers: Mathematics education and technology.* Berlin: Springer

Dreyfus, T. (in press). Didactic design of computer based learning environments. In C. Keitel, & K. Ruthven (Eds.), *Learning from computers: Mathematics education and technology.* Berlin: Springer.

Dugdale, S. (1992). The design of computer-based mathematics instruction. In J. Larkin & R Chabay (Eds.), *Computer assisted instruction and intelligent tutoring systems: Shared issues and complementary approaches* (pp. 11-45). Hillsdale, NJ: Erlbaum.

Hillel, J., Lee, L., Laborde, C., & Linchevski, L. (1992). Basic functions through the lens of computer algebra systems. *Journal of Mathematical Behavior, 11*(2), 119-158.

Kaput, J. (1989). Supporting concrete visual thinking in multiplicative reasoning. *Focus on Learning Problems in Mathematics, 11*(1), 35-47.

Karian, Z. (Ed.). (1992). Symbolic computation in undergraduate mathematics education. Mathematical Association of America, *MAA Notes series* (24).

Koedinger, K. (1992). Emergent properties and structural constraints: Advantages of diagrammatic representations for reasoning and learning. In H. Narayanan (Ed.), *Proceedings of the AAAI Spring Symposium on Reasoning with Diagrammatic Representations.* Stanford, CA.

Mascarello, M., & Winkelmann B. (1992). Calculus teaching and the computer: On the interplay of discrete numerical methods and calculus in the education of users of mathematics. In B. Cornu & A. Ralston (Eds.), *The influence of computers and informatics on mathematics and its teaching* (pp. 108-116). Science and technology education document series 44. Paris: UNESCO.

Nesher, P. (1986). Are mathematical understanding and algorithmic performance related? *For the Learning of Mathematics, 6*(3), 2-9.

Pea, R. (1985). Beyond amplification: Using the computer to reorganize mental functioning. *Educational Psychologist, 20*(4), 167-182.

Schwarz, B., & Dreyfus, T. (in press). Measuring integration of information in multirepresentational software. *Interactive Learning Environments.*

Shama, G., & Dreyfus, T. (in press). Visual, algebraic and mixed strategies in visually presented linear programming problems. *Educational Studies in Mathematics.*

Thompson, P. (1985). Experience, problem solving and learning mathematics: Considerations in developing mathematics curricula. In E. Silver (Ed.), *Teaching and learning mathematical problem solving* (pp. 189-236). Hillsdale, NJ: Erlbaum.

Yerushalmi, M., & Chazan, D. (1990). Overcoming visual obstacles with the aid of the Supposer. *Educational Studies in Mathematics, 21*(3), 199-219.

Yerushalmi, M. & Schwartz, J. (1989). *Visualizing algebra: The function analyzer* [computer program]. Pleasantville, NY: Educational Development Center and Sunburst Communications.

INTELLIGENT TUTORIAL SYSTEMS

Gerhard Holland
Gießen

1. INTRODUCTION

The following is an attempt to contribute to the topic of *intelligent tutorial systems* (ITS) as an object of research in mathematics education and development. In the debate in mathematics education about the use of advanced software for mathematics instruction, tutorial systems have only a low status beside mathematical tools like DERIVE and mathematical microworlds like Cabri géométre. There are at least two reasons for this:

1. As far as ITS are available, very few will run on school computers, are adaptable to the requirements of countries and school systems other than those for which they were developed, and are offered additionally at prices within the reach of schools.

2. Because of negative experience with *programmed instruction* in the 1960s, and subsequently with simple and low-yield drill and practice programs for simple skills, many mathematicians have a general distrust toward tutorial systems.

My contribution will have met its goal if it succeeds in initiating a qualified debate about the significance of tutorial systems for mathematics instruction and for research into mathematics education. After explaining the classical architecture of intelligent tutorial systems (section 2), the system HERON for solving word problems (by K. Reusser) is presented as an example (section 3). Subsequently (section 4), the paradigm of ITS as a private teacher is contrasted with the concept of a mathematical microworld with tutorial support. Finally, I give an extensive presentation of a general concept that can be used to subsume a large number of (potential) tutorial systems for mathematics instruction and is intended to contribute toward reducing the development cost for ITS (section 5).

2. INTELLIGENT TUTORIAL SYSTEMS

The primary theoretical motive in using methods of artificial intelligence (AI) to develop "intelligent" tutorial systems, which yield the same performance as a private teacher, has been an objective for more than 10 years in advanced research in the still recent field of *artificial intelligence and education.* This, however, is unaffected by the illusion of revolutionizing the

R. Biehler, R. W. Scholz, R. Sträßer, B. Winkelmann (Eds.),
Didactics of Mathematics as a Scientific Discipline, 213-223.

school system by means of ITS that can be implemented, on school computers, thus making the teacher superfluous. The cognitive psychologist J. R. Anderson, however, already hopes that the comparably low level of the US-American school system can be raised by developing intelligent tutorial systems (Anderson, 1992). To justify these hopes, he refers to empirical studies that furnish the (not very surprising) evidence that having a student taught by a private teacher is much more efficient than collective teaching in the classroom.

The requirements addressed to an ITS that is to take over the functions of a private teacher are derived from the qualifications asked from a human private teacher.

1. The teacher must be an *expert* on the subject in question. In this function, the teacher must be able to answer student questions pertaining to the discipline, to solve tasks put to the student, and to analyze student answers for bugs and misconceptions.

2. The teacher must know how to *present* the subject matter in an appropriate way and which tools must be placed at the student's disposal in order to free teaching from unnecessary ballast.

3. The teacher must have an idea of each student's knowledge and skills and be able to adapt his or her own hypothetical *student model dynamically* to the student's learning progress.

4. The teacher must have knowledge about the *curriculum* (subject matter, learning goals, etc.), and have methodological knowledge and a repertoire of tutorial strategies at his or her disposal in order to be able to intervene tutorially in an optimal way at any point.

These four requirements allow us to comprehend the classical architecture of an ITS as an integrated information-processing system with an expert module, an *environmental module*, a module for *student modeling*, and a tutor module (Wenger, 1987).

While research is far advanced in some fields, achieving results that are significant from a mathematics education perspective as well (e.g., the analysis of systematic bugs and their causes in written subtraction, the transformation of algebraic terms, and linear equations), there is as yet no ITS for teaching in school that meets the high requirements of an ITS in all four components and can additionally be run on hardware available in schools. In spite of rapid progress in the development of hardware and software, the two requirements can hardly be reconciled at present for technical reasons alone. And the immense cost in time required to develop an ITS reduces the probability of much change in the present situation, if there is no success in developing shells, authoring systems (Lewis, Milson, & Anderson, 1987), or, at least, transferable architectures for individual modules of certain classes of intelligent tutorial system.

3. HERON, AN *ITS* TO SOLVE WORD PROBLEMS

As an example for an ITS, I shall present the system HERON developed by the Swiss cognitive psychologist K.Reusser to solve word problems. HERON has the following features in common with the geometry tutor developed by J. R. Anderson (Anderson, Boyle, & Yost, 1985) and frequently discussed in the literature, but not presented here for reasons of space:

1. The system exists not only as a prototype but also as a user-friendly software that can be run on school computers and has already been tested with students. (Results of testing Anderson' geometry tutor are reported in Wertheimer 1990.)
2. The subject-matter field is highly relevant for mathematics education.
3. The development of HERON is based on convincing principles of cognitive psychology and pedagogy.
4. The tutor does not support individualized tutorial strategies.

The founding principles, however, express diverging views of the two researchers concerning the function of an ITS. Anderson developed the analysis modules of his tutors (geometry tutor, Lisp tutor) primarily as cognitive student models within the framework of his own cognitive (ACT*) theory. He thus sees his theory confirmed where his tutors perform in practice. In contrast, K. Reusser considers that the demand addressed to an ITS of replacing an intelligent and adaptive teacher by a cognitive student modeling alone is a possible long-term objective whose desirability must also be questioned (Reusser, 1991). According to Reusser, "intelligence" should not be concentrated in the computer, but rather be spread out across the entire pedagogical setting, with the learner at its center. Not the computer, but the learner assisted by the computer should establish diagnoses, set goals, and make plans (Reusser, 1991).

3.1 Method of Solution and Problem Solving in Dialogue With HERON

HERON supports all word problems that can be solved with the so-called simplex method used in many German school textbooks. We shall explain how the simplex method is applied in the tutorial system HERON with an example taken from Reusser (1991; see Figure 1). The lower right-hand window contains the word problem. The student solves the problem in dialogue with HERON by forward chaining in the following steps:

Analyzing text, producing situation units.

1. The student uses the mouse to mark those text sections containing relevant quantitative information.
2. For each information marked in this way, HERON produces a graphic *situation unit* consisting of three fields, and the student enters the numerical value into the lower left-hand field.

3. The student enters the unit of measurement into the lower right-hand field, and a textual label into the upper right-hand field, for example, "content of father's can." The latter can be taken from a menu, the student only having to decide which of the phrases offered in the menu belongs to the situation.

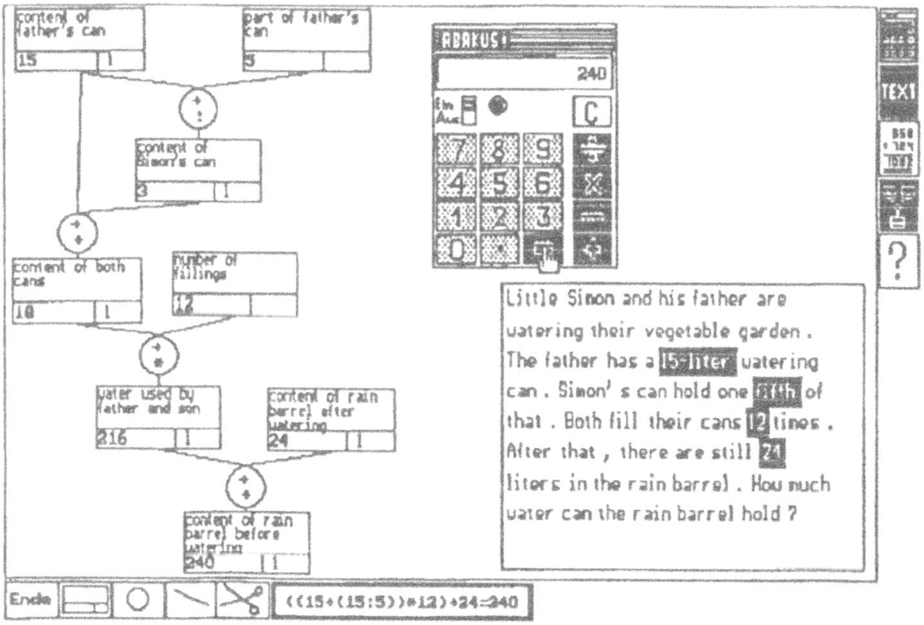

Figure 1. Screen display from HERON.

Producing a relational scheme.

4. The student selects two situation units from which a third quantity can be calculated (e.g., "content of father's can" and "part of father's can"). He or she uses the mouse to place these at a suitable spot on the monitor (e.g., the upper left-hand corner), selects the appropriate calculating operation from a menu, and links the circular operator node produced by the system by means of edges to the two situation units.

5. The system produces an empty subgoal node that is constructed according to the same principles as the situation units.

6. The student fills in the three fields of the subgoal node ("content of Simon's can"). Unit of measurement and label can be selected from a menu. The triplet of situation units is called a *relational scheme.*

Producing a tree structure.

7. The procedure is continued until the goal node representing the word problem's solution has been generated. The respective two starting units can be either situation units or goal nodes.

It should be noted that HERON also supports steps of backward chaining. For instance, the first relational scheme to be generated could be that which contains the goal node. In this case, the two parent nodes are not situation units, but unsolved subgoal nodes.

3.2 Supervision and Tutorial Support

HERON supervises the students' problem-solving process and gives feedback based on error analysis.

Besides the support the system gives by offering a menu to select for a large number of steps, help can be asked at any stage of the problem-solving process.

4. MATHEMATICAL MICROWORLDS WITH TUTORIAL COMPONENTS

In the research field of *Artificial Intelligence and Education*, the concept of microworld stands for a type of educational tool that differs significantly from the paradigm of an ITS. As microworlds are treated extensively elsewhere in this volume (see D. Tall's contribution), there is no need to define the concept of microworld here. ITS and microworlds differ mainly in their educational style. The latter are determined by the constraints the learning environment and the tutor exercise on the learner (Elsom-Cook, 1988) – or, positively, by the degree of freedom given to the learner to personally shape his or her own learning process. If this dimension is illustrated by a scale (Elsom-Cook, 1988), a traditional ITS like Anderson's geometry tutor is at one pole of the scale, while a microworld like that of Papert's LOGO is located at the other pole. That microworlds are more readily accepted by mathematics educators than ITS is most probably due principally to their preference for a teaching scenario that simultaneously enhances the learner's self-guidance of his or her learning process while not infringing on the teacher's role. However, a comparison of ITS and microworlds for mathematics instruction must not overlook the general differences in the goals for which they have been developed. While ITS primarily serves to enhance skills in applying knowledge of mathematical theorems and rules, mathematical microworlds (like the mathematical microworld MOTION; Thomson, 1987) have been developed mainly to promote conceptual knowledge.

As learning mathematical concepts cannot occur without any external guidance on given tasks, the developers of microworlds are confronted with the question of to whom the student should turn if he or she gets into difficulties when trying to solve a problem. A teacher rotating from work place to work place will soon be overburdened in this function. While this problem is significantly reduced if students work in pairs at the computer, it will nevertheless persist in principle. It is thus no wonder that there is an observable tendency today to equip mathematical microworlds with intelligent tutorial components (Holland, 1991; Laborde & Sträßer, 1991; Thomson, 1987). An interesting example of a microworld with tutorial components is the project "shopping on Mars" developed under the lead of T. O'Shea (Hennessy, Evertsz, & Floyd, 1989). Nonetheless, the developers of intelligent tutorial systems tend to give as much scope as possible to the self-shaping of the learning process and to metacognitive activities. One example is the system HERON presented in section 3. To close, some studies have followed the concept of "guided discovery learning" in an attempt to develop tutorial systems that are able to practice different teaching styles according to demand (Elsom-Cook, 1988, 1990).

5. TASK-ORIENTED *ITS* FOR MATHEMATICS INSTRUCTION

On a world scale, quite a number of ITS for mathematics instruction have been developed during the last decade. However, only a few have currently progressed beyond the prototype stage. As to subject matter, they can be assigned to almost all fields of school mathematics. Their favorite topics are: arithmetics, written arithmetics, algebraic term transformations, equations and equation systems, word problems, combinatorics, trigonometry, geometric proof, and differential and integral calculus. It is remarkable that the overwhelming majority of these systems is not intended to promote acquisition of knowledge of concepts, but rather serves to affirm skills in applying mathematical knowledge of theorems and rules. This, however, does not come as a surprise, because it seems to be much easier to develop ITS for mathematics skills than for the acquisition of mathematical concepts. A typical example for an ITS that can be used to train a demanding mathematical skill is Anderson's above-mentioned geometry tutor (Anderson, Boyle, & Yost, 1985).

The following will attempt to use the concept of *task-oriented* ITS to describe a common architecture for an extensive class of tutorial systems suited to learn and exercise the application of mathematical knowledge of theorems and rules in the context of intramathematical problem tasks (Holland, 1992). The ensuing possibility of developing some of the modules domain-independently should be used to reduce the enormous development cost for an ITS – just as Anderson's Teacher's Apprentice Project intended to develop an author system for ITS (Lewis, Milson, & Anderson, 1987). At the Institute for Didactics of Mathematics at the University of Gießen, three task-oriented ITS have been developed up to now and have been tested to some extent with university students – a tutor for geometric tasks of proofs and computation, a tutor for geometric construction tasks, and a tutor for transforming functions (the first two yet without a module for selecting tasks; cf. section 5.1, stage 19).

5.1 Characterization of Task-Oriented ITS

The following is a listing of the essential features of task-oriented ITS. A comparison with Anderson's tutors shows that the concept of task-oriented

ITS integrates principles that Anderson postulated for his own tutorial systems.

Educational goals and system requirements.

1. The global educational goal supported by the tutor is operationalized by an ideal *problem class,* that is, students are meant to be able to solve all the tasks belonging to this class after tutorial training.

2. The tasks are not one-step tasks of application (of a theorem or a rule), but *problem tasks* consisting of several steps that are solved by successively applying suitable *operators* (theorems and rules).

3. There is no deterministic method of solution, that is, there is generally more than one applicable operator for each step in the solution process. Hence, there are, in general, several solution plans or solutions for each task. (This is why tutorial systems for written methods of arithmetics are not among the systems considered here.)

4. The students *know* which operators are required or permissable for solving the task (transformation rules for transforming terms or equations, geometric theorems for tasks of geometric proof, rules for geometric loci for geometric construction problems). What is to be exercised here is the skill to *apply* the operators in the context of a problem solution consisting of several steps.

5. Educational goals are thus: (a) The students should be able to apply the relevant operators of the problem class in the context of a problem containing several steps. (b) The students should know and be able to apply heuristic methods to solve problems (e.g., working forward and working backward in problems of proof).

Global tutorial strategy.

6. The global educational goal is attained by solving problems of the problem class. A growth of learning occurs both through ITS feedback in case of faulty or unsuitable operator applications and through assistance that the students can ask for at any time. It should be noted that task-oriented ITS satisfy the demand formulated by J. R. Anderson that learning should take place within the context of problem-solving (Anderson, Boyle, Farrell, & Reiser, 1984).

ITS expert.

7. The ITS expert is a problem solver operating on a knowledge base in which knowledge about the applicability and effect of operators is represented as rule-based knowledge.

8. For each problem of the problem class the expert finds solutions that are appropriate to the knowledge state of the students.

9. The expert is able to check a student solution for correctness and quality. It is able to classify errors as they occur.

10. The expert is "transparent," that is, it uses only knowledge and methods the student is supposed to learn and use (it could not perform Stages 8 and 9 otherwise). It should be noted that subject-matter fields like geometric

proof, geometric constructions, algebraic term transformations, combinatorics, or integral calculus require the ITS to be equipped with a high-performance problem solver. The task-oriented ITS ability to provide the student with an informative error analysis justifies its being called an "intelligent" system, and this is at the same time the main difference to nonintelligent CAL systems of computer assisted learning (Lewis, Milson, & Anderson, 1987).

For J. R. Anderson, the expert in his tutorial systems is the model of an "ideal student" represented by a system of production rules. Real students are represented by *deviations* from the ideal student, that is, by omitting the rules not yet learnt and/or by adding buggy rules. With this, Anderson intends to attain a cognitive student modeling on the basis of his ACT* theory. As task-oriented ITS do not pursue the demanding goal of a cognitive student modeling, the costly and inefficient modeling by a production system can be dispensed with here.

Environment module.

11. For the dialogue between student and tutor, there is as little input with the keyboard as possible. Instead, menus and graphic input tools like mouse and graphic tablet are used in the sense of "direct manipulation." This should meet Anderson's requirement of liberating the short-term memory (Anderson, Boyle, Farrell, & Reiser, 1984).

12. For representing problem states and solution, a representation is chosen that makes the goal structure explicit (Anderson, Boyle, Farrell, & Reiser, 1984) and supports the planning of the solution (Collins & Brown, 1988). This purpose is served, in particular, by a two-dimensional representation of and/or trees, proof graphs, and algebraic term structures (Burton, 1988).

Monitoring by the ITS tutor.

13. The tutor monitors each step the student makes toward a solution. For this, he or she makes use of the expert (see 9).

14. The student may choose from several tutor modes for the tutor's response to errors. These are distinguished according to the scope they leave to the student in case of an erroneous or unfavorable operator application. Feedback after each false suboperation prevents the student from deviating from a solution path, but does not give the student an opportunity to find the error him or herself. To counter this, feedback is given only after completing work on the problem in order to exclude the risk of aimless error search. It should be noted that for his initial tutors (geometry tutor, Lisp-tutor), Anderson advocated and realized the principle of immediate feedback (Anderson, Boyle, Farrell, & Reiser, 1984). In the later tutors of the Teachers' Apprentice Project (Lewis, Milson, & Anderson, 1987), however, he also accepts other tutorial strategies.

Stepped tutor help.

15. At any stage in the problem-solving process, the student may call for help. This is offered by the tutor (using the expert module) in the form of hierarchically graded help. Help begins with general heuristic hints and ends with prescribing the very step toward a solution the expert would have chosen in this situation.

Student modeling.

16. While the student works on the problem, a *local student model* is established that refers only to the solution of the current problem (errors made, help called for). The local student model serves the feedback (see 14) and the dynamic backup of the *global* student model (see 18).

17. The diagnostic technique used to establish local student models is that of Model Tracing (Anderson, Boyle, & Yost, 1985; VanLehn, 1988). At each further stage of the problem-solving, the student's (false or correct) operator application is compared to the potential application of the expert. Model Tracing is possible, because the student is not allowed to chain operators (e.g., entering in the final result in case of a term transformation).

18. The global student model is backed up after each operation performed on the problem by means of the local student model. In a task-oriented ITS, it has the sole function of enabling the tutor to select suitable problems for the student (see VanLehn, 1988, p. 56). This can be achieved by simple bookkeeping of the problems hitherto worked on, and by additionally generating a hypothesis on the degree of its availability for each operator. At the beginning of each training, the global model does not contain any information.

Selecting the problems.

19. On the basis of the information provided by the global student model and the most recent local student model, the tutor selects a suitable problem from a *prestructured* problem collection.

It should be noted that "suitable" means that a particular student attains the global educational goal according to his or her own knowledge and skills by working on the smallest number of problems.

As yet, there are only few contributions on the general problem of advancement in a prestructured curriculum, and, in particular, on problem selection. For Anderson's tutors, the problem is not even mentioned.

The worth of selecting problems by the tutor using a global student model may be questioned if the structure of the task sequence is transmitted to the student for selecting an appropriate problem by him or herself.

6. CONCLUSION

Within the larger research field of *cognitive science*, the new research field *Artificial Intelligence and Education* has been established by regular conferences and periodicals during the last decade. Its objective is to develop flexible and adaptable tutorial systems for all imaginable fields of education and subject matter. One of the tasks of mathematics education is to participate in the development and testing of high-performance cognitive tools that support mathematical processes of learning. These will be either mathematical microworlds with intelligent tutorial components or intelligent tutorial systems for solving problem tasks, depending on whether the focus is on acquiring mathematical concepts and structures or on applying mathematical concepts, theorems, and rules. The guiding principle for developing such systems should always be that the learner's own shaping of his or her process of learning should be supported, while, at the same time, protecting the learner from unproductive errors and offering appropriate help in any situation.

REFERENCES

Anderson, J. R., Boyle, C. F., Farrell, R., & Reiser, B. (1984). Cognitive principles in the design of computer tutors. In P. E. Morris (Ed.), *Modelling cognition*. London: Wiley.

Anderson, J. R., Boyle, C. F., & Yost, G. (1985). The geometry tutor. *Proceedings of the International Joint Conference on Artificial Intelligence* (pp. 1-7). Los Altos, CA: Morgan Kaufmann.

Anderson, J. R. (1992). Intelligent tutoring and high school mathematics. In C. Frasson, G. Gauthiers, & G. I. McCalla (Eds.), *Intelligent tutoring systems* (pp. 1-10). Berlin: Springer.

Burton, R. R. (1988). The environment module of intelligent tutoring systems. In M. C. Polson & J. J. Richardson (Eds.), *Intelligent tutoring systems* (pp. 109-142). Hillsdale, NJ: Erlbaum.

Collis, A., & Brown, J. S. (1988). The computer as a tool for learning through reflection. In H. Mandl & A. Lesgold (Eds.), *Learning issues for intelligent tutoring systems*, (pp. 114-137). Berlin: Springer.

Elsom-Cook, M. T. (1988). Guided discovery tutoring and bounded user modeling. In J. A. Self (Ed.), *Artificial intelligence and human learning* (pp. 165-178). London: Chapman and Hall.

Elsom-Cook, M. T. (1990). Guided discovery tutoring. In M. T. Elsom-Cook (Ed.), *Guided discovery tutoring: A framework for ICA research* (pp. 3-23). London: Paul Chapman.

Holland, G. (1991). Tutorielle Komponenten in einer Lernumgebung zum geometrischen Konstruieren. In R. Sträßer (Ed.), *Intelligente tutorielle Systeme für das Lernen von Geometrie*. Occasional Paper 124, Universität Bielefeld/IDM.

Holland, G. (1992). Aufgabenorientierte tutorielle Systeme für den Mathematikunterricht. In U. Glowalla & E. Schoop (Eds.), *Hypertext und Multimedia. Neue Wege in der computerunterstützten Aus- und Weiterbildung*. Berlin: Springer.

Hennessy, S., O'Shea, T., Evertsz, R., & Floyd, A. (1989). An intelligent tutoring system approach to teaching primary mathematics. *Educational Studies in Mathematics, 20*, 273-292.

Laborde, J. M., & Sträßer, R (1990). Cabri-Geométre: A microworld of geometry for guided discovery learning. *Zentralblatt für Didaktik der Mathematik, 22*, 171-177.

Lewis, M. W., Milson, R., & Anderson, J. R. (1987). The teacher's apprentice: Designing an intelligent authoring system for high school mathematics. In G. Kearsley (Ed.), *Artificial intelligence and instruction: Applications and methods* (pp. 269-302). Reading, MA: Addison-Wesley.

Reusser, K. (1991). Tutoring systems and pedagogical theory: Representational tools for understanding, planning and reflection. In S. Lajoie & S. Derry (Eds.), *Computers as cognitive tools* (pp. 143-177). Hillsdale, NJ. Erlbaum.

Thomson, P. W. (1987). Mathematical microworlds and intelligent computer-assisted instuction. In G. Kearsley (Ed.), *Artificial intelligence and instruction, applications and methods* (pp. 83-110). Reading, MA: Addison-Wesley.

VanLehn, K. (1988). Student modeling. In M. C. Polson & J. J. Richardson (Eds.), *Intelligent tutoring systems* (pp. 109-142). Hillsdale, NJ: Erlbaum.
Wertheimer, R. J. (1990). The geometry proof tutor: An intelligent computer-based tutor in the classroom. *Mathematics Teacher, 83,* 308 - 317.
Wenger, E. (1987). *Artificial intelligence and tutoring systems*. Los Altos, CA: Morgan Kaufmannn.

CHAPTER 5

PSYCHOLOGY OF MATHEMATICAL THINKING

edited and introduced
by
Roland W. Scholz
Bielefeld / Zürich

Psychological research on mathematical learning, thinking, and instruction has accompanied the rise of didactics of mathematics as a scientific discipline since its very beginnings. In 1910, the German experimental psychologist David Katz (1913) produced the volume *Psychologie und mathematischer Unterricht* (Psychology and Mathematical Instruction) commissioned by the ICME. Obviously, this research project had been initiated by Felix Klein. Chapters of Katz's book deal with topics like the development of the concept of space and number.

The interest of mathematics teachers both in the nature of mathematical thinking, learning, and instruction and the methods psychologists use is also reflected by the Leipziger Lehrerverein (Leipzig Teacher Association) who founded and financed the "Institut für experimentelle Pädagogik and Psychologie" in 1906. One of the main outcomes of this institute is Freeman's (1910) volume on children's and adults' conception of numbers. Note that Freeman's studies used rigorous laboratory and experimental procedures.

As is well-known, many mathematicians also theorized on *mathematics*

R. Biehler, R. W. Scholz, R. Sträßer, B. Winkelmann (Eds.),
Didactics of Mathematics as a Scientific Discipline, 225-230.

as a human activity. Using the method of introspection and referring to his own experience of creating mathematics, Poincaré (1910, 1914) and many other mathematicians dealt with psychological questions like *insight* or *modes of thought* in mathematical thinking. Thus, traditionally, one may find a close relation between epistemology and the theory of mathematical cognitions.

The work of the psychologist who is most strongly associated with research on mathematical thinking, that is, Jean Piaget, was strongly influenced by Klein and Poincaré. For instance, when dealing with the question "What exactly is meant by geometrical intuition?" (Piaget, 1948/1963, p. 447), he discusses various definitions of intuition and intuitive thinking made by mathematicians.

In some respects, research on mathematical thinking attained a new quality through the constitution of the "International Group of Psychology in Mathematics Education" during ICME 3 in 1976 at Karlsruhe. Psychology in Mathematics Education (PME) was predominantely initiated by Ephraim Fischbein, Richard Skemp, and Hans Freudenthal in order to promote the exchange of scientific information in the field. Through approaching mathematical thinking from different perspectives, the research work of all three founders of PME was concerned with *understanding qualities of mathematical thinking*.

The object of understanding qualities of mathematical thinking and their dependance on types of (contextual) framing and representations, is still a main issue of current reseach in the PME community (see Goldin, 1992; Vergnaud, 1990). The relation between external and internal representation is, in some respects, the core linkage that brings cognitive psychology into mathematics education. Historically, there is a close relationship between the psychology of thinking and epistemology; thus one will find many cognitive issues being addressed in chapter 8 on *history and epistemology of mathematics and mathematics education*. When analyzing symbol schemata, technologies, and media, the researcher in mathematics education at least implicitly deals with topics of cognitive psychology (cf., e.g., *Kaput,* this volume).

There are some more links to chapters of this book. Whereas this chapter concentrates on the individual's acquisition of mathematics, most of the psychological approaches in chapter 3 on *interaction in the classroom* include a cognitive and social-psychological perspective. Naturally, many explanations of existing or nonexisting differences between different groups or populations of learners are cognitively founded (see, e.g., Lorenz & Hanna, this volume). Thus many results of the psychology of mathematical thinking are applied in chapter 6 on *differential didactics*. Last but not least, as Fey stresses in the first paper of chapter 1 on *preparing mathematics for the students,* curriculum developers have learned a lot in the last three decades from psychological theories of the child's cognitive development.

Ephraim Fischbein's contribution on *the interaction between the formal, the algorithmics and the intuitive components in a mathematical activity* provides a thorough model of mathematical activity, its genesis, growth, concepts, and qualities. Thus Fischbein, who himself incorporates both mathematics and psychology, approaches the cognitive foundation of mathematical thinking when distinguishing between the *formal aspect* (e.g., axioms and theorems), the *algorithmic* aspect, and the *intuitive* way of mathematical reasoning. He demonstrates that all three aspects are necessary for mathematical understanding. Though very often intuitions or certain skills may enhance each other, Fischbein reveals that primitive intuitions, Gestalt features, and algorithmic skills may also serve as obstacles and barriers in acquiring new mathematical knowledge. He stresses that these intuitive and primitive models tacitly influence the formal reasoning process, and reveals that Piaget, who was interested in separating stages of cognitive development, obviously was not attracted by this interplay of qualitatively different knowledge sources within the subject. When starting from different stages of mathematical thinking, Fischbein provides some examples for epistemological obstacles and interferences of different representations or models tied to different Piagetian stages. Fischbein himself applies and refers to a multitude of research methods. Using theoretical analysis, introspection, attentive observations, case studies, and experimental research, he illustrates how the interference of the formal, the algorithmic, and the intuitive components may promote and hinder each other.

Gerhard Steiner considers himself as a scholar of Piaget in the second generation. In the first part of his paper *From Piaget's constructivism to semantic network theory: Applications to mathematics education - a microanalysis,* he critically examines ideas and concepts of the Geneva School that are currently used in cognitive psychology. As we know, concepts like assimilation, accomodation, or schema are taught in many teacher-training programs and may be used actually and potentially for an academic understanding of the child's mathematical learning. In contrast, the INCR concept, for instance, is currently mentioned only occasionally.

While taking a close and inside look at Piagetian modeling, Steiner reveals that Piaget already anticipated the current "standard differentiation" of conceptual and procedural knowledge in his concepts of schema and systems of schemata. Both processes, accomodation and assimilation, take place in the learning of mathematics. Whereas assimilation is considered mostly as an active adjustment and integration of information into existing schemata, accomodation denotes the change of the individual's cognitive structure when being confronted with information that necessitates an enlarged or revised internal representation. When introducing the Piagetian concepts of "lecture des données" and "mise en relation," Steiner demonstrates how Piaget's theory provides access to an "internalization of connections according to an organizational plan" that has been abstracted from former actions. In the language of modern psychology, Piaget thus – in another terminology – was dealing with the formation and change of semantic networks.

In order to understand and to model how students organize, modify, and enlarge their mathematical knowledge, Steiner introduces the concept of an *algebraic mathematical network*. This concept allows for a microanalysis of algebraic-mathematical thinking. It provides an approach for preparing mathematical problems in such a way that the student's schema is actively modified. Steiner's goal is to foster a learner's autonomy in tackling algebraic problems when applying the Piagetian schema concept and progressive network analysis. Through a sequence of tasks prepared by the teacher, the student is influenced progressively and thus introduced to a freshly created and activated micronetwork. This progression of new (accomodated) networks provides an elaboration of the algebraic mathematical network.

How algebraic mathematical network analysis may be applied in the classroom is demonstrated by a pilot study on secondary school students. Thus Steiner shows how Piagetian theory may be used for the derivation of didactical practice in dealing with trinominals. The methodological difficulties of judging and measuring the change of mathematical network analysis are briefly discussed.

Joachim Lompscher is one of the collaborators and scholars of Galperin, Davydow, and Rubinstein. One may say that Rubinstein (1958) developed the philosophical basis of Soviet Psychology (cf. Goldberg, 1978). He demonstrated that, during the transition from an act's connection with practical experience to its association with theoretical thought, a reorientation occurs. That is, practical activity is an extremely important stimulus for the formation of thought. By combining these ideas with those from the Geneva School and with that of the Sociohistorical School of Leont'ev and Vygotsky, the classroom experience is conceived of as a part of the social relation of the student and a constituent of the subject-object relation for both, that is, for cognitive development and for teaching.

Due to the current fundamental changes in political and national systems in Eastern Europe and the former Soviet Union, the further development of this theory in these countries is questionable. The selected contributions and learning-teaching experiments reviewed by Joachim Lompscher were discontinued in the late 1980s. Three branches of the Sociohistorical School are concisely described and discussed. In Lompscher's paper on *the sociohistorical school and the acquisition of mathematics,* the didactical experiments of *Galperin* provide an interpretation and application of Vygotsky's concept of internalization or interiorization. According to this approach, the solving of tasks has to be organized on various levels of activity in order to become internalized. Starting from material activity, the learner should proceed by verbalizing for others via verbalizing for oneself and end up with a nonverbal mental level. Thus, Galperin provides sequences of proximal development for the learner. The core idea of *Davydow's* interpretation is the principle of ascending from the abstract to the concrete. In his teaching experiments, students start working with symbols and graphical models, thus recognizing the general structure and relationships, and finally may apply them to the concrete mathematical object, for instance, natural numbers. In his own series of studies, *Lompscher* has investigated the course of discovery of connections in the representation of verbal statements on real situations. In his teaching experiments, he leads students through different stages of activities in coping with structures of text problems ending up with an independent coping with objects of learning as a result of goal formation, information and strategy sampling, and so forth.

Richard Lesh and Anthony E. Kelly are committed to the research approach that most strongly influenced the psychology of mathematical thinking of North America during the last two decades, that is, *constructivism.* From a constructivist point of view, reflective ability is considered to be the major source of knowledge on all levels of mathematics (cf. von Glasersfeld, 1991, p. xviii). Thus, as Lesh and Kelly conclude in their contibution on *action-theoretic and phenomenological approaches to research in mathematics education*, constructivism is not simply a perspective on children's thinking but rather more a theory on thinking. Thus constructivism is considered to be the essential and fundamental feature of thinking. As Lesh and Kelly state, the student makes sense of the terms, words, and signs. They presume that students are permanently inventing, testing, rejecting, and revising models in order to interpret and understand their environment. When looking for general concepts of system change, they introduce the concepts of *evolution, generation and mutation, selection, adaptation,* and *accomodation* that clearly rely heavily on the framework of the Geneva School, that is, *genetic epistemology.* Lesh and Kelly briefly sketch three *teaching experiments in conceptually rich environments* in which the process of model revision may be traced.

Thus, at least with respect to the four contributions on the *psychology of mathematical thinking*, in some respects, Piaget seems to be everywhere. As Lompscher's contribution shows, the role of the cultural tradition represented by the teaching subject as emphasized by Vygotsky (1978) may be regarded not only as complementary (see Bartolini-Bussi, this volume) but also as a constructive integration of the social-psychological framework to the principles of cognitive development. Nevertheless, I shall end with another remark on Piagetian research, which is highly significant for an understanding of the child's acquisition of mathematics and hence for a development of didactics of mathematics, that is, *developmental psychology.* Note that all four contributions in this chapter do not refer to the well-known Piagetian theory of developmental stages but rather to general concepts like schema or accomodation. The qualitative change in the cognitive structures was modeled in the comprehensive and closed theory of cognitive stages. In general, the main results of Piaget's theory were replicated completely successfully, and, today, neo-Piagetian models like *Siegler's rule assessment approach* (Siegler, 1986) may be considered as updates of Piagetian theory within the *language of the information-processing approach that* shaped cognitive psychology in the late 1970s and 1980s.

REFERENCES

Freeman, F. N. (1910). *Untersuchungen über den Aufmerksamkeitsumfang und die Zahlauffassung bei Kindern und Erwachsenen*. Leipzig: Veröffentlichungen des Instituts für experimentelle Pädagogok und Psychologie des Leipziger Lehrervereins.

Goldberg, J. G. (1978). Psychological research into mathematics learning and teaching in the U.S.S.R. and Eastern Europe. In F. Swetz (Ed.), *Socialist mathematics education*. Southhampton, PA: Burgundy Press.

Goldin, G. A. (1992). On developing a unified model for the psychology of mathematical learning and problem solving. In W. Geeflin & K. Graham (Eds.), *Proceedings of the 16th PME Conference* (Vol. 3, pp. 235-261). Durham, NH: University of New Hampshire

Glaserfeld, E. von (1991). Introduction. In E. von Glasersfeld (Ed.), *Radical constructivism in mathematics education* (pp. xiii-xx). Dordrecht: Kluwer.

Katz, D. (1913). *Psychologie und mathematischer Unterricht*. Leipzig: Teubner.

Piaget, J. (1968). *The child's conception of space*. London: Routledge & Kegan Paul. [Original work published in 1948]

Poincaré, H. (1910). *Der Wert der Wissenschaft*. Leipzig: Teubner.

Poincaré, H. (1914). *Wissenschaft und Methode*. Leipzig: Teubner.

Polya, G. (1954). *How to solve it*. Princeton, NJ: Princeton University Press.

Rubinstein, S. L. (1958). *Grundlagen der allgemeinen Psychologie*. Berlin: Volk und Wissen.

Siegler, R. S. (1986). *Children's thinking*. Englewood Cliffs. NJ: Prentice-Hall.

Vergnaud, G. (1990). Epistemology and psychology of mathematics education. In P. Necher & J. Kilpatrick (Eds.), *Mathematics and cognition*. Cambridge: Cambridge University Press.

Vygotsky, L. S. (1978). *Mind and society: The development of higher psychological processes*. Cambridge, MA: Harvard University Press.

THE INTERACTION BETWEEN THE FORMAL, THE ALGORITHMIC, AND THE INTUITIVE COMPONENTS IN A MATHEMATICAL ACTIVITY

Efraim Fischbein
Tel Aviv

1. INTRODUCTION

Essentially speaking, mathematics should be considered from two points of view: (a) *mathematics as a formal, deductive rigorous body* of knowledge as exposed in treatises and high-level textbooks; (b) *mathematics as a human activity.*

The fact that the ideal of a mathematician is to obtain a strictly coherent, logically structured body of knowledge does not exclude the necessity to consider mathematics also as a creative process: As a matter of fact, we want students to understand that mathematics is, essentially, a human activity, that mathematics is invented by human beings. The process of creating mathematics implies moments of illumination, hesitation, acceptance, and refutation; very often centuries of endeavors, successive corrections, and refinements. We want them to learn not only the formal, deductive sequence of statements leading to a theorem but also to become able to produce, by themselves, mathematical statements, to build the respective proofs, to evaluate not only formally but also intuitively the validity of mathematical statements.

In their exceptional introductory treatise, "What is mathematics?" Courant and Robbins have written:

> Mathematics as an expression of the human mind reflects the active will, the contemplative reason, and the desire for aesthetic perfection. Its basic elements are logic and intuition, analysis and construction, generality and individuality. Though different traditions may emphasize different aspects, it is only the interplay of these antithetic forces and the struggle for their synthesis that constitute the life, the usefulness and supreme value of mathematical science. (Courant & Robbins, 1941/1978, p. I).

In the present paper, I would like to consider the interaction between three basic components of mathematics as a human activity: the formal, the algorithmic, and the intuitive.

1. *The formal aspect.* This refers to axioms, definitions, theorems, and proofs. The fact that all these represent the core of mathematics as a formal

R. Biehler, R. W. Scholz, R. Sträßer, B. Winkelmann (Eds.),
Didactics of Mathematics as a Scientific Discipline, 231-245.

science does not imply that, when analyzing mathematics as a human process, we may not take them into account.

Axioms, definitions, theorems, and proofs have to penetrate as active components in the reasoning process. They have to be invented or learned, organized, checked, and used actively by the student.

Understanding what rigor means in a hypothetic-deductive construction, the feeling of coherence and consistency, the capacity to think propositionally, independently of practical constraints, are not spontaneous acquisitions of the adolescent.

In Piagetian theory, all these capabilities are described as being related to age – the formal operational period. As a matter of fact, they are no more than open potentialities that only an adequate instructional process is able to shape and transform into active mental realities.

2. *The algorithmic component.* It is a mere illusion to believe that by knowing axioms, theorems, proofs, and definitions as they are exposed formally in textbooks, one becomes able to solve mathematical problems. Mathematical capabilities are also stored in the form of solving procedures, theoretically justified, which have to be actively trained. There is a widespread misconception according to which, in mathematics, if you understand a system of concepts, you spontaneously become able to use them in solving the corresponding class of problems. We need *skills* and not only *understanding*, and skills can be acquired only by practical, systematic training. The reciprocal is also sometimes forgotten. Mathematical reasoning cannot be reduced to a system of solving procedures. The most complex system of mental skills remains frozen and inactive when having to cope with a nonstandard situation. The student has to be endowed with the formal justification of the respective procedures. Moreover, solving procedures that are not supported by a formal, explicit justification are forgotten sooner or later.

Certainly, there is a problem of age, of the order of what to learn first and how to teach. But, finally, I expect that students, who learn the basic arithmetical operations, for instance, are taught sooner or later not only the algorithms themselves but also *why* they do what they do. This profound symbiosis between *meaning* and *skills* is a basic condition for productive, efficient mathematical reasoning.

3. *A third component of a productive mathematical reasoning is intuition*: intuitive cognition, intuitive understanding, intuitive solution.

An intuitive cognition is a kind of cognition that is accepted directly without the feeling that any kind of justification is required. An intuitive cognition is then characterized, first of all, by (apparent) *self-evidence*. We accept as self-evident, statements like: "The whole is bigger than any of its parts." "Through a point outside a line one may draw a parallel and only one to that line." "The shortest way between two points is a straight line."

Being apparently self-evident, intuitively accepted cognitions have a coercive impact on our interpretations and reasoning strategies. Intuitive cognitions may sometimes be in accordance with logically justifiable truths, but sometimes they may contradict them. Consequently, intuitions may play a facilitating role in the instructional process, but, very often, contradictions may appear: Intuitions may become obstacles – epistemological obstacles (Bachelard) – in the learning, solving, or invention processes.

2. HISTORICAL EXAMPLES

Some historical examples may help to clarify this statement. How can we explain why Euclidian geometry – which is true mathematics despite all its imperfections – had been developed in Antiquity, while non-Euclidian geometries appeared only in the 19th century, 2,000 years later? If mathematics is a closed domain with regard to reality, if mathematics is essentially a logical construction, what makes the difference? There is a fundamental difference: Euclidean geometry is based on intuitively accepted statements (including the famous fifth postulate) and "common notions." All of them are intuitively acceptable. As one knows, Aristotle distinguished between axioms (or common notions) and postulates (see Boyer & Merzbach, 1989, p. 120). This was, in fact, the idea. Building deductively, one has to start from some basis that can be accepted without proof. Playing with axioms that contradict our intuition would mean to accept certain statements without proof and without the *direct* feeling of their certainty. Non-Euclidian geometries do not hurt logic but they are counterintuitive. The entire conception of mathematics had to be changed in order to feel free to accept, as axioms, statements that contradict intuition.

A similar situation happened with infinity. Let us first recall the distinction between *potential* and *actual* infinity. A process is said to be potentially infinite if one assumes that it can be carried out without ever stopping it. *Actual* infinity refers to infinite sets of elements considered in their totality. The process of division of a geometrical segment is potentially infinite, while the totality of natural, rational, or real numbers constitute examples of actual infinity. It has been shown that even 11- to 12-year-olds are able to accept intuitively the potentially infinite extension of a line segment (Fischbein, 1963) or its potentially infinite division.

On the contrary, actual infinity is a counterintuitive, abstract concept. Our intelligence is adapted to finite magnitudes and, consequently, reasoning with infinite magnitudes leads to apparent paradoxes. As an effect, great philosophers, scientists, and mathematicians like Aristotle, Gauss, or even Poincaré rejected the use of the concept of actual infinity.

It was only in the 19th century, with Cantor, that actual infinity became accepted as a mathematical concept as a result of a complete change of perspective.

In the following, I will refer specifically to various types of interaction between the formal, the algorithmic, and the intuitive components of a mathematical activity.

3. OPERATIONS AND INTUITIVE MODELS

What has been said about the role of intuitive acceptance in the history of science may be claimed also with regard to the learning process. The relationship between the formal and the intuitive aspects of mathematical reasoning in learning, understanding, and solving processes is very complex. Sometimes there is a certain congruence, but, very often, conflictual phenomena may appear that lead to misconceptions, systematic mistakes, and epistemological obstacles. Especially sensitive to such conflicts are the domains related to infinity and probability, but, as a matter of fact, in every branch of mathematics, one may encounter concepts, statements, and operations that are difficult to understand and accept because of such contradictory relationships between the formal and the intuitive constraints.

Let me mention a few examples. A very widespread misconception is that "multiplication makes bigger" and "division makes smaller." That misconception has been encountered not only in elementary school students (Fischbein, Nello, & Marino, 1985) but also in the preservice teacher (see, e.g., Tirosh, Graeber, & Glover, 1986). A systematic analysis revealed a world of psychological problems.

Let us consider the following two problems:

1. From 1 quintal of wheat, you get 0.75 quintals of flour. How much flour do you get from 15 quintals of wheat?
2. 1 kilo of a detergent is used in making 15 kilos of soap. How much soap can be made from 0.75 kilos of detergent?

These are two examples from a set of questions given to 628 5th-, 7th-, and 9th-grade students from 13 different schools in Pisa, Italy. The students were asked to choose only the solving operation without effectively performing the computation. We quote the percentages of correct answers, according to grades (see Fischbein, Nello, & Marino, 1985, p. 10):

Problem 1: 79 (Gr. 5); 74 (Gr. 7); 76 (Gr. 9)
Problem 2: 27 (Gr. 5); 18 (Gr. 7); 35 (Gr. 9)

For both problems, the solution consists in the multiplication 15 x 0.75. Formally and procedurally the solution is the same. What makes the difference?

As one may observe by reading the two problems carefully, in the first problem, the operator is a whole number (15), while, in the second, the operator is a decimal. From a formal point of view, this should not make any difference: Multiplication is a commutative operation. But *intuitively* things look totally different.

Let us imagine that behind the operation of multiplication lies an intuitively acceptable model (and, in fact, taught in elementary classes): Multiplication is repeated addition. The model is adequate, but only as long as one deals with whole numbers. Three times five, means, in this interpretation, 5 + 5 + 5 = 15.

But what would 0.75 times 5 mean? *Formally*, "0.75 times 5" and "5 times 0.75" lead to the same result. But intuitively, they do not. 0.75 times 5 does not have an intuitive meaning. It *cannot* be represented in the terms of the repeated addition model.

In a multiplication *A* x *B*, verbally expressed as "A times B," *A* is the operator and *B* the operated. If Operator *A* is a decimal, the multiplication has no intuitive meaning. As a consequence, when addressing a multiplication problem in which the operator is a decimal, the student will not grasp the solving procedure directly, that is, intuitively. The "repeated addition model" operating behind the scenes will prevent the right solution instead of facilitating it. As an effect of this situation (the influence of the "repeated addition" model for multiplication applicable to whole numbers), the student is led to believe intuitively that "multiplication makes bigger" and "division makes smaller." These statements are true, are intuitively acceptable, but only as long as the operator is a whole number.

4. ALGORITHMS AND INTUITIVE MODELS

4.1 Example: The Operation of Subtraction

One knows, today, that students make various systematic mistakes in performing subtraction, and many such "bugs" have been identified. I do not intend to enter into details. I only want to specify that at least a number of these bugs might be predicted from the primitive model of subtraction.

If you have in a container a number *A* of objects, (e.g., marbles) and you want to take out a number of them, *B* (the primitive model of the operation of subtraction), you can do it only if $B < A$. If $B > A$, the student will tend to reverse the operation $B - A$. For instance (Resnik, 1983, p. 73):

$$\begin{array}{r} 326 \\ \underline{-\ 117} \\ 211 \end{array}$$

Another possibility, derived from the primitive model, is just to consider, when $B > A$, that you take out as much as you can from the container and the container remains empty. For instance (Resnik, 1983, p. 73):

$$\begin{array}{r} 542 \\ \underline{-\ 389} \\ 200 \end{array}$$

If the student has learnt the patent of "borrowing," several situations may occur. The most typical difficulty appears when the student has to "borrow" from 0. If $B > A$, you borrow from the next container, but if this container is empty, then you may write 0, or you may borrow from the bottom, or you may skip over the empty container and try a third one.

Borrow from bottom instead of zero:	702 - 368 454
Borrow across zero:	602 - 327 225

(With regard to misconceptions in subtraction, see also Maurer, 1987; Resnik, 1983.)

5. CONCEPTS AND INTUITIVE REPRESENTATIONS

5.1 The Concept of Set

Linchevski and Vinner (1988) have analyzed a number of misconceptions held by elementary school teachers concerning the mathematical concept of set. They have identified the following misconceptions: (a) Subjects consider that the elements of a set must possess a certain explicit common property. (b) A set must be composed of more than one element. The idea of an empty set or of a singleton is rejected. (c) Repeating elements are considered as distinct elements. (d) An element of a set cannot be an element of another set. (e) To these we may add a fifth common misconception, that is, that two sets are equal if they contain the same number of elements.

A very simple interpretation may account for all these misconceptions. If the model one has in mind, when considering the concept of set, is that of a *collection of objects*, all these misconceptions are predictable. An empty collection, or a collection containing only one object, are obviously nonsense. We never constitute classes of objects that are absolutely unrelated conceptually (your name, a pair of old shoes, and the imaginary number *i*). In every practical situation, two identical elements that, nonetheless, have a separate existence (e.g., two dimes) are counted separately. The same object cannot be in two different containers at the same time. Two collections of objects are considered equal if they contain the same number of elements.

I do not affirm that students identify, explicitly and consciously, the mathematical concept of set with the notion of a collection of concrete objects. What I affirm is that, while considering the mathematical concept of set, what they have in mind – *implicitly* but *effectively* – is the idea of a collection of objects with *all* its connotations. There is no subjective conflict here. The intuitive model manipulates from *behind the scenes* the meaning, the use, and the properties of the formally established concept. The intuitive model seems to be stronger than the formal concept. The student simply forgets the formal properties and tends to keep in mind those imposed by the model. And the explanation seems to be very simple: The properties imposed by the concrete model constitute a coherent structure, while the formal properties appear, at least at first glance, rather as an arbitrary collection. The set of formal properties may be justified as a coherent one only in the realm of a clear, coherent mathematical conception.

In my opinion, the influence of such tacit, elementary, intuitive models on the course of mathematical reasoning is much more important than is usually acknowledged. My hypothesis is that this influence is not limited to the preformal stages of intellectual development. My claim is that even after individuals become capable of formal reasoning, elementary intuitive models continue to influence their ways of reasoning. The relationships between the concrete and the formal in the reasoning process are much more complex than Piaget supposed. The idea of a tacit influence of intuitive, primitive models on a formal reasoning process does not seem to have attracted Piaget's attention. In fact, *our information-processing machine is controlled not only by logical structures but, at the same time, by a world of intuitive models acting tacitly and imposing their own constraints.*

5.2 The Concept of Limit

Moving to a higher level of mathematical reasoning, we may find very beautiful examples of the complexity of the relationship between its formal, algorithmic, and intuitive components. Without understanding these relationships, it would be difficult, in fact, rather impossible, to find the right pedagogical approach.

In order to make sure that psychological comments are not mere speculation, I consider it to be useful to quote genuine mathematicians. I am referring to "What is mathematics" by Courant and Robbins (1941/1978).

I have chosen the concepts of *limit* and *convergence,* because they play a central role in mathematical reasoning. At the same time, the interplay between the formal, the algorithmic, and the intuitive aspects is rich in psychological and didactic implications.

But let us quote from the text of Courant and Robbins:

> The definition of the convergence of a sequence a_n to a may be formulated more concisely as follows: *The sequence $a_1, a_2, a_3, \ldots$ has the limit a as n tends to infinity if, corresponding to any positive number ε no matter how small, there may be found an integer N (depending on ε) such that:*
>
> $$|a - a_n| < \varepsilon \text{ for all } n \geq N.$$
>
> This is the abstract formulation of the notion of the limit of a sequence. Small wonder that when confronted with it for the first time, one may not fathom it in a few minutes. There is an unfortunate, almost snobbish attitude on the part of some writers of textbooks, who present the reader with the definition without a

> thorough preparation as though an explanation were beneath the dignity of a mathematician . . . .
>
> There is a definite psychological difficulty in grasping this precise definition of limit. Our intuition suggests a "dynamic" idea of a limit as the result of the process of "motion": We move on through the row of integers 1, 2, 3, . . . n, . . . and then observe the behavior of the sequence a_n. We feel that the approach $a_n \to a$ should be observable. But this "natural" attitude is not capable of clear mathematical formulation. To arrive at a precise definition we must *reverse* the order of steps; instead of first looking at the independent variable n and then at the dependent variable a_n, we must base our definition on what we have to do if we wish actually to check the statement $a_n \to a$. In such a procedure, we must first choose an arbitrarily small margin around a and then determine whether we can meet this condition by taking the independent variable n sufficiently large. Then, by giving symbolic names, ε and N, to the phrases "arbitrarily small margin" and "sufficiently large n" we are led to the precise definition of limit. (Courant & Robbins, 1941/1978, pp. 291-292)

Intuitively, it is relatively easy to understand, as Courant and Robbins say, the concepts of limit and convergence. Intuitively, one may consider a sequence of numbers a_n that come closer and closer to a certain number a as n tends to ∞. The number a is then the limit of the sequence a_n, and the sequence is said to converge to a. If one adds also an example, things become totally clear *intuitively*. For instance, one may consider the sequence whose nth term is $a_n = 1/n$: The series

$$1, 1/2, 1/3, \ldots 1/n, \ldots$$

has the limit 0, for increasing n: $1/n \to 0$ as $n \to \infty$.

But we cannot go directly from the intuitive representation to the formal, rigorous definition. The formal definition reverses the order of ideas, contradicts the natural, dynamic representation of the process. And this makes the definition of limit, as a matter of fact, counterintuitive, difficult to grasp. We do not start by describing the process of approaching a by a sequence of numbers a_n. We start by mentioning, strangely enough, a positive number ε "no matter how small," and *afterwards* we introduce N and $n \geq N$. That is, it is not ε that depends on N (as happens in reality) – the interval $|a_n - a|$ becomes smaller as we go on increasing N (respective n) – but, in the formal definition, we make N "dependent on ε." We reverse the natural order of the thinking process.

As a matter of fact, the formal definition above is not entirely "purified" from every intuitive element. The term "tends" ("the sequence $a_1, a_2, a_3, \ldots$ has the limit a as n *tends* to infinity . . .") is not a purely abstract term. We continue to keep in mind, tacitly, an intuitive model. The term "tends" has a psychological, not a mathematical or a physical meaning. People "tend to," are "inclined" to. "Tend to" has a connotation of desire, of aspiration. Numbers do not *tend*. They exist or do not exist. The term "tends to" is what remains from the initial intuitive, dynamic interpretation of the concepts of convergence and limit. It expresses the potential infinity that is intuitively acceptable. I suppose that mathematicians have felt intuitively that, by trying to eliminate completely any intuitive residual (in this case, in which the processuality is essential), they would have made the formal product meaningless. The term "tends to" is a compromise between the dynamic of the primitive, intuitive representation of convergence and the need to freeze an infinite given set of elements in a formal definition. When one "tends," one does not move, but one does not stay totally rigid either.

As an effect of this conflictual relationship between the formal definition and the intuitive representation of the concept of limit, various misconceptions may appear. Shlomo Vinner (1991) asked 15 gifted students in a prestigious high school to define the concept of limit (after the concept had been taught). Only one student gave a formulation that could be accepted, though incomplete. The other 14 students exhibited some typical misconception. Shlomo Vinner mentions the following main misconceptions:

> 1. A sequence "must not reach its limit" (thus the sequence 1, 1, 1, . . . would be said not to converge to a limit).
>
> 2. The sequence should be either monotonically increasing or monotonically decreasing. Thus, for instance, the sequence whose nth element is given by
>
> $$a_n = 1 + \frac{(-1)^n}{n}$$
>
> does not tend to a limit.
>
> 3. The limit is the "last" term of the sequence. You arrive at the limit after "going through" infinitely many elements. (Vinner, 1991, p. 79)

As Cornu (1991) has shown, the term "tends to" possesses various primitive meanings in the student's mind, and these interact with the formal concept. "Tends to" may mean:

> to approach (eventually staying away from it)
> to approach . . . . without reaching it
> to approach . . . . just reaching it"
> to resemble (. . . such as "this blue tends towards violet") (Cornu, 1991, p. 154)

The interpretation the student will confer on the term "tends to" in relation to the concept of limit will then depend on his or her intuitive model. The student who does not accept that the sequence 1, 1, 1, . . . does converge to a limit (which is, in fact, 1) holds, intuitively, that "tends to" implies: (a) that the intervals between the successive terms of the sequence and the limit have to become smaller and smaller, and (b) that the limit is never reached. Both conditions are never fulfilled in the above example (for a discussion of the epistemological obstacles related to the concept of limit, see Cornu, 1991).

As a matter of fact, the concept of limit is a contradictory one (in the dialectical, Hegelian, sense) because our mind is *naturally* not adapted to the conceptualization of actual infinity.

Another example: The idea that the area of a circle is the limit of sequences of polygons cannot, in fact, be grasped intuitively: It is a contradictory one. When we have the circle, we have no more polygons. *Intuitively*, a polygon has a number of sides, maybe a very great number of sides. A "something" that is simultaneously circle *and* polygon has no meaning at an intuitive level. The contradiction may be eliminated only at a pure, formal level. But the pure, formal level, is, itself, psychologically impossible. We tend to it in mathematics, but, as a matter of fact, we never reach it psychologically.

As an effect, we get the epistemological obstacles of the students concerning the notions of limit and continuity, that is, the various partial interpretations we may find in students (the limit is never reached or the limit is always reached).

The same types of obstacle may be identified in the history of mathematics. Some mathematicians (like Robins, 1679-1751, see Cornu, 1991, p. 161) claimed that the limit can never be attained. Others, like Jurin (1685-1750) said that the "ultimate ratio between two quantities is the ratio reached at the instant when the quantities cancel out" (cited in Cornu, 1991).

These contradictory attitudes gave birth to the concept of "infinitesimals" or "arbitrary small numbers" that express the effort to conceptualize *a process* intuitively seen as endless.

Let me add another example. In a study devoted to measuring the degree of intuitiveness of a solution (Fischbein, Tirosh, & Melamend, 1981), the following question has been addressed:

> Given a segment *AB* = 1m. Let us suppose that another segment *BC* = $^1/_2$ m is added. Let us continue in the same way, adding segments of $^1/_4$ m, $^1/_8$ *m*, etc. What will be the sum of the segments *AB* + *BC* + *CD* . . . (and so on)? (Fischbein, Tirosh, & Melamed, 1981, p. 494, 495)

The following categories of answers have been recorded:

1. Sum = 2 (5.6%) (correct)
2. Sum = infinite (51.4%)
3. "The sum is smaller than 2" or "The sum tends to 2" (16.8%). (Fischbein, Tirosh, & Melamed, 1981, p. 499)

As one can see, only a very small percentage of students gave the correct answer (S = 2). The explanation is that, as we mentioned above, actual infinity is counterintuitive. In order to accept that the sequence 1 + $^1/_2$ + $^1/_4$ + . . . = 2, one has to grasp *intuitively* the entire actual infinity of the sequence. Because this does not happen, the students easily forget the correct answer (S = 2) and consider the infinity of the sequence as a *potential* infinity (the sum *tends* to 2, or the *sum is smaller than* 2).

Asking high school or college students to find the decimal equivalent of $^1/_3$, they willingly write $^1/_3$ = 0.333 . . . On the other hand, they would hardly accept that 0.333 . . . *equals* $^1/_3$. As in the above example, they claim that 0.333 . . . *tends* to $^1/_3$. We encounter here the same type of intuitive obstacle as above. In addition, one has to emphasize the following aspect:

If a student accepts that $^1/_3 = 0.333\ldots$, he or she should accept also that $0.333\ldots = {}^1/_3$: The relation of equality is symmetrical. In reality, as it has been shown (see Kieran, 1981), the intuitive, tacit model associated with the equality sign is usually that of an input → output process that is not symmetrical!

6. THE IMPACT OF A RIGID ALGORITHM ON AN INTUITIVE REPRESENTATION

In a series of interviews with preservice mathematical teachers, the following type of problem has been presented:

> Five kilos of apples cost 15 shekels. How much will 7 kilos of apples cost?

It is a classical elementary problem of proportionality. Some solved the problem by determining the price of one kilo (15 : 5 = 3) and, after multiplying by 7, they got: 3 x 7 = 21. Some students wrote directly the proportion $^{15}/_5 = {}^x/_7$.

A second problem has been posed:

> Seven workers finish a certain piece of work in 28 days. In how many days will five workers finish the work?

The students affirmed that there was also a problem of proportion and wrote: $^7/_{28} = {}^5/_x$ They found that $x = 20$, and this was their result.

They were then asked to analyze the answer: If seven workers finish the piece of work in 28 days, less workers (that is, five), will finish the work in less days. The students understood that they made a mistake. They have applied a schema automatically, blindly; and thus the intuitive, direct interpretation, which would have been useful, did not function.

Sometimes, the intuitive background manipulates and hinders the formal interpretation or the use of algorithmic procedures. But, sometimes, it is the blind application of schemas that leads to wrong solutions, although the appeal to a direct, intuitive interpretation would have prevented the solver from giving an erroneous answer.

7. THE INTERACTION BETWEEN THE FORMAL CONSTRAINTS AND SOLVING ALGORITHMS

Solving procedures, acting as overgeneralized models, may sometimes lead to wrong solutions in disregard of the corresponding formal constraints. Let me consider some examples.

It has been found that students often would write $\sin(a+b) = \sin a + \sin b$, or $\log(a+b) = \log a + \log b$. Obviously, the property of distributivity of multiplication over addition $[m(a+b) = ma + mb]$ does not apply in the above situations. Students forget that one deals with a formal property of multiplication and addition. They transform it in a solving model and, by external similarity, it becomes a solving procedure.

The same type of common mistake, in which a solving technique does not obey the formal rules and is thus wrongly applied, appears in the following example:

$$\frac{an}{an+by} = \frac{\cancel{an}}{\cancel{an}+by} = \frac{1}{by}$$

Such categories of mistakes are well-known to teachers. Maybe, what is less understood is that, in order to overcome such errors, the student needs to gain a fuller understanding of the relationships between the formal and the algorithmic components in mathematics. The student has to understand, in my opinion, the formal basis (definitions and theorems) that justifies an algorithm. It is the blind learning of algorithms that leads to these types of misuse. In the absence of a clear understanding of the formal frame and justification, the superficial similarity of problems leads to wrong generalizations.

8. THE FIGURAL CONCEPTS

A most interesting situation with regard to the interaction between the figural (intuitive) and conceptual aspects occurs in the domain of geometry.

Psychology textbooks usually distinguish between concepts and images as the two basic components of a thinking activity. But geometrical figures occupy a special position. What is a line, a triangle, a sphere, or a cube? Certainly they are images. They possess a certain shape. But, in the flux of a geometrical reasoning they are not mere images in the usual sense. (I am not referring to drawings. I am referring to geometrical, mathematical entities.) They are ideal, abstract entities. They possess a kind of universality that characterizes only concepts. *Every* property of a geometrical figure is derived from the definition of the respective figure, from the axiomatic structure to which it belongs. Consequently, one may claim that geometrical figures, though spatial images, possess qualities that characterize only concepts: ideality, abstractness, universality, definition dependence, a kind of purity and perfection that does not exist in nature. In geometrical reasoning, we deal with figures that are not mere images, but idealized mental entities completely subordinated to axiomatic constraints. We may then claim that a geometrical figure is a mental object that is not reducible to usual concepts or images. It is not a mere concept, because it is a spatial representation. A concept is an *idea* that, strictly speaking, does not possess figural qualities. On the other hand, a geometrical figure is not a mere image, because *all* its properties are strictly, rigorously imposed by a definition. A geometrical figure is, *at the same time*, figure and concept. The drawing of a circle or a triangle is a graphic model of a geometrical figure, not the geometrical figure itself.

But that total symbiosis between figural (intuitive) and conceptual properties in a geometrical figure is usually only an ideal situation. Very often, the formal constraints and the figural ones interact and conflict among themselves, and such conflicts may influence the flow of geometrical reasoning.

It is difficult for children to accept that a square is a rectangle, a rhombus, or even a parallelogram, even if they know the respective definitions. The figural, the Gestalt particularities are so strong that they annihilate the effect of the formal constraints.

Alessandra Mariotti (1992) reports the following example: A 16-year-old student, Alessia (Grade 11) has been given the following problem.

How many angles do you see in Figures 1a and 1b? (see Figure 1)

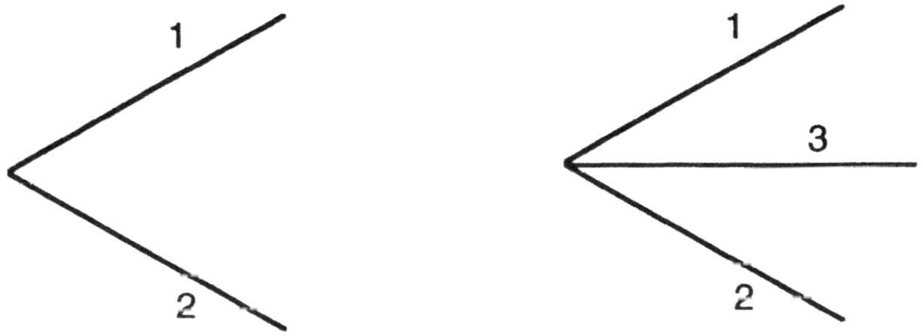

Fig. 1 a Fig. 1 b

Alessia: Whenever I see two lines that intersect, I know that the space between the lines is an angle. I think that in both figures there is only one angle, even if, at first, I thought that in the second figure there were two angles. I can explain my supposition. First I thought that in this representation, Line 1 and Line 2 form one angle and Line 2 and Line 3 form a second angle. However, now I think that there is only one angle formed by crossing lines (1,2) and that Line 3 is the bisector of this angle. (Marrioti, 1992, p. 11)

Alessia's difficulty is generated by the fact that the concept is unable to control the figure. And this, not because she does not possess the concept correctly but because the figure still carries with it Gestalt features inspired by practice. As a matter of fact, the complete symbiosis discussed above does not yet exist; if you cut a piece of cake into two halves, you get two pieces of cake, not three (Alessia's first interpretation). If Line 3 is the bisector of the angle it cannot belong, at the same time, to two other angles (the second interpretation). In the above example, the concept of angle does not yet control totally the intuitive, figural properties and their interpretation. In the interaction between the formal and the intuitive constraints, it is the intuitive constraints that are, in this example, decisive.

8. SUMMARY

The main claim of the present paper is that, in analyzing the students' mathematical behavior, one has to take into account three basic aspects: the formal, the algorithmic, and the intuitive.

The formal aspect refers to axioms, definitions, theorems, and proofs. The algorithmic aspect refers to solving techniques and standard strategies. The intuitive aspect refers to the degree of subjective, direct acceptance by an individual of a notion, a theorem, or a solution. Sometimes these three components converge. But, usually, in the processes of learning, understanding, and problem-solving, conflictual interactions may appear. Sometimes a solving schema is applied inadequately because of superficial similarities in disregard of formal constraints. Sometimes, a solving schema, deeply rooted in the student's mind, is mistakenly applied despite a potentially correct, intuitive understanding.

But, usually, it is the intuitive interpretation based on a primitive, limited, but strongly rooted individual experience that annihilates the formal control or the requirements of the algorithmic solution, and thus distorts or even blocks a correct mathematical reaction.

The interactions and conflicts between the formal, the algorithmic, and the intuitive components of a mathematical activity are very complex and usually not easily identified and understood. Theoretical analyses, attentive observations, and experimental research have to collaborate in revealing the multiple sources of mistaken attitudes in a mathematical activity. This implies that the intimate collaboration between psychology and didactic experience represents a basic condition for the progress of mathematics eduction.

REFERENCES

Boyer, C. B., & Merzbach, U. C. (1989). *A history of mathematics*. New York: Wiley.

Cornu, B. (1991). Limits. In D. Tall (Ed.), *Advanced mathematical thinking* (pp. 153- 165), Dordrecht, Netherlands: Kluwer.

Courant, R., & Robbins, H. (1978). *What is mathematics? An elementary approach to ideas and methods.* Oxford: Oxford University Press.

Fischbein, E. (1963). *Conceptele Figurale* [in Roumanian]. Bucuresti: Editura Academiei, R.S.R.

Fischbein, E., Deri, M., Nello, M. S., & Marino, M. S. (1985). The role of implicit models in solving verbal problems in multiplication and division. *Journal of Research in Mathematics Education, 16*(1), 3-17.

Fischbein, E., Tirosh, D., & Melamed, U. (1981). Is it possible to measure the intuitive acceptance of a mathematical statement? *Educational Studies in Mathematics, 12,* 491-512.

Kieran, C. (1981). Concepts associated with the equality symbol. *Educational Studies in Mathematics, 12,* 317-326.

Linchevski, L., & Vinner, Sh. (1988). The naive concept of sets in elementary teachers. *Proceedings of the Twelth International Conference, Psychology of Mathematics Education* (Vol. 2.) Vezprem, Hungary.

Mariotti, M. A. (1992). Imagini e concetti in geometria. *L'Insegnamento della Matematica e delle Scienze Integrate, 15*(9), 863-885.

Maurer, S. B. (1987). New knowledge about errors and new views about learners: What they mean to educators and what more educators would like to know. In A. H.

Schoenfeld (Ed.), *Cognitive science and mathematics education* (pp. 165-188). Hillsdale, NJ: Erlbaum.

Resnik, L. B. (1983). Procédure et compréhension en arithmétique élémentaire. *Seminaire de Didactique de Mathématiqe 1982-1983*. Grenoble: IMAG.

Tirosh, D., Graeber, A. D., & Glover, R. M. (1986). Preservice teachers' choice of operation for multiplication and division word problems. *Proceedings of the Tenth International Conference, Psychology of Mathematics Education* (pp. 57-62). London: University of London Institute of Education

Vinner, Sh. (1991). The role of definitions in the teaching and learning of mathematics. In D. Tall (Ed.), *Advanced mathematical thinking* (pp. 65-79). Dordrecht, Netherlands: Kluwer.

FROM PIAGET'S CONSTRUCTIVISM TO SEMANTIC NETWORK THEORY: APPLICATIONS TO MATHEMATICS EDUCATION - A MICROANALYSIS

Gerhard Steiner
Basel

1. FROM PIAGET'S "STRUCTURES D'ENSEMBLE" TO "SEMANTIC NETWORKS" AND MORE OF THESE CONCEPTUAL TRANSITIONS

Many discussions have been led on whether or not Piaget's theory has substantially contributed to school education: to planning, implementing, and evaluating both instruction and learning. It is, indeed, not self-evident that Piaget's developmental or epistemological concepts lead to a better understanding of academic learning and achievement. While some of his concepts are still used vividly and do have a long-lasting influence on educational activities (assimilation and accommodation, schema, schema construction), others have undergone some kind of a metamorphosis in the new "psychotope" of current cognitive psychology (structure d'ensemble, mise en relation), and still others have been abandoned or even forgotten (e.g., the INRC group).

The "schema" concept and the concept of "schema construction" remained almost unchanged, although Rumelhart and Norman (1973, 1976; Rumelhart, 1978) have elaborated the schema concept and have tried to give it more processual precision. The concept of "schema" in Piaget's sense cannot be discussed without mentioning in parallel the concept of "operation." When working with children during the transition from preoperational to concrete-operational stages, it becomes clear that higher developmental structures (recognizable on the grounds of particular combinations of schemata) result from what Piaget called "abstraction à partir de l'action" (abstraction from one's own actions). This gives the action a particular significance in the context of this chapter, especially with regard to mathematical thinking, for example, when poor math students are trying to solve problems just by manipulating mathematical symbols. Actions of an individual may be internalized, becoming optimally reversible and flexible in their use; this transforms them, according to Piaget, into "operations" that, in turn, do not exist as isolated processual units but are organized into

R. Biehler, R. W. Scholz, R. Sträßer, B. Winkelmann (Eds.),
Didactics of Mathematics as a Scientific Discipline, 247-261.

wholistic systems of operations. "Abstraction" then refers to the organizational aspect of the generated system of operations. It is not easy to define the difference between systems of operations and combinations (i.e., systems) of schemata. It seems to me that the difference refers to a certain early anticipation by Piaget of what are, in modern terms, the conceptual and procedural aspects of the same knowledge structure. As far as the combinations (or systems) of schemata are concerned, they correspond to what Piaget called "structure d'ensemble." In modern terms, we would call them parts of semantic or other (e.g., arithmetic) networks. Thus, Piaget's "structures d'ensemble" are no longer formal and generalized structures, but have obtained, all of a sudden, the colorful face of semantic networks; but, nevertheless, the action as well as the internalization aspects should not be lost from sight.

Internalization has to do with one of the most prolific concepts of Piaget's, the "mise en relation," that is, the counterpart of "lecture des données" ("reading" from the information given). Having children look at, for example, arithmetic material leads them to process surface features such as colors, numbers, shapes, and so forth. This is "lecture des données," whereas connecting certain judgments about lengths, numbers, or positions of the material without just "reading off" what they look like is what Piaget calls "mise en relation" (Steiner 1974b, 1983) and what Bruner (1957, 1973) refers to as going "beyond the information given." This process corresponds to internally connecting the elements of reasoning and internally operating on the items of a task. Therefore, "mise en relation" leads per se to an internalization of connections according to an organizational plan that has been abstracted from the former actions executed with and on the material at hand. "Mise en relation" includes a connecting process that equals the connecting process through a "named relation" as stated by recent semantic network theory (cf. Lindsay & Norman, 1972). Thus, Piaget's concepts of "structure d'ensemble" as well as "mise en relation," seen as theoretical entities, have become parts of current semantic network or schema theories, although under new terms.

Some of Piaget's concepts have proved not to be of great importance for educational activities during elementary school grades and later. Astonishingly enough, this is true for, for example, the famous "stage" concept including the "décalage" problem (i.e., the time shift in the acquisition of structurally identical systems of operations on materials that differ in certain aspects of content or situational presentation). Juan Pascual-Leone (1970, 1976) has dealt with both these concepts and the corresponding behavioral phenomena and provided the scientific community with an interesting "neo-Piagetian" mathematical model for the transition from one developmental stage to the next one indicating the crucial variables that influence the equilibration processes taking place during these transitions. Pascual-

Leone's contribution is a theoretical one to developmental theory, not so much to teaching and education. For this reason I shall not go into it here.

Quite different considerations stem from a somewhat older disciple of Piaget's: Hans Aebli, one of the very first PhD students and later critics. As early as 1963, he focused on the stage concept and the "décalage" problem showing that many factors other than just the structural organization are responsible for the developmental level (stage) of a child: the complexity of the material to be learned, its concreteness, the time spent with the material, the number of repetitions in dealing with the items, as well as the motivation to cope with one problem or another. All these factors are of utmost importance for preparing learning situations not just in developmental experiments but above all in classrooms. Following this line of reasoning, Steiner, a student of Aebli's and, thus, of Piaget's in the second generation, attempted to integrate Piaget's structural with Bruner's representational approach to development in order to avoid further problems with décalage-like shifts in development or problems in teaching and learning due to different aspects of materials or situations the child has to deal with (Steiner, 1974a).

A further and highly remarkable elaboration of Piaget's theory, another version of neo-Piagetian thinking, was presented by Robbie Case (1978, 1985), who started from similar questions to those that Aebli was asking years before. In Case's view, development is the result of a continuing reorganization of executive strategies that a child uses in tackling problem situations that transcend former ones in complexity. Similar to Aebli's considerations of factors affecting the child's operational level, Case stresses the complexity and perceptual organization of a task and the individual's affective disposition (Aebli was focusing on motivation). But Case relies particularly on two factors: (a) the *M* power (already focused by Pascual-Leone), defining a child's short-term memory capacity, and (b) his or her cognitive style (mainly the independence from distracting stimuli in the surrounding environment). He used these two factors to emphasize the individual's contribution to processing the information given in the problem situations. With these factors in mind, it becomes possible to plan teaching as well as learning processes that correspond to the operational level of the child. However, two points in task analysis have to be observed strictly by the planning experimenter or teacher: enhancing the salience of particular parts of the task or the problem presented to the learning child and reducing task complexity. (For an elaborated treatment of task complexity or "cognitive load," see, also, Chandler & Sweller, 1991; Sweller, 1988.)

Back now to Piaget! His way of describing structural change in development by means of formal and rather highly specialized mathematical structures such as groupings, groups, or even higher ones such as lattices (Inhelder & Piaget, 1955; Piaget, 1947) has disappeared from any educational discussion. These structures have been criticized for their restricted usefulness or rigidity in describing real behavioral development and change and have been replaced by content-specific descriptions of development or learning processes, respectively (Aebli, 1978, 1987).

As can be seen, there is a strong conceptual shift from Piaget's terms toward current cognitive terms (and the corresponding view of the behavioral phenomena) that fit in with the requirements of both educational learning and instructional theories, particularly in regard to math education. Therefore, we shall discuss the following problems mainly in terms of modern schema theory or network theory, respectively; but, from time to time, the reader will be aware of the heritage of Piaget's theoretical approach.

2. SEMANTIC NETWORK THEORY AND SCHEMA THEORY FOR MATH EDUCATION

It is a well-known statement that the use of schema theory in teaching is of utmost importance (see, e.g., Glaser, 1984). Let me first clarify what I have in mind when using the concept of "schema," what its relations to "semantic networks" are, and, in particular, what "schema" means in mathematics education.

Following the classical interpretation by Norman and Rumelhart (cf. Rumelhart, 1978; Rumelhart & Norman, 1973, 1976), a schema is an activated part of a semantic network. "Semantic network," in turn, is the cognitive psychologist's metaphor about how human knowledge is stored in and can be accessed from memory. Thus, a schema is always a representational, permanently modifiable unit, a meaning structure of a particular (although restricted) scope that represents actions, operations (these latter ones as systems of internalized actions in Piaget's sense), or concepts.

Within an individual's semantic network, which contains his or her world knowledge, there are certain domain-specific parts of knowledge such as arithmetic or algebraic-mathematical knowledge. The nodes of the corresponding algebraic-mathematical networks are the domain-specific concepts such as the several kinds of numbers but also concepts like fraction, equation, function, and many others, while the relations that connect the conceptual nodes are defined by mathematical operations from simple additions up to, for example, logarithmic operations. An algebraic-mathematical schema is, accordingly, an activated part of the corresponding algebraic-mathematical network (AMN).

As far as the relations between schemata and rules or schemata and algorithms are concerned, one could say that the schema contains (a) activated conceptual knowledge from a certain part of an AMN, and this in two possible formats: symbolic or iconic (in Bruner's, 1966, sense); and (b) rules or algorithms that constitute the corresponding procedural part of that same schema knowledge. To obtain a complete knowledge of such rules, it is necessary, according to Sweller and Cooper (1985), to acquire a large number of schemata incorporating those rules, a statement, by the way, that I do not agree with. I shall come back to this.

3. MICROANALYSIS OF ALGEBRAIC-MATHEMATICAL THINKING

3.1 Three Preliminary Remarks

1. The choice of factorizing trinomials and, very briefly, functions for micronanalysis is due to the fact that these areas offer themselves for demonstrating several characteristics of math learning as well as the nature of AMN.

2. The following microanalyses do not try to simulate school situations, but allow a close look through the glasses of a cognitive psychologist working in educational psychology – after having taught himself for many years on all levels.

3. Several authors have dealt with the analysis of algebra learning and mathematical reasoning processes. Sweller and Cooper (1985), for example, had their students construct schemata to transform equations by thinking move by move through already solved problems, so-called "worked examples," instead of having them waste a lot of time by hunting for problem-solving techniques. Zhu and Simon (1987), on the other hand, trained their Chinese students in detecting the production systems (or rules) for factorizing elementary trinomials. The focus of these studies was on finding rules or constructing a sequence of schemata. What is still missing according to my view is an attempt to perform a careful application of semantic network theory – here in the form of AMN theory – to algebra problems.

If ever the close connectedness of knowledge is crucial in regard to retrieving information, use of knowledge, problem-solving, and so forth (and many studies, e.g., the ones using the expert/novice paradigm, support this view), then we have to apply AMN theory very systematically and stringently within the specific domain of mathematics learning.

3.2 Factorizing Trinomials

While tutoring our subjects, we always started from a mathematical situation including some operations that the student was already able to master, for example, from the following trinomial (that, by the way, comes close to the ones used by Zhu & Simon, 1987):

$$x^2 + 8x + 16$$

The knowledge for grasping the meaning of this trinomial (mentally represented in what we call a schema) includes a complex compound of subschemata that can be represented by the following graph (see Figure 1).

The first remarkable fact is that the trinomial as such is viable only in connection with the to-be-squared binomial or, more generally, the to-be-multiplied binomials.

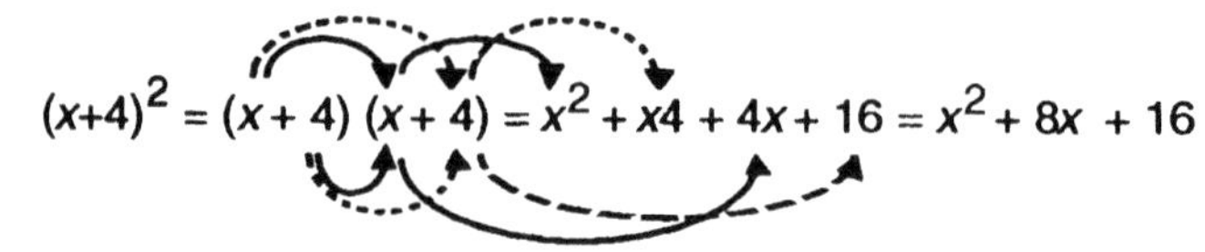

Figure 1. Scheme for representing student's knowledge in solving trinominals.

Focusing from a theoretical point of view on the schema "factorizing trinomials," one can recognize that it implies two complementary parts: the direct part of "multiplying binomials" (marked by the arrows in the graph), and the reverse part of "factorizing trinomials," both parts with a corresponding set of subschemata: multiplications, additions, multiplicative as well as additive decompositions, and some knowledge that is often overlooked concerning commutativity laws (stemming from the partial multiplications of $x4$ and $4x$, respectively).

The conceptual knowledge of the "multiplying binomials/factorizing trinomials" schema involves the above-mentioned subschemata as well as their functional reversals in their full interplay. Such knowledge obviously contains much more than just the procedural or algorithmic knowledge part of the schema, which, in turn, often gives rise to plain manipulation of the mathematical symbols at hand.

It was said that the schema is an activated part of the AMN. But of which one? The answer is: Of the one on which the schema is instantiated. This reveals the prototype character of a schema that enables individuals to interpret one instance that they are faced with out of a set of other possible instances. Applying algebraic network theory in this context means instantiating schemata in a way of systematically enlarging the corresponding AMN. This will be performed by progressive transformation (see Figure 2).

Let us now progressively transform the trinomial and ask the student what will happen to the left-hand side of the equation as a result of the respective transformation. It should be noted that our example corresponds to a slightly advanced level of handling trinomials, but not to the exact teaching in a lesson since it is heavily abbreviated.

In classical math education in secondary schools, the problems to be solved would typically look different: After a first problem, a second, a third, a fourth one, and so forth would be exposed (written in the math work book), each problem having its own alphanumerical appearance and its operational structure, and would be solved by the execution of the appropriate algorithms. Each of the problems would map in the student's mind a certain microstructure basically isolated from the other ones within the AMN.

The situation is totally different with progressive transformations: Each transformation leads to a freshly created equation, the corresponding activated micronetwork of which is, metaphorically, a "neighbor" of the foregoing one and, thus, leads to a systematic elaboration of the AMN.

Equation 1 to start with:

$$(x+4)^2 = (x+4)\,(x+4) = x^2 + 8x + 16$$

$$\downarrow$$

Transformation 1: $(x+\quad)\,(x+\quad) = x^2 + \square x + 12$

Equation 2 to continue: $(x+3)\,(x+4) = x^2 + 7x + 12$

$$\downarrow$$

Transformation 2: $(\quad+\quad)\,(\quad+\quad) = x^2 + \square x + 24$

Equation 3 to continue: $(x+6)\,(x+4) = x^2 + 10x + 24$

$$\downarrow$$

Transformation 3: $(\qquad)\,(\qquad) = x^2 + 10x - 24$

Equation 4 to continue: $(x+12)\,(x-2) = x^2 + 10x - 24$

$$\downarrow$$

Transformation 4: $(\qquad)\,(\qquad) = 2x^2 + 20x - 48$

Figure 2. Progressive steps in the transposition of trinominals.

3.3 Cognitive Learning Requirements

The most important requirement to be fulfilled by the student is to carefully anticipate the changes on the left-hand side of the equation before just initiating some operational algorithm. Anticipations are, thus, the core processes in handling transformations. This procedure implements the notion that a schema is a source of prediction, an internal model to be instantiated: Its function is to provide the student with the ability to interpret the situation he or she faces. Good teaching helps the student to generate predictions, hypotheses, or anticipations, which are tested finally by backward multiplications that provide the student with feedback or debugging information when errors occur or when insecurities dominate the reasoning process – which is often the case in math learning. Such a procedure would impede a poorly understood plain manipulation of algebraic symbols.

Anticipations usually include several sub- or microprocesses such as comparisons as well as inferences as, for example, in Transformation 1: Comparisons of the constant terms of the two right-hand sides of Equations

1 and 2: 16 as 4 x 4, 12 possibly as 3 x 4, and inferences regarding the consequences of the multiplications for the coefficient of the linear term. By means of a backward multiplication, the student may check whether or not the anticipations were correct. As far as Transformation 2 is concerned (with Equation 3 as transient result), the comparison microprocesses reveal that doubling the numerical term of one of the binomials (6 instead of 3) has the characteristic effect of changing not just the constant term but also the linear term of the trinomial. To recognize this means to assimilate the interplay of the subschemata involved.

The student proceeding in this way is far from passively receiving disconnected ideas or retrieving rote-learned facts, but is, instead, actively involved in moving mentally within the algebraic-mathematical micronetwork (AMMN) that is activated by each transformation. The comparisons back and forth from one side of the equation to the other or from the former equation to the latter involved in the anticipatory activities may remind us of the "oscillating comparisons" between partial and final goals suggested by Scardamalia and Bereiter (1985), although in a different learning context.

Whereas Transformations 1 and 2 are gradual in kind, just changing the numerical size of some terms, Transformation 3 is quite different, rather essential in kind, and the respective anticipations are much more complex than in the foregoing examples: What remains unchanged? And where do changes occur — at the surface rather than in the depth? Superficially, "$10x$" remains the same, but the deep structure, in other words, the "operational anatomy" changes remarkably. It is from the anticipation of two different signs with the binomials (in the brackets) that the composition of the "$10x$" may be anticipated.

It is with such anticipatory steps that the rules of the particular constructions of both the linear and the constant terms are derived. I am returning now to the aforementioned problem (tackled, as I said, by Sweller & Cooper, 1985) of how many problems have to be solved or how many schemata have to be instantiated to derive a rule: In my view, it is not a question of the number of solved problems or schemata used, but rather a question of the quality of the connections in the interplay of the respective subschemata that are established by means of the anticipatory microprocesses that go on in handling the transformation.

Let us have a look at Transformation 4 and ask a question concerning long-term math learning goals with the progressive transformation's approach:

Equation 5 to continue: $(\quad)(\quad) = 2x^2 + 20x - 48.$

Equation 5 might be transformed spontaneously at a certain moment by the students themselves to:

$$(\quad)(\quad) = 2(x^2 + 10x - 24)$$

and be factorized by means of the now familiar schemata. With a systematic progressive transformational treatment, the student becomes accustomed to a new approach in handling complex cases of factorizations. He or she will dare to tackle it, starting again (in the following example) by isolating a common factor, and handling the trinomial according to an appropriate schema use:

$$1/2\, x^2 + 5x - 12$$

$$1/2\, (x^2 + 10x - 24).$$

Progressive transformation does not only lead to new instantiations of schemata or to tightening the AMMN but also sharpens the student's focus for spontaneously finding possible transformations by which complex problems can be turned – at least temporarily – into more simple and transparent ones.

Thus, a long-term goal is to foster a learner's autonomy in tackling algebra problems: The use of schemata made flexible by progressive transformations and elaborated AMMN provides the student with the cognitive foundations as well as with the feeling of becoming mathematically more and more self-efficient (Bandura, 1982). Progressive transformation always leads to a motivational "optimal match" (Heckhausen, 1969).

3.4 AMMNs as Parts of AMNs

In well-constructed mathematics curricula, schemata from AMMN will be integrated into more encompassing networks. For instance, the factorization schema as an activated part of a micronetwork in its advanced form will become an integrated part in schemata for understanding and handling functions. To give an example:

Equation 6: $y = 2x^2 + 4x + 2.$

In the present functional Equation 6, the "factorizing schema" is an integrated part of the "function schema," which, in turn, is the condition for understanding the equation as well as for representing it in a graph; the former allowing, after Transformation 6, the factorization of the right-hand side of the equation (following Thaeler, 1985, p. 238).

Transformed equation 6: $y = 2(x^2 - 2x + 1) = 2(x - 1)^2$

$$\downarrow$$

Transformation *n*: $y = -2(x - 1)^2$

Figure 3. Factorizing scheme.

The schema-bound knowledge or, more precisely, the conceptual symbolic knowledge that is expressed in Transformation *n* may be matched with the corresponding iconic representation, the graph of the function, which can be generated easily if the student has been led through several progressive transformations of which the following line gives one possible example:

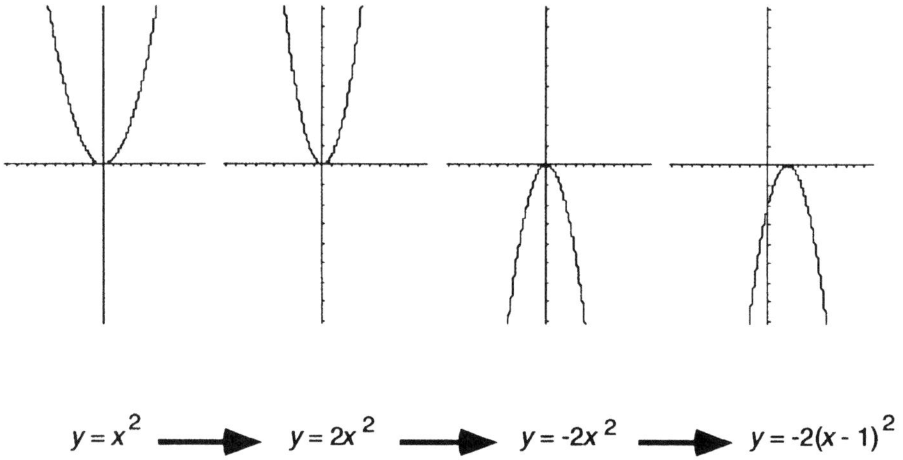

Figure 4.

(The present "cumulative" representation of the graphs tries to illustrate the progressive-transformative character of our instructional procedure.)

Combining factorizing schemata and function schemata including iconic function representation knowledge shows a substitution of the one schema under the other or, in other words, an integration of one AMMN into a more encompassing AMN.

4. SOME PRELIMINARY EVIDENCE OF EFFECTS OF "PROGRESSIVE TRANSFORMATION" – REPORT ON A PILOT STUDY

4.1 Method

Subjects. Twelve poor mathematics achievers in the 10th grade of a Basel senior high-school (9 females, 3 males; mean age 17;1) volunteered for a pilot study.

Procedure. The main structure of the pilot study was a pretest - treatment - Posttest 1 - Posttest 2 - procedure. Each test contained measurements of motivation toward mathematics, algebra test achievement, individual preliminary assessments of task difficulty, as well as – after having solved it – predictions about the correctness of the solutions.

The 12 subjects were assigned to three treatment groups: generative, transformative, and conservative. The conservative treatment corresponded to ordinary high-school-style mathematics education; the transformative treatment was derived from the "progressive transformation" approach; and so was the generative treatment, except for the fact that students were to generate the transformations by themselves instead of receiving the suggestions from the tutor.

Treatments included – after the pretest – six lessons each within two weeks. Posttest 1 was administered one day after treatment; Posttest 2, six weeks after Posttest 1 to control for long-term treatment effects.

Algebra tasks tested and trained during treatments. The focus was on fractions, factorizations, and combinations thereof with an increasing degree of complexity and therefore of difficulty.

Hypotheses. The generative as well as the transformative treatments as opposed to the conservative one were expected to lead to:

1. better algebra test results;
2. qualitatively different algebraic reasoning;
3. more confidence in problem-solving;
4. more accuracy in judging task difficulties;
5. more ease in predicting the correctness of the problem solutions.

No hypotheses were formulated about changes in motivation toward mathematics learning, although we hoped for an increase in motivation scores.

The study was mainly an elaborated single-case study with the goal to test, to a certain degree, the theoretical approach regarding, whether or not 10th graders were an adequate sample for such research questions and the mathematical content of these; furthermore, to learn from the particular observations in those single case studies, to formulate further research questions, and to control for the appropriateness of the instruments used (mainly for motivation measurements).

4.2 Results

Data analyses. In all three algebra tests (pre- and posttests), correct solutions, number of errors, as well as not tackled tasks were scored. Qualitative error analyses were performed by using thinking aloud protocols. Scores also included estimated task difficulties as well as predicted correctness of solutions. The scores of all three treatment groups were compared over the duration of the three tests (approximately 2 months). Thinking aloud protocols were recorded after all three tests while students were solving critical test items in order to find qualitative changes in the students' algebraic reasoning style before and after treatment.

Particular results.

1. All three treatments led to better algebra test results as far as the number of errors was concerned. There was no qualitatively salient effect of the generative and the transformative treatments as opposed to the conservative one. Thus, Hypothesis 1 could not be confirmed.

2. Contrary to the number of errors due to carelessness, which rather grew in the generative and transformative groups, the number of systematic errors (e.g., missing the interplay of operations; not responding to a slight hint

from the tutor in the thinking aloud interview) declined over both posttests for the generative as well as the transformative treatments. This latter result was very strong in both the algebra test results and the thinking aloud protocols. Thus, Hypothesis 2 could be confirmed.

3. A similar result was obtained for the number of not tackled problems: The number of these declined drastically over the two posttests for the generative and transformative treatments; this was not the case for the conservative treatment. We interpret such a result as a confirmation of Hypothesis 3, which addressed individual confidence in tackling problems at all.

4. The results referring to the students' estimations of task difficulty as well as the predictions of correctness of solutions are somewhat contradictory as yet, and do not permit either confirmation or falsification of the corresponding hypotheses.

5. Small gains in motivation to handle algebra tasks and cope with sometimes difficult mathematical problems were distributed fairly evenly across all three treatment groups.

4.3 Conclusions

If systematic errors are essentially schema-bound (in the sense of the first parts of this chapter), then a decline of systematic errors indicates a positive treatment effect as does the increased number of problems tackled over the three tests. Fewer systematic errors means theoretically better AMMNs or at least a more adequate use of the accessible networks; this, in turn, may explain the higher degree of confidence when faced with difficult problems.

The troubles students have when forced to estimate the difficulty of each task or their certainty regarding the correctness of a worked out solution might be due to a long-lasting attitude, particularly in poor math achievers, of observing the single tasks mainly in terms of their surface structure.

It is concluded from the results that:

1. The effective treatments should be offered over more than just six lessons.

2. Instead of trying to repair poor AMN at 10th-grade levels, we should start earlier, probably with 8th graders, to foster both the very first construction and the elaboration of the schemata required for the particular algebra tasks.

3. The study was working exclusively with poor mathematics students. It is not known what effects the generative and the transformative treatments would have with bright or even highly gifted students. So it is necessary to control for a possible aptitude-treatment interaction, especially in regard to progressive transformations.

5. DIDACTICAL IMPLICATIONS FOR AN IMPLEMENTATION OF THE "PROGRESSIVE TRANSFORMATION'S" APPROACH

Since transformative treatment is, as we have to be aware of, not a content but definitely a cognitive, process-bound procedure, the application of the progressive transformation type of teaching as well as a possible generative teaching for gifted students has to take place with any algebraic-mathematical content from the very beginning of arithmetic teaching (Steiner 1974a, b, 1983, 1988) up to the highest forms of mathematics education in secondary schools and colleges.

I suppose that an equilibrium has to be established between systematic use of transformational teaching procedures and consolidating procedures such as practicing, rehearsing, applications to everyday problems (in a way that fulfills the "situated learning" requirements), and further embedding the mathematical structures into texts, and so forth. (By the way, there is good reason to apply progressive transformations to text problems as well as to other science problems, e.g., in physics or statistics!)

Much of the success of the use of the progressive transformation paradigm will depend on the mathematically adequate construction of transformation sequences that systematically lead to the elaboration of the AMN. A cooperation between mathematicians, educational or cognitive psychologists, curriculum planners, math textbook authors, and teachers is urgently needed.

One problem that does not resolve itself is the measurement of the effects of progressive transformations. Since the approach involves mainly procedures and not so much products, measurements by means of test results are indirect and tend to miss the actual reasoning and learning processes. Teachers have to encourage the students' thinking aloud to obtain more process-oriented results that can be evaluated. Exams should include, besides the problem solutions, attempts and approaches to anticipations. All this would be part of the development of widely restructured curricular units including students' work books and other materials.

Of very special merit would be (this is just a concluding idea) to develop a process-oriented prognostic instrument based on the progressive transformation's approach to predict students later mathematics achievements.

This chapter was first of all cognitive in kind: a partial application of Piaget-derived and adapted schema theory or algebraic-mathematical network theory. It is more than just a vision, since preliminary results support the theoretical direction of research and implementation as well. It might open up a path to a new culture of mathematical reasoning and learning.

REFERENCES

Aebli, H. (1963). *Über die geistige Entwicklung des Kindes.* Stuttgart: Klett.

Aebli, H. (1978). Von Piagets Entwicklungspsychologie zur Theorie der kognitiven Sozialisation. In G. Steiner (Ed.), *Piaget und die Folgen. Die Psychologie des 20. Jahrhunderts* (Vol. 7, pp. 604-627). München: Kindler.
Aebli, H. (1987). Development as construction: Nature and psychological and social context of genetic constructions. In B. Inhelder, D. de Caprona, & A. Cornu-Wells (Eds.), *Piaget today* (pp. 217-232). Hillsdale, NJ: Erlbaum.
Bandura, A. (1982). Self-efficacy mechanisms in human agency. *American Psychologist, 37*(2), 122-147.
Bruner, J. S. (1957). Beyond the information given. In J. M. Anglin (Ed.), *Jerome S. Bruner: Beyond the information given. Studies in the psychology of knowing* (pp. 218-238). New York: Norten. [Reprint]
Bruner, J. S. (1966). *Toward a theory of instruction.* Cambridge, MA: Harvard University Press.
Bruner, J. S. (1973). *Beyond the information given.* New York: Norton.
Case, R. (1978). Piaget and beyond: Toward a developmentally based theory and technology of instruction. In R. Glaser (Ed.), *Advances in instructional psychology* (Vol. 1, pp. 167-228). Hillsdale, NJ: Erlbaum
Case, R. (1985). *Intellectual development: Birth to adulthood.* Orlando, FL: Academic Press.
Chandler, P., & Sweller, J. (1991). Cognitive load theory and the format of instruction. *Cognition and Instruction, 8*(4), 293-332.
Glaser, R. (1984). Education and thinking. *American Psychologist, 39*(2), 93-104.
Heckhausen, H. (1969). Förderung der Lernmotivierung und der intellektuellen Tüchtigkeiten. In H. Roth (Ed.), *Begabung und Lernen* (pp. 193-228). Stuttgart: Klett.
Inhelder, B., & Piaget, J. (1955). *De la logique de l'enfant à la logique de l'adolescent. Essai sur la construction des structures opératoires formelles.* Paris: Presses Universitaires de France.
Lindsay, P. H., & Norman, D. A. (1972). *Human information processing.* New York: Academic Press.
Pascual-Leone, J. A. (1970). A mathematical model for the transition rule in Piaget's developmental stages. *Acta Psychologica, 63,* 301-345.
Pascual-Leone, J. A. (1976). A view of cognition from a formalist's perspective. In K. Riegel & J. Meacham (Eds.), *The developing individual in a changing world.* Hague: Mouton.
Piaget, J. (1947). *Traité de logique. Essai sur la logique opératoire.* Paris: Colin.
Rumelhart, D. E. (1978). Schemata: The building blocks of cognition. In R. Spiro, B. Bruce, & W. Brewer (Eds.), *Theoretical issues in reading comprehension.* Hillsdale, NJ: Erlbaum.
Rumelhart, D. E., & Norman, D. A. (1973). Active semantic networks as a model of human memory. *Proceedings of the Third International Joint Conference on Artificial Intelligence.* Stanford, California.
Rumelhart, D. E., & Norman, D. A. (1976). *Accretion, tuning, and restructuring: Three modes of learning.* San Diego, CA: University of California, Technical Report # 63.
Scardamalia, M., & Bereiter, C. (1985). Fostering the development of self-regulating in children's knowledge processing. In S. F. Chipman, J. W. Segal, & R. Glaser (Eds.), *Thinking and learning skills. Vol. 2: Research and open questions* (pp. 563-577). Hillsdale, NJ: Erlbaum.
Steiner, G. (1974a). On the psychological reality of cognitive structures: A tentative synthesis of Piaget's and Bruner's theories. *Child Development, 45,* 891-899.
Steiner, G. (1974b). Kind und Mathematik. *Zeitschrift für Pädagogik, 20*(5), 677-702.
Steiner, G. (1983). Number learning and constructing coherent networks by using Piaget-derived operative principles. In M. Zweng, T. Green, J. Kilpatrick, H. Pollak, & M. Sydam (Eds.), *Proceedings of the Fourth International Congress on Mathematical Education.* Boston: Birkhauser.
Steiner, G. (1988). *Lernen. Zwanzig Szenarien aus dem Alltag.* Bern: Huber.
Sweller, J. (1988). Cognitive load during problem solving: Effects on learning. *Cognitive Science, 12,* 257-285.

Sweller, J., & Cooper, G. A. (1985). The use of worked examples as a substitute for problem solving in learning algebra. *Cognition & Instruction, 2*, 58-89.

Thaeler, J. S. (1988). Input-output modification to basic graphs: A method of graphing functions. In A. F. Coxford & A. P. Shulte (Eds.), *The ideas of algebra, K-12. Yearbook of the National Council of Teachers of Mathematics* (pp. 229-241). Reston, VA: The Council.

Zhu, X., & Simon, H. A. (1987). Learning mathematics from examples and by doing. *Cognition & Instruction, 4*, 137-166.

THE SOCIOHISTORICAL SCHOOL AND THE ACQUISITION OF MATHEMATICS

Joachim Lompscher
Berlin

1. INTRODUCTION

Every kind of didactics is based – in a more or less explicit and differentiated way – on psychological theories, concepts, and facts, in particular, on those of developmental and learning psychology. One of the psychological concepts that is, at present, increasingly discussed internationally is the so-called sociohistorical school, which is particularly tied to the names of Vygotsky, Luria, and Leont'ev. Ideas and results of the Geneva School (and of others as well) inspired them to critical retorts, but also to constructive integration (e.g., Elkonin, 1960, 1978; Leont'ev, 1966/1978; Leont'ev & Tichomirov, 1963; Obuchova, 1972; 1981, Vygotsky, 1964).

After characterizing the theoretical conception of this school in theses, some examples will be used to show its potential for the acquisition of mathematics.

2. THE DEVELOPMENTAL AND ACQUISITIONAL CONCEPTION OF THE SOCIOHISTORICAL SCHOOL

The individual's development takes place under concrete sociohistorical conditions, which consist, in the most general sense, in that a human being (as a member of the species and as an individual in this framework) assures his or her own existence and growth by activity. This means the interplay between human beings and the world, characterized by its social, material, active, purposeful, conscious character, and in which human beings set themselves as subjects with regard to sections of the world, making the latter their object. Subject-object relationships are mediated via direct or indirect relationships to other subjects, while subject-subject relationships are mediated via relationships to objects.

In interplay and communication, human beings shape and reshape their natural and social bases of existence, continuously experiencing feedback from nature and society in doing so. The means, conditions, and objects developed by and for the activity of previous generations – that is, human culture – must be appropriated by subsequent generations in order to enable

R. Biehler, R. W. Scholz, R. Sträßer, B. Winkelmann (Eds.),
Didactics of Mathematics as a Scientific Discipline, 263-276.

them to participate in the life of society and in shaping it – at least partially – and developing it further. This occurs in social interaction, that is, in joint activity of coordinating action, exchanging information, and developing forms of cooperation in order to satisfy needs and to realize goals.

The psychological functions serving to regulate this activity occur virtually twice in their developmental history: initially shared between different persons, as a social relationship and division of functions ("interpsychical functions" according to Vygotsky), and then increasingly as psychological, internal ("intrapsychical") functions proper. This is based on the unity and interplay of interiorization and exteriorization, the activity becoming increasingly a mediated one: as means to attain ends, which are initially simple and directly available, although, later, more and more complex and specially produced things are used. From operating with objects develops operating with signs as bearers of signification, which (may) have different meanings. Means of acting upon objects and other subjects become means of influencing oneself and of self-awareness.

The explanation of psychological novelties and particularities of an evolutionary step and of the transition from one level of development to another must begin by analyzing the social situation of development, which is mainly determined by the respective social position and by the changing social relationships and conditions of activity, by the emerging internal and external contradictions, and by the strategies and forms of coping with these. In this framework, a concrete level of development will be characterized by two zones of development: the "zone of actual performance," comprising everything the child (or adolescent) can already cope with independently on the basis of previous development and acquisition. At the same time, this contains a potential of performance beyond this status that the individual cannot yet realize alone, but with guidance, support, by imitation, and so forth. This "zone of proximal development" can be transformed into activity and guided on to a next "zone of actual performance" in social interaction and cooperation (with adults, older children, or peers), this leading to a next "zone of proximal development." Stages of cooperation thus alternate with stages of independence, while contents, forms, and levels, both of cooperation and of independence, change and grow.

An instruction intended to enhance development must be applied at the respective social situation of development and at the zone of proximal development, which means to concentrate not simply on the developmental stage just reached, but rather on the developing, not fully formed psychological functions and to actively encourage their growth by creating and offering conditions, contents, and means of activity that help students to cope with demands belonging to their next zone of development.

This conception, which has been presented here in an extremely abbreviated form (for more details, see, e.g., Elkonin, 1989; Galperin, 1980; Jantzen, 1986, 1991; Lektorski, 1990; Leont'ev, 1964, 1979; Luria, 1982,

1987; Saporoshez, 1986, 1990; Vygotsky, 1985/1987; see, also, Rubinstein, 1958, 1962, 1963, who was in a way related to the sociohistorical school, but made important contributions to the conception of activity and acquisition; see, also, Abulchanowa-Slawskaja & Bruschlinki, 1989), has been applied and extended in various directions and subject matter fields. It is also the basis of the conception of learning activity and its development (e.g., Davydov, 1969, 1977, 1986, 1988 a, b; Davydov, Lompscher, & Markova, 1982; Elkonin & Davydov, 1962, 1966; Engeström, 1987, 1990; Lompscher, 1975, 1988, 1989 a, b, in press). Here, learning is considered, on the one hand, as result and prerequisite of any activity ("learning by activity"), and, on the other hand, as a specific activity ("learning as an activity," "learning-activity"). Learning activity is directly or indirectly interrelated with teaching activity, is oriented toward acquiring social knowledge and skills by individually reproducing these, and assumes that a desire to learn arises, which concretizes in learning goals and motives to learn whose realization requires engaging in activities that are adequate to the object, the conditions, and to the goals. Learning activity develops within the process of personality development from other activities and again contributes essentially to personality development – dependent on how it is acquired itself, is developed into a specific activity. In this case, it can become a dominating activity that determines the individual's position within the system of social relationships, influencing personal growth quite intensely (particularly at young and middle school age). Sooner or later, the activity of learning cedes this position to other kinds of activity within the framework of an individually emerging, increasingly differentiated, and, at the same time, concentrating activity system of a person in which lifelong learning takes a significant place. Among the essential novel psychological structures, which are created mainly in and by learning activity, are communicative and cooperative competence; learning strategies; cognitive motivation; theoretical reasoning; scientific systems of concepts, rules, and methods; and so forth.

3. SELECTED CONTRIBUTIONS TO ACQUIRING MATHEMATICS

One branch of the sociohistorical school found its expression in Galperin's theory of stagewise formation of mental activities (e.g., Galperin, 1967, 1968, 1982; Galperin & Talysina, 1968, 1972; Lompscher, 1967, 1973; Podolsky, 1987; Salmina, 1981, 1988; Talysina, 1969, 1975, 1988), which, as the name says, placed activity at the center of theoretical and empirical studies, analyzing, in particular, the specific role of the orientation base and stagewise interiorization for the formation and acquisition of activities. The focus of these studies was on conditions of the acquisition of basic knowledge and skills that should be available to all students. In the field of mathematics, various areas were worked on.

3.1 Acquisition of Elementary Mathematical Concepts

In one of their first studies in 1957, Galperin and Talysina (1967) were able to show that poorly performing students, no matter whether they had had geometry lessons before (6th and 7th grade), could be enabled in learning experiments to correctly form and use concepts like straight line, line segment, perpendicular, angle, bisectrix, to differentiate them from similar concepts, to identify figures in any position, to orient themselves consistently toward the verbal formulation of a task even if the drawing diverged from it, and to independently apply the acquired concept-forming strategy to new concepts (e.g., from the concept of straight line and angle to bisectrix, to adjacent angle, etc.). The demand here was one of classification (assigning concrete phenomena to definite classes on the basis of certain features). The instructional strategy was:

1. To explicitly formulate the features necessary and sufficient for assignation, giving them to the students as a basis of their activity.
2. To present tasks of different degree of difficulty whose solution required applying these features in a definite order.
3. To organize the solving of such tasks on various levels of activity: (a) as a material activity on the object or as a materialized activity on the basis of a written basis of orientation; (b) as a verbalizing activity "for others," which contained all the steps and features according to the basis of orientation; (c) as a verbalizing activity "for oneself," which only verbalized certain turning points; and (d) as a nonverbal or mental activity, which became increasingly reduced and automatized.
4. To make the transition to the next higher stages of interiorization dependent on the degree of mastery of the activity on the respective level, and to use recourse to previous stages to overcome errors or difficulties.

As the basis of orientation comprised the necessary features, the activity could in a way be carried out from the very outset – in principle, without error and with conscious orientation toward what was essential in each case and thus appropriated. As soon as the strategy had been formed using some of the concepts, it could be applied by the students independently to acquiring other concepts without having to go through all the stages of interiorization. Realization of this conception in experiments involving individuals and groups led to a significant increase in efficiency and quality of the learning processes.

According to the same principles and with analogous results, training programs for proving geometrical theorems, for acquiring geometrical transformations, and so forth were elaborated and tested. The transformation activity's components (e.g., determining the points in the initial object that permit its reconstruction; determining the objects in relation to which the transformation shall take place; determining the transformation itself – turnings, shifts, and so forth; realizing the transformation activity; and analyzing the transformed object) were at first analyzed and trained separately and then integrated stagewise to a holistic activity by means of an appropriate basis of orientation. Similarly, the number concept, the number value system, and basic calculus, as well as solving text or word problems, were trained (see below).

Galperin's and his colleagues' conception inspired numerous studies and applications in various subject matter fields. Transferring the principles and methods of analysis and training was sometimes formal and superficial. Correct applications of the conception yielded high learning results as a rule. This was mainly due to the fact that orienting the learners toward the respectively essential features and relations in the object of learning was not left to chance and how things would go, but was strictly guided. Experiments that succeeded in realizing the orientation in a generalized way for a large number of objects or events, in a way enabling the learners to establish (and realize) the basis of orientation for subclasses or concrete cases themselves, proved particularly efficient. Stagewise training and interiorization of the respective activities in solving tasks appropriate for this purpose (as a unity of application and acquisition) was an essential condition for increasing independence of the learners in coping with complex and novel learning demands.

The potential of this conception, however, at the same time indicates its limitations, the focus being, as a rule, on an individual, sometimes complex activity (and on training for it). How it can be integrated into superordinate contexts of the learners' activity is a question that remains underdeveloped. Questions of motivation and defining goals thus play a subordinate role. The emphasis is on acquiring individual concepts and skills, or closely defined complexes of the latter, and less on the structure and the system of entire subjects or courses. Above all, the conception is mostly oriented – in spite of the high status of activity – to presenting what is to be acquired ("transmission strategy"), to strictly guiding the process of acquisition, and hence on determining learning from without.

These limitations were overcome by the conception of learning activity and its formation, which was mainly developed by Elkonin and Davydov. They opposed a strategy of activity and formation to that of transmission. The theory of the stagewise formation of mental activity was integrated as an essental component into a larger context – that of activity. This is what will be shown in the next sections.

3.2 Formation of the Number Concept

In the frame of Galperin's theory, several studies have already been carried out with preschoolers and elementary school students on the formation of the number concept and on operating with numbers. As an alternative to the traditional orientation toward sets of individual objects (or counting, etc.), measuring and the relation between quantities and units of measure and hence the relativity of the number (dependence on the measure that is laid to a quantity) and the adequancy of the measure and of measuring was made the basis of developing the number concept and operating with numbers (Galperin & Talysina, 1968, pp. 72-134; Talysina, 1969, pp. 107-120).

Davydov (1962, 1966, 1969) chose the same starting point. In teaching experiments extending over several years with entire classes, he developed and realized a training course (1st to 3rd grade) following the teaching strategy of ascending from the abstract to the concrete (e.g., Davydov, 1977, 1986, 1988 a, b; Lompscher, 1989 a, b; Seeger, 1989). His intention was to shape and form learning activity so as to ensure that elementary theoretical reasoning – as a novel psychological formation in the zone of proximal development – occurs in younger school children from first grade on. In mathematics, the children were to acquire a full-sized concept of number, which requires profound abstraction of the feature of quantitiveness from all other features of the objects. Measuring proved to be a practical activity suitable for that purpose. In order to be able to study and consciously grasp features of the number and of operating with numbers, the children must be given opportunities to detach themselves from the objective. This is achieved by working with symbols and graphical models if the basic features and relations obtained by manifold practical-objective activities can be fixed in them in a general form, and if they can be used to operate.

On this basis, students learn to reconcretize general relationships and theorems, to form terms, equations, and word problems themselves, and to solve them; the transitions between the abstract and the concrete being at first realized in deployed activity, then slowly reduced. The natural numbers and calculating with them then appears to the children as a concretization, as a special case of general mathematical features and relations. Abstractions are obtained and analyzed by practical-objective activity of their own and they are applied to various concrete phenomena, or the latter are derived from them. In any case, the emphasis is on deriving, founding, arguing, and on other cognitive operations. Calculating and the training of calculating skills is being based on an understanding of the general laws of numbers and on the relationships between them. Activities that were first unfolded are reduced, interiorized, and automatized stepwise and stagewise.

The introduction of younger students to the world of numbers occurred – very briefly – in several steps:

1. Within the context of most different situations, objects are compared with regard to certain features (length, breadth, height, weight, area, etc.) while introducing the concept of equal, larger, and smaller, which are assigned the appropriate symbols, and the respective quantities are designed as *A*, *B*, and so forth.

2. Where direct comparison of quantities (by juxtaposition, superposition, etc.) is difficult or impossible, possibilities of indirect comparison are sought and found – under guidance – in measuring: A measure is used to

establish how much larger or smaller (or equal) a quantity is as compared to another, that is, how many times the measure (a thread or a stick) can be laid on, or how many times it is contained (a cup in a container of liquid). This multiplicative relationship is first fixed with scratches, small sticks, and the like, then with digits. Fixings like $A/c = 5$, $B/c = 4$ will result (the measure c being contained 5 times in Quantity A, 4 times in Quantity B); $5 > 4$, follows $A > B$, or – in generalized form – $A/c = K$, $B/c = M$; $K > M$, $A > B$.

3. Not only different quantities are measured with the same measure but also equal quantities with different measures (liquids with cups of various sizes, lengths with threads of various lengths, areas with panels of different size, etc.): $A/c = K$, $b > c$, follows $A/b > K$, and so forth. For children, the concept of number is not reduced to quantities or to counting individual objects, but is represented by the general formula of $A/c = N$.

4. The measure can be varied in different ways, for example, as a sequence rising multiplicatively. Even first graders will thus acquire the concept of a number having several digits, and are getting to know different number value systems (the dual, the decadic one, etc.). "Experimenting" with measure also leads to cases in which it is smaller than the quantity to be measured ("is not quite contained in the quantity") – this leads to the concept of fractioned number. In work with directed quantities, the concept of positive and of negative number is formed.

5. Comparing and measuring quantities also provides occasions of transforming equality and inequality into one another, thus studying and exercising addition and subtraction (as focal points for first graders). While working with line segments and other objects, the children will recognize, for instance, that when two quantitities are given, a third is virtually given as well (or can be derived or calculated): When drawing the equation $a + b = c$ as a line segment, for instance, the children realize that the third quantity is definitely established by the two others, and independent of which order they choose. The general form of the operation is varied accordingly ($a = c - b$, etc.) and transformed in different concrete tasks. Word problems, too, are formed and varied by the children themselves on this basis, studying reversability and other features of addition and subtraction in doing so. The exercises necessary to form these skills are contained in this framework of activities. Multiplication and division are elaborated in a similar way in second grade. Geometrical concepts and operations are developed in connection with arithmetical ones.

Although amount and demand level of the mathematical subject matter were significantly increased as compared to widespread teaching practices in elementary school, the forming of mathematical concepts and skills and the development of mathematical reasoning (e.g., in solving word and factual problems and transfer problems for theoretical generalization) could be raised to a significantly higher level as compared to the control population (belonging, in part, to higher grades), as has been shown in numerous studies.

3.3 Solving Word and Fact Problems

Discovering mathematical connections in the verbal representation of real situations and working on such situations with mathematical means contains significant potential for the acquisition of mathematics, but, at the same time – as numerous studies and the international debate have shown – considerable difficulties for the majority of the students. This topic also has had, and still has, a significant role in the framework of the sociohistorical tradition (cf., e.g., Davydov, 1969, 1986; Nikola & Talysina, 1972; Talysina, 1969). I shall confine myself here to an example taken from my own studies (Lompscher, 1989 a, b, 1990).

In elementary school, students are confronted with innumerable text and factual tasks – as a rule, these are presented mainly as a field of application and exercise for the mathematical topics respectively treated. In this work, the students grow accustomed to routines that rather more impede than enhance the solving of such problems. As they "know" beforehand which mathematical context the problems will be about, they confine themselves in most cases to a superficial analysis, which is mainly attuned to establishing which indications (text, indicative terms) are present in the text and which operations can be carried out with these. This strategy is even encouraged by many teachers and even textbooks by corresponding questions, hints, and so forth, but it often leads – as soon as a task has real problem character for the students – to inadequate, superfluous, or simply nonsensical operations and solutions (as in the so-called "captain tasks") or to no approach or solution being found. In most cases, students will not even realize their mistakes or the nonsensical character of their answer, as self-control is little developed and they do not relate the result obtained to the problem text or to the question. The difficulties of text and factual problems usually do not lie in executing the calculating operations, but rather in coping with the cognitive demand of an adequate problem representation and analysis, in uncovering the mathematical structure underlying the real situation, and in deriving mathematical operations from it.

In teaching experiments with fourth graders, but also with fifth and third graders, we checked whether and how the students can be reoriented toward another strategy (which, for them, means to re-learn), the teaching strategy of ascending from the abstract to the concrete being taken as a basis. The students first had to form a concept of the general structure of mathematical word and factual problems and then to concretize the latter with various task structures. This required that the students profoundly analyze different problem structures. Goal-means-condition analysis seemed to be the suitable heuristic strategy for this: First the goal is analyzed as to which means might serve its realization in order to look from this aspect for the statements made to this purpose in the text, or which are the conditions of looking for these and deriving them. The starting point thus is the analysis of the unknown, the goal from which subgoals are derived "proceeding backwards," which then can be realized on the basis of the available and required data "proceeding forward." In the actual process of solving, these two methods of proceeding – global and local strategy – are, of course, merged. The success of this method depends significantly on knowledge about functional relationships between quantities, and from a profound analysis of the real situation presented, in order to be able to discover the mathematical structure concealed in it, to recognize its elements and relations, and to use these for solving the problem: for finding what is sought. The most important stages and conditions of the training process were the following:

1. A relatively substantial problem text containing statements relevant and irrelevant for the solution led to various, but, as a rule, unsuccessful, student attempts at solution. By this, solving "such difficult problems" became a specific learning goal that was subdivided, in the process of learning-teaching, into subgoals.

2. In joint activity, a general structure was discovered in different problems and fixed in a graphical learning model (Figure 1), the analysis of what was sought forming the starting point.

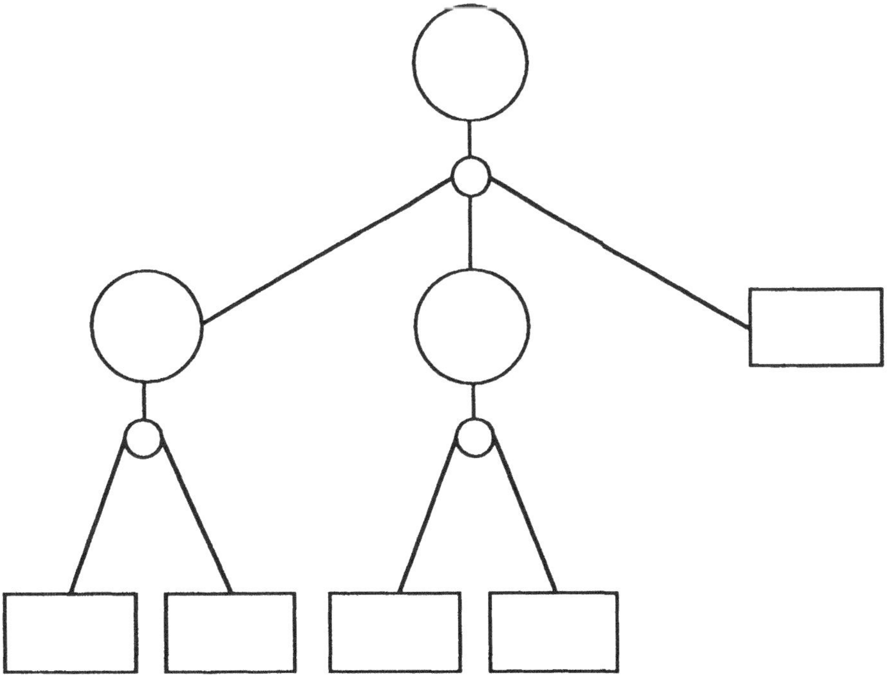

Figure 1. Structure model for solving word problems (according to Reinhold).

3. The functional relationships between different quantities (e.g., starting time, duration, finishing time, price per item/number/sum) were analyzed

systematically by practical-objective and mental activities (real and imagined change of a quantity, checking its effect on others) and generalized. For establishing the structural model corresponding to a problem, first, the respective general concept (size, price, etc.), then its concretization was used. The functional relationships served to justify the mathematical operations: if ... and ... are given, ... can be calculated by ... ; if ... is unknown, I need ... and ... to calculate it.

4. Using various problem structures, by formulating and reformulating texts, transforming things known into things unknown and vice versa, changing the quantitative data of the various quantities, or transforming problems into questions and questions into problems, the subactivities necessary to solve word and factual problems of (a) grasping the goal (formulating what is sought); (b) grasping the essential quantities and the relationships between them; (c) establishing adequate mathematical equations; (d) solving the equations; (e) checking and evaluating the solution path and the numerical result found; and (f) formulating an answer referring to the goal or question were established and integrated into a holistic, flexible activity of problem-solving oriented toward uncovering and working on the respective structure, verbalizing and justifying the method selected, first extensively, then increasingly briefly as the students grew accustomed to systematic, founded methods, and toward conscious use of the relevant mathematical concepts and operations. Thoughtless, routine "solving" was prevented by the fact that each problem, in principle, presented, in some aspects, different demands and a different problem character to the children. A differentiated analysis of the mathematical demand structures enabled us to vary the demands on the children's mental activity in manifold ways, increasing them slowly but systematically.

To record and analyze the learning results, we used various methods, which, as a whole, showed a high superiority of the experimental classes as compared to the control classes. One example is given in Figure 2: Students in the experimental class were able to discover a problem's mathematical structure even if they were less familiar with the contents of the real situation presented than with other tasks (Problem b was about liquids – an unexpected object for the students – while the facts, operations, and text structure were analogous to Problem a). Even poorly performing students (Group III) were able to cope with the demand relatively well, while the average students in the control classes (Group II) were mostly overtaxed. Similar results were obtained with transfer problems, which yielded a significantly higher level of development in abstract reasoning.

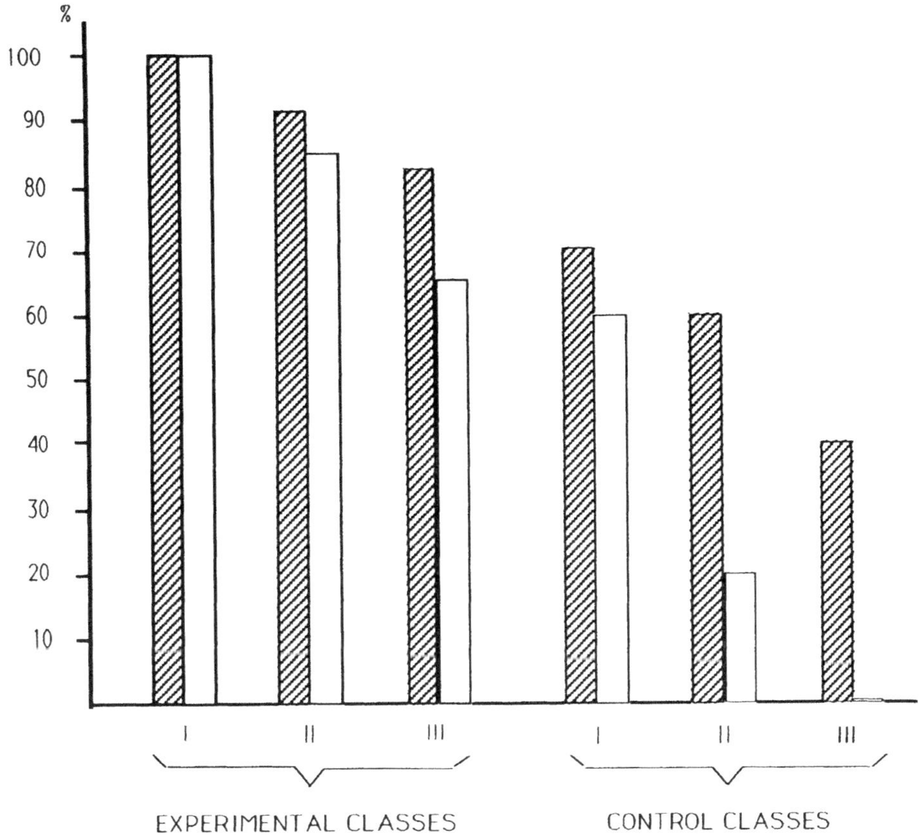

Figure 2. Percentages of correct solutions in two structurally equivalent word problems with familiar (unshaded columns) and unfamiliar (shaded columns) contents (see Reinhold, 1988).

4. CONCLUSION

The contributions of the psychology oriented toward sociohistorical and activity theory to the field of acquiring mathematics have been presented here only briefly and in small sections. It must at least be pointed out that learning activity and its formation is not considered as a purely individual process, but that a significant status is allotted to the joint activity of the children in analyzing and in looking for connections and solutions, in planning and in justifying, in realizing activities, and in checking and evaluating their results (including the analysis of errors). Joint activity is the genetically original one, and individual cognition and competence develops from the very process of interaction, communication, and cooperation in coping with situations containing unknowns – problem situations – requiring much space. Independent reasoning, applying one's knowledge and skills to unfamiliar situations, recognizing and evaluating novel, useful activity in uncertain situations does not develop, or only in a limited scope, if the students are accustomed primarily to receiving and reproducing ready-made insights. Learning activity means active, increasingly independent coping with objects of learning, and increasingly self-determined learning as a result of systematically enabling the students to form learning goals; to select and use learning strategies adequate to the objects, the conditions, and goals; to responsibly use learning aids and learning time; and so forth.

The interest in the sociohistorical tradition and conception of activity has grown on an international scale within recent years (e.g., Bol, Haenen, & Wolters, 1985; Engelsted, Hedegaard, Karpatschof, & Mortensen, 1993; Engeström, 1987, 1990; Hedegaard, Hakkarainen, & Engeström, 1984; Hildebrand-Nilson & Rückriem, 1988; Moll, 1990; Säljö, 1991; Van Oers, 1990; Wertsch, 1985 a, b). The scope and variety of theoretical and empirical work in this direction has increased significantly. It will be able to make a productive contribution to solving problems of acquiring mathematics in the future.

REFERENCES

Abulchanowa-Slawskaja, K. A., & Bruschlinski, A. W. (1989). *Filosofsko-psichologiceskaja koncepcija S. L. Rubinsteina* [S. L. Rubinstein's philosophical-psychological conception]. Moskva: Mysl.

Bol, E., Haenen, J. P. P., & Wolters, M. (Eds.). (1985). *Education for cognitive development.* Den Haag: SVO.

Davydov, V. V. (Ed.). (1969). *Psichologiceskie vozmoznosti mladsich skolnikov v usvoenii matematiki* [The mental potentials of younger school students in the acquisition of mathematics]. Moskva: Prosvescenie.

Davydov, V. V. (1977). *Arten der Verallgemeinerung im Unterricht.* Berlin: Volk und Wissen.

Davydov, V. V. (1982). The psychological characteristics of the formation of elementary mathematical operations in children. In T. P. Carpenter, J. M. Moser, & T. A. Romberg (Eds.), *Addition and subtraction: A cognitive perspective* (pp. 234-238). Hillsdale, NJ: Erlbaum.

Davydov, V. V. (1986). *Problemy razvivajuscego obucenija* [Problems of developmental teaching]. Moskva: Pedagogika.

Davydov, V. V. (1988a). Problems of developmental teaching. The experience of theoretical and empirical psychological research. Excerpts. *Soviet Education, 30*(8), 15-97; *30*(9), 3-83; *30*(10), 3-77.

Davydov, V. V. (1988b). Learning activity: The main problems needing further research. *Multidisciplinary newsletter for activity theory, 1,* 29-36.

Davidov, V. V., Lompscher, J., & Markowa, A. K. (Eds.). (1982). *Ausbildung der Lerntätigkeit bei Schülern.* Berlin: Volk und Wissen.

Elkonin, D. B. (1960). *Detskaja psichologija* [Child psychology]. Moskva: Ucpedgis.

Elkonin, D. B. (1980). *Psychologie des Spiels.* Berlin: Volk und Wissen.

Elkonin, D. B. (1989). *Izzbrannye psichologiceskie trudy* [Selected psychological studies]. Moskva: Pedagogika.

Elkonin, D. B., & Davidov, V. V. (Eds.). (1962). *Voprosy psichologii ucebnoj dejatelnosti mladsich skolnikov* [Issues in the psychology of the learning activity of younger school students]. Moskva: Izd. APN.

Elkonin, D. B., & Davidov, V. V. (Eds.). (1966). *Vozrastnye vozmoznosti usvoenija znanij* [Age-stage-related potentials in the acquisition of knowledge]. Moskva: Prosvescenie.

Engested, N., Hedegaard, M., Karpatschof, B., & Mortensen, H. (Eds.). (1993). *The societal subject.* Aarhus: Aarhus University Press.

Engeström, Y.(1987). *Learning by expanding*. Helsinki: Orienta-Konsultit Oy.
Engeström, Y. (1990). *Learning, working and imaging*. Helsinki: Orienta-Konsultit Oy.
Galperin, P. (1967). On the notion of internalization. *Soviet Psychology, 5*(3), 28-33.
Galperin, P. (1969). Stages in the development of mental acts. In M. Cole & I. Maltzman (Eds.), *A handbook of contemporary Soviet psychology* (pp. 249-273). New York: Basic Books.
Galperin, P. (1980). *Zu Grundfragen der Psychologie*. Berlin: Volk und Wissen.
Galperin, P. (1982). Intellectual capabilities among older preschool children: On the problem of training and mental development. In W. W. Hartup (Ed.), *Review of child development research* (Vol. 6, pp. 526-546). Chicago, IL: The University of Chicago Press.
Galperin, P., & Talysina, N. F. (Eds.). (1968). *Formirovanie znanij i umenij na osnove teorii poetapnogo usvoenija umstvennych dejstvij* [Formation of knowledge and skills on the basis of the step-wise acquisition of mental actions]. Moskva: Izd. Moskovskogo universiteta.
Galperin, P., & Talysina, N. F. (Eds.). (1972). *Upravlenie poznavatelnoj dejatelnostju ucascichsja.* [Guiding the cognitive activity of students] Moskva: Izd. Moskovskogo universiteta.
Hedegaard, M., Hakkarainen, P., & Engeström, Y. (Eds.). (1984). *Learning and teaching on a scientific basis.* Aarhus, Denmark: Aarhus universitet.
Hildebrand-Nilshon, M., & Rückriem, G. (Eds.). (1988). *Proceedings of the 1st international congress on activity theory.* Berlin: System Druck.
Jantzen, W. (1986). *Abbild und Tätigkeit. Studien zur Entwicklung des Psychischen.* Solms-Oberbiel: Jarick Oberbiel.
Jantzen, W. (1991). *Psychologischer Materialismus, Tätigkeitstheorie, Marxistische Anthropologie.* Hamburg: Argument.
Lektorski, W. P. (Ed.). (1990). *Dejatelnost': Teorii, metodologija, problemy* [Tätigkeit: Theorien, Methodologie, Probleme]. Moskva: Politizdat.
Leont'ev, A. N. (1964). *Probleme der Entwicklung des Psychischen.* Berlin: Volk und Wissen.
Leont'ev, A. N. (1966/1978). Predislovie [Preface] In P. Fraisse & J. Piaget (Eds.), *Traité de psychologie experimentale* (Vols. 1-6). Moskva: Progress.
Leont'ev, A. (1979). *Tätigkeit, Bewußtsein, Persönlichkeit.* Berlin: Volk und Wissen.
Leont'ev, A. N., & Tichomirow, O. K. (1963). Posleslovie [Postscript to Russian translation of] Inhelder, B., & Piaget, J.: *La genese des structures logiques elementaires.* Moskva: Izd. inostrannoj literatury.
Lompscher, J. (Ed.). (1967). *Probleme der Lerntheorie.* Berlin: Volk und Wissen.
Lompscher, J. (Ed.). (1973). *Sowjetische Beiträge zur Lerntheorie. Die Schule P. Galperins.* Köln: Pahl-Rugenstein.
Lompscher, J. (Ed.). (1975). *Theoretische und experimentelle Untersuchungen zur Entwicklung geistiger Fähigkeiten.* Berlin: Volk und Wissen.
Lompscher, J. (Ed.). (1988). *Persönlichkeitsentwicklung in der Lerntätigkeit.* Berlin: Volk und Wissen.
Lompscher, J. (Ed.). (1989a). *Psychologische Analysen der Lerntätigkeit.* Berlin: Volk und Wissen.
Lompscher, J. (1989b). Formation of learning activity in pupils. In H. Mandl, E. De Corte, N. Bennett, & H. F. Friedrich (Eds.), *Learning and instruction. European research in an international context* (Vol. 2.2, pp. 47-66). Oxford: Pergamon Press.
Lompscher, J. (1990). *Aufsteigen vom Abstrakten zum Konkreten im Unterricht – Versuche zu einer alternativen Lehrstrategie.* Berlin: Akademie der Pädagogischen Wissenschaften.
Lompscher, J. (in press). Learning activity. In *The international encyclopedia of education* (2nd ed.). Oxford: Pergamon Press.
Luria, A. (1982). *Sprache und Bewußtsein.* Berlin: Volk und Wissen.
Luria, A. R. (1987). *Die historische Bedingtheit individueller Erkenntnisprozesse.* Berlin: Deutscher Verlag der Wissenschaften.
Moll, L. C. (Ed.). (1990). *Vygotsky and education. Instructional implications and applications of sociohistorical psychology.* Cambridge: Cambridge University Press.

Nikola, G., & Talysina, N. F. (1972). Formirovanie obscich priemov resenija arifmeticeskich zadac [Formation of general strategies for arithmethical problem solving]. In P. Galperin & N. F. Talysina (Eds.), *Upravlenie poznavatelnoj dejatelnostiju ucascichsja* [Guiding the cognitive activity of students] (pp. 209-261). Moskva: Izd. Moskovskogo universiteta.
Obuchowa, L. F. (1972). *Etapy razvitija detskogo myslenija* [Developmental stages of childhood thought]. Moskva: Izd. Moskovskogo universiteta.
Obuchowa, L. F. (1981). *Koncepcija Zana Piaget: za i protiv* [For and against the conception of Jean Piaget]. Moskva: Izd. Moskovskogo universiteta.
Podolski, A. I. (1987). *Stanovlenie poznavatelnogo dejstvija: naucnaja abstrakcija i real-nost'* [The etiology of cognitive activity: Scientific abstraction and reality]. Moskva: Izd. Moskovskogo univeriteta.
Rubinstein, S. L. (1958). *Grundlagen der allgemeinen Psychologie*. Berlin: Volk und Wissen.
Rubinstein, S. L. (1962). *Sein und Bewußtsein*. Berlin: Akademie-Verlag.
Rubinstein, S. L. (1963). *Prinzipien und Wege der Entwicklung der Psychologie*. Berlin: Akademie-Verlag.
Salmina, N. G. (1981). *Vidy i funkcii materializacii v obucenii* [Types and functions of materialization in teaching]. Moskva: Izd. Moskovskogo universiteta.
Salmina, N. G. (1988). *Znak i simvol v obucenii* [Signs and symbols in teaching]. Moskva: Izd. Moskovskogo universiteta.
Saporoshez, A. W. (1986). *Izbrannye psichologiceskie trudy* [Selected psychological studies]. Moskva: Pedagogika
Saporoshez, A. W., Sintschenko, W. P., Owtschinnikowa, O. W., & Tichomirow, O. K. (Eds.). (1983). *A. N. Leont'ev i sovremennaja psichologija* [A. N. Leont'ev and modern psychology]. Moskva: Izd. Moskovskogo universiteta.
Säljö, R. (Ed.). (1991). Culture and learning. *Learning and Instruction, 1*(3).
Seeger, F. (1989). *Davydov's theory of generalization: Theoretical thinking and representation in learning and teaching algebra*. IDM Occasional Paper 117. Bielefeld: Universität Bielefeld.
Talysina, N. F. (1969). *Teoreticeskie problemy programmirovannogo obucenija* [Theoretical problems of programmed teaching]. Moskva: Izd. Moskovskogo universiteta.
Talysina, N. F. (1975). *Upravlenie processom usvoenija znanij* [Guiding the process of knowledge acquisition]. Moskva: Izd. Moskovskogo universiteta.
Talysina, N. F. (1988). *Formirovanie poznavatelnoj dejatelnosti mladsich skolnikov* [Training the cognitive ability of younger students]. Moskva: Prosvescenie.
Van Oers, B. (1990). The development of mathematical thinking in school: A comparison of the action-psychological and information-processing approaches. *International Journal of Educational Research, 14*(1), 51-66.
Vygotsky, L. S. (1964). *Denken und Sprechen*. Berlin: Akademie-Verlag.
Vygotsky, L. (1985/1987). *Ausgewählte Schriften* (Vols. 1-2). Berlin: Volk und Wissen.
Wertsch, J. V. (Ed.). (1985). *Vygotsky and the social formation of mind*. Cambridge, MA: Harvard University Press.
Wertsch, J. V. (Ed.). (1985). *Culture, communication and cognition: Vygotskyan perspectives*. Cambridge: Cambridge University Press.
Zaporozec, A. V. (1990). *Entstehung und Aufbau der Motorik. Eine tätigkeitspsychologische Studie*. Berlin: Deutscher Verlag der Wissenschaften.

ACTION-THEORETIC AND PHENOMENOLOGICAL APPROACHES TO RESEARCH IN MATHEMATICS EDUCATION: STUDIES OF CONTINUALLY DEVELOPING EXPERTS

Richard Lesh and Anthony E. Kelly
Princeton / New Brunswick

1. ASSUMPTIONS ABOUT STUDENTS' THINKING

We begin with the assumption that students actively construct meaning. They are not tabula rasa upon which teachers "write" knowledge. Each student makes sense of the world in terms of the understandings of the world that he or she brings to it. These understandings or models of the world are constantly being revised, and are never in a final state. Thus, we are in general accord with the precepts of what has become known as constructivism.

2. MODELS

We do, however, pay particular attention to models. By a model we mean a *structural metaphor* or a *pattern* that provides thinkers with the ability to describe, predict, and control the behavior of complex systems. A model allows them to make informed decisions on the basis of a subset of the total available cues. It allows them to "filter" information intelligently, to suggest information that may fill in "holes" in their understanding of a task, and to recognize superfluous information. Models may contain, but are not limited to, facts and procedural rules. Rather, they serve to organize facts and rules into systems for understanding and for action. Models tend to be multidimensional and unstable. Consequently, they are often revised or restructured depending on the conditions and purposes that exist in a given situation.

2.1 The Characteristics of Models and How They Develop

When we study children and teachers, we find that both groups propose models that are tested, rejected, revised, or revisited, *all without any clear notion of exactly what an expert response might look like for a given problem.* How is it that people perceive the need to develop beyond the constraints of their own current conceptualizations of their experiences? How is it that they so often develop in directions that are generally better

R. Biehler, R. W. Scholz, R. Sträßer, B. Winkelmann (Eds.),
Didactics of Mathematics as a Scientific Discipline, 277-286.

without a preconceived notion of best? How do models evolve? We will illustrate this process with three examples: (a) a study of teachers tutoring students; (b) a study of teachers designing authentic assessment tasks; and (c) a study of teachers designing scoring rubrics for authentic assessment tasks.

2.2 Evolution

The models that underlie the interpretation phases of mathematical problem-solving evolve in a manner similar to how other types of organisms or systems evolve. We invite the reader to indulge our use of this analogy, because we feel that the perspective that it provides is more important than whether the correspondence is tight and "correct" at every single juncture. The processes that students and teachers engage in can be described as involving generation and mutation, selection, adaptation, reorganization, differentiation, and accumulation.

2.3 Generation and Mutation

In the tutoring study (Lesh & Kelly, 1991), for example, students proposed a variety of different ways to think about a problem. In the early stages, they suggested several models based on additive relationships, subtractive relationships, fractions, or proportions. These models were expressed in a variety of different ways: as numbers, as verbal arguments, as graphs, as sketches, and so forth. As the students explored a relationship through a given representation, they oftentimes pursued features of the representation that, in turn, suggested the pursuit of an alternative relationship. In this way, the models were dynamic, unstable, and subject to mutation.

In the same study, teachers began by suggesting several ways to improve tutoring for a given problem: revising the problem statement, focusing on the required procedural skills, focusing on the mathematical structure, focusing on the student's affective response, or focusing on the student's mathematical response. Each of these generations is, of course, intimately connected to the others. As teachers explored one of them, their thinking often mutated in ways parallel to the students'. For example, revisions of the problem statement often led to discussions about skills and their importance; the idea of importance would sometimes lead to questions of how students responded to the problems *affectively*; and so on.

In the problem-design study (Lesh, Hoover, & Kelly, 1993), teachers began by collecting a wide variety of *stimuli* for context-setting for mathematical problems: state lotteries, stock reports, housing costs, political cartoons, recipes, even bungee-jumping. They also attempted to design into the tasks a wide number of *implicit demands* on students to generate models for addition, subtraction, fractions, graphing, or logical argument. Mutation was seen for these suggestions, for example, in scenarios about stock reports, which raised questions about students' prior knowledge; or problems involving graphs were queried as to which procedures were being assessed and their importance.

In the assessment-design study, teachers were able to suggest a variety of ways of evaluating a student's solution: length of answer, "density" of answer, presence or absence of numbers, accuracy of calculations, structure of the argument, or its effectiveness as a communication. Discussions of each of these generations again would lead to a proliferation of variants on themes: How important was length of answer if the calculations were inaccurate? How important was the accuracy of a calculation if it did not have supporting representations (e.g., graphs) that made its reasoning clearer?

The models in each of the above situations were presented as tentative, unstable, temporary, and "fuzzy." The expectation was that some might flourish and others perish. The driving force for mutation appears to be an attempt to address at each modeling cycle what are seen as the complexities of the task demand. As the solution models become more mature, complexities, which are first seen as independent and disjointed, are later subsumed into or seen as irrelevant to the solution of the problem.

2.4 Selection

Not all models survive all task/student/environmental demands. Several mechanisms appeared to be involved in selection: (a) *trial by consistency* that is, teachers and students asked themselves whether each new idea "made sense" based on their own current conceptions and experiences; (b) *trial by ordeal:* In the tutoring study, teachers' ideas for improvement were "field tested" in a tutoring session with a real student and the ideas either did or did not enhance learning. In the problem-design study, suggested problems measured up to a set of guiding principles or failed to do so (Lesh et al., 1993). In the assessment design study, proposed scoring rubrics measured up to a set of guiding principles or failed to do so; (c) *trial by jury*: In each of the three studies, each group of teachers compared and contrasted their suggestions against those in consensus-forming discussions. In these discussions, arguments were forwarded for which suggestions to keep, which to weed out, how they should be organized and prioritized, and how they might form a coherent conceptual system.

Overall, the driving force in selection appeared to be the result of the resolution of various cognitive conflicts: task-interpretation mismatches, interpretation-representation mismatches, environment-model mismatches, model-tool mismatches, specific-general mismatches, and procedure-logic mismatches.

2.5 Adaptation

As a positive consequence of mutation, demands from the environment (challenges to proposed models) cause problem solvers, collectively and individually, to modify or extend existing ideas in one or more of the following ways: reorganization, coordination, differentiation, and integration.

Reorganization. We wanted to provide experiences that would encourage teachers and students to switch to some completely new ways to *think about* their suggestions. Sometimes these reorganizations occurred when *cognitive conflicts* were presented. We noted some of the pertinent mismatches above. Other reorganizations occurred when "wild ideas" (e.g., metaphors and analogies drawn from brainstorming sessions) were considered that suggested a reinterpretation of a direction or approach.

Coordination, or the "building up" process. Over time, teachers and students gradually constructed more *flexible* and *stable* conceptual systems for interpreting their suggestions. Sometimes we encouraged alternating between situations in which attention was focused on the *constituent parts* of complex acts, other times on situations in which the focus was on the flexibility and coordination of the *systems-as-a-whole.*

To help teachers gradually coordinate and refine their tutoring systems-as-a-whole, we gradually increased the complexity of the contexts in which the learner was to perform – while preserving the basic structure of the task. For example, the complexity of tutoring sessions increased naturally as teachers gradually noticed new types of relevant factors ranging from mathematical issues, to psychological issues, to pedagogical issues; and, tutoring activities also became more complex as we introduced ways to use graphics (other computer-based tools) as parts of hints, feedbacks, or follow-up questions.

In the problem-design and assessment-design projects, we raised concerns such as how well a given problem statement would draw upon the students' experiences, or how well it documented students' work. Alternatively, we asked if a scoring rubric that appeared satisfactory for teachers was of equal value for parents or for the students themselves.

Differentiation: The "splitting" process. Conceptual systems do not simply get "built up" (or constructed) in a bottom-up manner; models also get "sorted out." Teachers and students discriminate among alternative models: those they have constructed and those that they have been given. The differentiation process sometimes means that students and teachers temporarily lose sight of the "large picture" when they pay attention to details of a model. Alternatively, when the focus is on a single model, they lose sight of others. We have found it to be a useful intervention with teachers and students to redirect their attention to larger issues or components of their models or alternative models that they are neglecting.

2.6 Accumulation

When models are developing, the problem solver does not start from scratch each time. The parts of the models that have served well in the past are retained and become part of a larger and more comprehensive solution. The accumulating model draws upon representations and notation systems that were helpful in the past. The emerging model helps guide the teacher in tutoring decisions, in designing tasks, and evaluating responses; it helps guide students in applying articulated models to new problem situations.

Arising out of the design of action-theoretic studies of emerging expertise among students and teachers, we have noted the following: In the tutoring study, teachers moved progressively away from rule-oriented mathematics and toward model-centered mathematics. Their interventions were reduced in number and changed in character. They became less concerned, for example, with procedural errors, and more interested in students' thinking: Were students using more than one representational system? Were the models they were constructing equal to the tasks that were set?

In the problem-design study, teachers moved away from designing problems that had a single correct answer (for which students' thinking and reasoning processes were not documented) toward ones in which the central goal of the task was to promote model building and model documentation.

In the response-evaluation study, teachers moved away from assigning holistic scores that covered many types of responses toward assessment procedures that considered the *conditions* of testing, student-related factors, task-related factors, and their curricular goals in mathematics. They then considered how to produce rich *descriptions* of the students' work. Finally, considering both the conditions and descriptions, they evaluated the students' work.

3. CONSEQUENCES OF CONSTRUCTIVISM FOR A RESEARCH METHODOLOGY

When researchers adopt a constructivist orientation toward thinking and learning, they must adapt their research methodology accordingly. Given the assumptions of constructivism, can researchers predict with confidence the state or level of construction of a concept that a student will reveal? If not, attempts to prescribe what constitutes an "expert" state, and what constitutes a "novice" state are open to question. As a corollary, pre- and posttests that reify these a priori codifications of expertise are also open to question. Further, research and teaching agendas whose goal is to bridge this hypothesized "gap" with *pre*scriptions may be misguided. Detailed observations of children's thinking make clear that students' thinking is often inadequately described by either the novice or expert prescriptions of researchers (e.g., Carpenter, Fennema, & Romberg, 1993; Maher, Davis, & Alston, 1991). Some children's thinking is haphazard, showing some "expert" characteristics and some "novice" characteristics. The thinking of other children frequently goes beyond the expectations of "expertise" that were assumed for them (Lesh, Post, & Behr, 1989).

Further, since children's thinking evolves in complex ways over protracted periods of time, the a priori timing of a prescribed instrument to be used as a "posttest" may be quite arbitrary and may or may not succeed in recording the changes in learning that it was designed to record. If children construct ideas complexly over a long period of time, then researchers must be willing to make continuous, rich, longitudinal observations of children.

Researchers must also focus on authentic tasks. Researchers in mathematics education should be primarily concerned about students' construction of real mathematics, not about drawing remote inferences about mathematical problem-solving based on scores from indices such as multiple-choice assessments of procedural knowledge. We should be concerned with mathematical problem-solving, not with surrogates of this process.

Finally, the constructivist approach suggests that researchers should pay particular attention to the environment in which children are learning. Some environmental factors (which include teachers and technology) will encourage and prompt children's thinking, others will constrain it. From this observation, we draw the lesson that the researchers themselves are part of the environment that is both studying children's thinking and eliciting it.

3.1 Children, Teachers, and Researchers

We wish to make the more general assertion that constructivism is not simply a statement about how *children* think; rather, it is a statement about the *nature* of thinking. A corollary of this premise is the principle that whatever characteristics we ascribe to children's thinking, we should be willing to ascribe both to teachers' thinking and to our own thinking as researchers. If we claim that children construct internal representations, or models, of the world, and that these models are incomplete, flawed, subject to revision, and evolve over a long period of time, then we must apply these principles equally to our study of the models of the world held by teachers and researchers.

We should not adopt uncritically the premise that mathematics teachers are "expert" at mathematics, teaching, or tutoring. As researchers, we equally should not entertain the conceit that we have an error-free metavision of the thinking of all our "subjects." Teachers and researchers construct models and often revise them. To use the language of Gadamer (1975, 1976), the *horizons* of children, teachers, and researchers are limited and contain what Heidegger would call many *blindnesses* (Heidegger, 1962). While each group has knowledge that the other does not yet possess, all of our models of the world are historical, incomplete, fractious, contain misconceptions and biases, and continually evolve.

3.2 Factors in a Research Methodology

Authentic performance: Tasks for students. We wish to elicit and develop the mathematical intuitions of students using authentic tasks. An authentic task for a student includes constructing mathematical models to gain leverage over general problems (the stage of model *construction*), explorations of the qualities of mathematical objects (the stage of model *exploration*), and application of mathematical models to new situations (the stage of model *application*). The type of authentic tasks that we have been developing and using are called model-eliciting problems. Model-eliciting problems are designed in accordance with the following principles (see Lesh et al., 1993):

1. *The Model Construction Principle*: Does the task create the need for a model to be constructed, or modified, or extended, or refined? Does the task involve constructing, explaining, manipulating, predicting, or controlling a structurally significant system? Is attention focused on underlying patterns and regularities rather than on surface-level characteristics?

2. *The Simple Prototype Principle*: Is the situation as simple as possible, while still creating the need for a significant model? Will the solution provide a useful prototype (or metaphor) for interpreting a variety of other structurally similar situations?

3. *The Model Documentation Principle*: Will the response require students to explicitly reveal how they are thinking about the situation (givens, goals, possible solution paths)? What kind of system (mathematical objects, relations, operations, patterns, regularities) are they thinking about?

4. *The Self-Evaluation Principle*: Are the criteria clear for assessing the usefulness of alternative responses? Will students be able to judge for themselves when their responses are good enough? For what purposes are the results needed? By whom? When?

5. *The Model Generalization Principle*: Does the model that is constructed apply to only a particular situation, or can it be applied to a broader range of situations?

6. *The Reality Principle*: Is the scenario of the problem contrived so that it would contradict students' knowledge of the scenario in a "real-life" situation? Will students be encouraged to make sense of the situation based on extensions of their own personal knowledge and experiences?

Model-eliciting tasks allow and encourage students to display and document their mathematical problem-solving.

Authentic performance: Tasks for teachers. We wish to study teachers in authentic tasks that include: (a) *writing* high quality instruction/assessment activities aimed at authentic performance in mathematics; (b) *evaluating* the scope and depth of clusters of authentic performance activities (to generate instruction-relevant interpretations of the results that students produce); (c) *making insightful observations* of students' behaviors during videotapes of the preceding activities; (d) *interpreting* and evaluating students' responses when multiple types and levels of correctness are possible; and (e) *tutoring* students using concrete materials and nondirective questioning to mold and shape their sense-making strategies during authentic performance activities.

3.3 Teaching Experiments

The teaching experiments that we emphasize can be characterized as longitudinal development studies in conceptually rich environments (Lesh, 1983). They focus on "real-life" classroom experiences, such as those that are involved when: (a) teachers use concrete materials to interview students to identify specific conceptual strengths and weaknesses; (b) teachers observe groups of students as they work on complex project-based activities; (c) teachers assess strengths and weaknesses of results that students produce during realistically complex "authentic" performance activities; or (d) teachers lead discussions and plan other instructional activities that focus on deeper and higher-order understandings of elementary mathematical ideas and processes. Thus, classroom activities for students provide contexts for encouraging the development of teachers' knowledge, beliefs, and abilities.

Throughout the experiments, activities for instruction and assessment are integrated: (a) Every assessment activity is designed to be a valuable learning experience for the individuals involved. Therefore, neither students nor teachers are required to take time away from instruction-focused activities to document key achievements and improving abilities; (b) Many activities used for instruction enable the learner automatically to produce documentation to verify the kind of learning that has occurred; (c) Assessment activities are aimed at generating information to facilitate wise instructional decision-making by teachers, students, and others whose primary aims are to encourage student development; (d) Assessment activities for *students* form the basis for instructional activities for *teachers*. Examples of such activities include: adapting existing project-based activities to focus on the "real-life" experiences of targeted groups of students (e.g., young women, specific minority populations); making insightful observations during the preceding classroom activities; or conducting follow-up interviews to diagnose the strengths and weaknesses of individual students. All of the assessment activities focus on "authentic performance" in the sense described in Lesh and Lamon (1992).

The teaching experiments provide us with:

1. *Insight about the nature of teachers' knowledge*: The teaching experiments enable us simultaneously to facilitate and investigate the development of teachers' knowledge about: (a) the nature of mathematics, (b) the nature of realistic problem-solving and decision-making situations in which mathematics is useful, and (c) the nature of "talents" that contribute to the preceding kinds of understandings and capabilities.

2. *Knowledge and prototypes to clarify the nature of high-quality authentic performance activities for students*: Participating teachers develop startup kits of field-tested authentic performance activities for students – while helping us to clarify key characteristics of "excellence" in such activities.

3. *Knowledge and prototypes to clarify the nature of high-quality authentic performance activities for teachers*: The materials we develop for our teaching experiments provide prototypes for some of the most important components of a field-tested, on-the-job, teacher education program in which teachers can simultaneously develop and document their mathematical/psychological/instructional knowledge and abilities.

3.4 Documentation of Change

In our teaching experiments, we wish to move away from state-focused documentation and toward progress-focused documentation of learning.

State-focused documentation aims mainly at: (a) evaluating (or assigning values to) states of development, (b) identifying deficiencies with respect to some (preconceived and static) standard, and (c) inferring that progress has been made by comparing one evaluation with another (using a subtraction-based model to describe differences).

Progress-focused documentation aims at monitoring progress in a direct manner by focusing on activities in which it is possible simultaneously to encourage and document development in directions that are increasingly "better" (without using pretests and posttests, which embody a fixed and final definition of "best"). In progress-focused documentation, relevant activities tend to contribute to both learning and the documentation of learning. Therefore, distinctions between instruction and assessment tend to blur and the roles of teachers become particularly important. Further, the quality of teachers' contributions tends to be strongly influenced by their own understandings about the nature of mathematics, mathematics learning, and mathematics problem-solving. Finally, the role of support and administrative personnel (e.g., parents and supervisors) is neither ignored nor diminished.

In practice, teaching experiments for students provide an ideal context for a teaching experiment for teachers. Teachers study the development of mathematical thinking of students; the development of mathematical thinking and instructional models of the teachers is studied, in turn, by our project staff.

3.5 The Role of the Researcher

The constructivist paradigm demands that we recognize that the models that we use to make sense of the performances of both teachers and students are themselves subject to revision and restructuring. For that reason, we find it difficult to maintain the belief that we can act as disinterested, objective observers of our "subjects." Our "subjects" are actual teachers and actual students engaged in meaningful learning. In a very real sense, we learn from them; they are co-collaborators in the search for knowledge about how to improve instruction, learning, and assessment. Our "data" are our models, our understandings, of the models that the teachers and students are using to make sense of the tasks that we have set them. Because we do not claim omnipotence about what constitutes "true" learning, instruction, and assessment, we remain open *during the teaching experiments* (and afterwards) to corrections to and revisions of our own models of these concepts.

4. SUMMARY

In summary, constructivist perspectives on learning have radical consequences for how we define knowledge for children *and* teachers *and* researchers. Constructivism demands authenticity in instruction and assessment and nonabsolutism in the design and interpretation of teaching experiments. The knowledge gained from teaching experiments remains historical, situated, and open to revision since it involves the application of the researchers' best current theoretical models to help understand the cognitive models of teachers who, in turn, are growing in their understanding of the cognitive models of their students.

REFERENCES

Carpenter, T., Fennema, E., & Romberg, T. (1993). *Rational numbers: An integration of research.* Hillsdale, NJ: Erlbaum.

Gadamer, H-G. (1975). *Truth and method.* (G. Barden & J. Cumming, Trans. and Eds.). New York: Seabury Press.

Gadamer, H-G. (1976). *Philosophical hermeneutics.* (D. E. Linge, Trans.). Berkeley, CA: University of California Press.

Heidegger, M. (1962). *Being and time.* (J. Macquarrie & E. Robinson, Trans.). New York: Harper and Row.

Lesh, R. (1983). Conceptual analyses of problem solving performance. In E. Silver (Ed.), *Teaching and learning mathematical problem solving* (pp. 309-329). Hillsdale, NJ: Erlbaum.

Lesh, R., Hoover, M., & Kelly, A. E. (1993). Equity, assessment, and thinking mathematically: Principles for the design of model-eliciting activities. In I. Wirszup & R. Streit (Eds.), *Developments in school mathematics education around the world* (Vol. 3). Reston, VA: National Council of Teachers of Mathematics.

Lesh, R., & Kelly, A. E. (1991). *Human simulation of computer tutors: Lessons learned in a ten-week study of twenty human mathematics tutors.* Paper presented at the International Psychology of Mathematics Education (PME) Conference, Assisi, Italy.

Lesh, R., & Lamon, S. (1992). *Assessing authentic performance in school mathematics.* Washington, DC: AAAS.

Lesh, R., Post, T., & Behr, M. (1989) Proportional reasoning. In M. Behr & J. Hiebert (Eds.), *Number concepts and operations in the middle grades.* Reston, VA: National Council of Teachers of Mathematics.

Maher, C. A., Davis. R. B., & Alston, A. (1991). Brian's representation and development of mathematical knowledge: A four-year study. *The Journal of Mathematical Behavior, 10*(2), 163-210.

CHAPTER 6

DIFFERENTIAL DIDACTICS

edited and introduced
by
Roland W. Scholz
Bielefeld / Zürich

Apparently, mathematics is learned by different populations that, at least phenomenologically, show variations in mathematical performance and in access to the acquisition of mathematical knowledge. The problem regarding how this variation in performance and perhaps ability may be explained is approached from different points of view. Two extreme positions may be identified with respect to the impact of *gender, socioeconomic status, social and ethnic minorities, culture, and personality* on the learning of mathematics.

One position assumes that there are no systematic fundamental differences between different groups, such as the genders, with respect to the learning of mathematics. According to this view, in some respects, everybody equals anybody like nobody is equal to someone else.

The other position postulates that there are systematic differences in the structure and dynamics of gender, socioeconomic status, race, culture, and personality with respect to learning mathematics. With reference to the subdiscipline of differential psychology (Anastasi, 1954), the branch of re-

R. Biehler, R. W. Scholz, R. Sträßer, B. Winkelmann (Eds.),
Didactics of Mathematics as a Scientific Discipline, 287-290.

search on mathematics education that systematically investigates differences with respect to these variables can be designated as *differential didactics.*

Clearly, the question as to which of the two positions is adequate is not just theoretical but must be answered by empirical research. Today, both positions can be found among researchers concerned with the possible impact of differential variables.

Any research that runs under the heading "differential didactics" should include:

1. a diagnosis encompassing a category or a variable (i.e., an independent variable) and a hypothetical or an empirically accessible outcome (i.e., a dependent variable) in order to describe what is different;
2. an explanation or model for the genesis of differences between groups;
3. a description of didactical efforts and their impact on the "outcomes" of mathematics education.

With respect to knowledge of the state of the art presented in the contributions of this chapter, some general conclusions for investigations of differential didactics can be suggested. For instance:

1. Standard tests are no suitable means to provide insight into the genesis of group differences in mathematical thinking and in learning mathematics.
2. Mathematical learning and thinking is a complex affair shaped by social experience, "personal life formation," and adaptation to cultural constraints. As a result, it is unusual to find a single variable or a monocausal explanation for these differences.
3. Presumably, many differences in mathematical performance are due to different internal representations and qualities of information processing (see chapter 5 on *psychology of mathematical thinking*).
4. Didactical efforts used to optimize mathematics education for different populations should vary in quality and not in quantity.

In most industrialized countries, about 5% of the population are currently considered to be mathematically (mentally) disabled in that they are referred to special schools where they receive a mathematics instruction that differs from that of the rest of their age cohort. Conversely, a small percentage of students has excellent access to mathematics.

The question whether highly gifted students or different genders should be promoted in special programs of mathematics education belongs to a set of "circularily reappearing" questions that are often dealt with controversially. H. G. Steiner (1986, pp. 280-282) has pointed out two issues that are important for a scientific understanding of debates on these topics: First, these topics define political paradigms (Dubiel, 1985): Any position and argumentation is defined by conceptions of humanity, cultural values, or social-philosopical theories. Second, didactics of mathematics as a scientific discipline has to rationalize these controversies by revealing the foundations of the different positions in the scientific knowledge and facts already attained.

Before I turn to the different papers, I want to point out a potential research mistake, which I shall call "hidden type I error," that may be inherent in any study on differential didactics.

The potential research mistake has been discussed nicely with respect to sex differences in laterality by Springer and Deutsch (1981). For instance, there are apparently more surveys reporting that male rats show significantly thicker right hemispheres than female rats. Does this mean that this finding is generally true? It is usually not known how many studies were run on an issue. Furthermore, there is a lack of information on which studies got published and which have remained unpublished. Thus there may be a hidden type I error due to the scientific community's convention that only significant results and/or findings that are related to hypotheses are published.

Does mathematics learning show a similar bias to that suspected in laterality? Clearly we have to admit that we cannot answer this question and hence we cannot exclude this possibility.

Jens Holger Lorenz, in his paper on *mathematically retarded and gifted students,* acknowledges that groups of individuals differ qualitatively in their mathematical thinking. He discusses various disciplinary approaches for explaining differences in arithmetic skills and in the acquisition of fundamental mathematical concepts both for mathematically retarded and for highly gifted students.

He reveals that, from certain perspectives (like psychodiagnostics or neuropsychology, although they are often used for assessing differences), no methodological-didactical measures can be derived, whereas, from other perspectives (e.g., cognitive psychology), one may provide some access for an understanding of both shortcomings and giftedness in the acquisition of mathematics. When pointing at the qualitative differences in information processing among groups of highly gifted students, he concludes that, because *of their individual styles of learning, mathematically highly gifted students – like retarded students – require a teaching method of their own.*

Whether or not a differential didactics exists or should be applied for males and females is a difficult question.

In her contribution, *Should girls and boys be taught differently? Gila Hanna* critically examines different bodies of research concerned with gender differences. As an expert on measurement and evaluation of studies in education, she elaborates that (given the published findings) there is no clear evidence for a general superiority of male mathematical achievement. She argues that potential structural differences in the gender's approaches are derived more from pronounced assertions than from from solid empirical evidence. Furthermore, when considering the last centuries, gender differences in mathematics achievement (see also Robitaille and Nicol, this volume) seem to be diminishing. If at all, boys seem to outperform girls in the field of problem-solving. Thus Hanna doubts whether differential didactics would be in the interest of women, as a presumed short-term advantage – in the long run – might reverse progress in levelling out female and male mathematics achievement.

Note that not only a hidden type I error may be found but also a hidden differential didactics that handicaps female students. First, research on textbooks has shown that examples of mathematization mostly come from the boys' range of experience. Second (as Hanna reports), boys receive more attention in the classroom than girls. Thus we should not aim to introduce but rather to eliminate a differential didactics for gender.

If one wants to differentiate mathematics instruction for different populations, one should know the communalities. Nearly every national school system provides different treatments, that is, different intensities and perhaps qualities of mathematical instruction by means of streaming, by different levels of the school system, or by different curricula for certain students. Clearly, determining difference often starts from common features. Thus, the topic of differential didactics leads directly to the question on which mathematics may and should be taught to all. Starting from a historical analysis on the spread of mathematics, *Zalman Usiskin's* paper on *from "mathematics for some" to "mathematics for all"* deals with a didactical analysis of mathematics to be taught to all in order to meet the needs of the individual and of society. As he elaborates, the accessibility of mathematics has not only increased dramatically, but also, because so much (cognitive) work has been taken over by the computer, a fundamental change of the core of mathematics education has to be expected.

There are various primary links to other chapters of this volume. As many misconceptions of mathematical problems underlie how low achievers cope with mathematical problems, both Fischbein's paper on *formal, algorithmic, and intuitive components in a mathematical activity* as well as Lompscher's contribution on the teaching experiments of the *sociohistorical school* provide insights into retarded students' mathematical thinking. Thus the contributions of chapter 5 on *psychology on mathematical thinking* are most close to this chapter. However, the paper of *David Robitaille and Cynthia Nicol* on *Comparative international research in mathematics education* or *Ubiratan D'Ambrosio's cultural framing of teaching and learning mathematics,* both from chapter 8, also deal with issues of differential didactics.

REFERENCES

Anastasi, A. (1954). *Contributions to differential psychology.* New York: Macmillan.

Dubiel, H. (1985). Was ist Neokonservatismus? Frankfurt: Suhrkamp.

Springer, S. P., & Deutsch, G. (1981). *Left brain, right brain.* New York: W.H. Freeman

Steiner, H. G. (1986) Sonderpädagogik für testsondierte "mathematisch hochbegabte" Schüler oder offene Angebote zur integrativ-differenzierenden Förderung mathematischer Bildung?" In *Beiträge zum Mathematikunterricht 1986* (pp.280-284). Bald-Salzdethfurth: Franzbecker.

MATHEMATICALLY RETARDED AND GIFTED STUDENTS

Jens Holger Lorenz
Bielefeld

1. MATHEMATICALLY RETARDED STUDENTS

1.1 The Problem

Research in the field of learning disabilities in arithmetic skills has not yet reached the stage that dyslexia research has achieved during the last 25 years. This is all the more surprising, as studies strictly basing their diagnosis of dyscalculia on a developmental lag of 2 years as compared to performance in other school subjects have shown that about 6% of students must be evaluated as showing extremely poor performance in arithmetic (Kosc, 1974), and that at least 15% must be considered to have such trouble with calculation that they need help (Lorenz, 1982). On the whole, it is stressed that dyscalculia occurs much more often in elementary school than problems with reading and orthography (Klauer, 1992). One of the reasons for this deficit is that attempts at explaining dyscalculia were made from very diverging fields of science and diversifying research approaches. Research was further impeded by the fact that older research approaches used definitions of dyscalculia that were oriented toward discrepancy models. While a case of dyscalculia may be assumed if an arithmetical substandard performance is present (a) in students showing at least average intelligence, or (b) partial underachievement at each level of intelligence, there has been so little proof in the past that this definition is feasible that it has now been rejected for dyslexia as well (Grissemann & Weber, 1982). Moreover, it seems to make little sense for pedagogical reasons to dismiss all students who do not fall under this definition of discrepancy, but are nevertheless in need of individual help in the field of arithmetics.

1.2 Research and Explanatory Approaches

Psychodiagnostics. Psychodiagnostics, which is oriented toward test methodology, emerged from the problem of selection, that is, the need to identify appropriate versus less appropriate candidates for a specific demand. It created the construct of "intelligence," which seemed to justify selection and assignation to certain school types. Although different cognitive abilities are considered to determine intelligence within the various intelligence models,

R. Biehler, R. W. Scholz, R. Sträßer, B. Winkelmann (Eds.),
Didactics of Mathematics as a Scientific Discipline, 291-301.

almost all of them have in common that components like spatial intuition, short-term and long-term memory, language factors, and calculatory competence must be included. However, in testing methodology, this creates the problem that the requirement of having factors of intelligence correlate only in a small way eliminates the very components that are responsible for calculatory competence, for the latter must prove to be independent of the other factors of intelligence. This impeded a clarification of mathematical ability by methodology alone. Thus, Ginsburg's assessment of this approach is devastating:

> By contrast, standard tests are of little value .... They yield unhelpful labels like "low mathematical aptitude," and worst of all, they fail to reveal children's strengths. The tests say nothing specific about what the child can do and about how instruction should proceed. All this is positively harmful to the child who has trouble learning. (Ginsburg, 1977, p. 149)

Due to the simplistic idea about the causes of dyscalculia, it was impossible to establish elaborate didactical-methodological approaches to the problem. Curricular aids were derived from test items in the vein of associationism, and appropriate exercises to improve arithmetical competence were developed. This kind of task analysis thus mostly led to a simple drill-and-practice unit that subdivided the subject matter to be learned into small steps. Resnick characterizes the behaviorist methods developed from the psychodiagnostic approach as follows:

> [Skinner] and his associates showed that "errorless learning" was possible through shaping of behavior by small successive approximations. This led naturally to an interest in a technology of teaching by organizing practice into carefully arranged sequences through which the individual gradually acquires the elements of a new and complex performance without making wrong responses en route. This was translated for school use into "programmed instruction" – a form of instruction characterized by very small steps, heavy prompting, and careful sequencing so that children could be led step by step toward the ability to perform the specific behavioral objectives. (Resnick, 1983, pp. 7-8)

Special education. Special education tackles some factors of ability relevant for the learning of young children and elementary school students in an isolated way in specific assistance programs. The genesis of problems with calculating is mainly seen in factors like (a) disturbance of the body schema; (b) visuomotor integration disturbances; and (c) spatiovisual weakness of grasp and representation (Johnson & Myklebust, 1971). This approach, however, remains attached to a defectological view and is insufficiently specific with regard to subject matter and content to be able to derive detailed statements about how arithmetical contents are learned.

Success is expected here of orthopedagogical exercise treatments concentrating on symptom clusters, which are made responsible as a basis for disturbing learning or processing and integrating information. Besides these measures, which, while focusing on the particular ability deficits, are not specific with regard to content, special education tries to respond organizationally to the problem of impeded achievement. Students are integrated into special classes or schools, absolving a slowed-down version of the elementary school curriculum. Special methodological approaches have not been developed for arithmetic, the only basis being a "weeding-out" hypothesis, that is, the subject matter curriculum is lightened and disburdened of "superfluous" contents. Underhill, one of the major representatives of American special education, considers only memory a factor relevant for calculus:

> Children with Low Mathematics Ability
>
> Some children master mathematics concepts more slowly than do the majority. They have difficulties conceptualizing mathematical ideas, understanding mathematical generalizations, and remembering computational procedures and basic facts when such ideas are presented in conventional ways. There are many reasons why children experience difficulties . . . suffice it here to say that learning styles, socioeconomic status, heredity, and teacher skill are but a few of the factors which are positively correlated with pupil sucess in mathematics.
>
> Characteristics: Children with low mathematics ability are usually characterized by low-normal intelligence, poor reading achievement, and poor memory. They are frequently from homes of lower-class culture and may have parents whose backgrounds reflect poor mathematics achievement and attitudes.
>
> A. Since intellectual development is correlated with mental age, expect slow children to perform at a less abstract level than bright and average children.
>
> B. Since low-ability students progress more slowly, allow them more time at the concrete and semiconcrete levels of concept development.
>
> C. Use a multi-sensory approach to accomodate various learning styles.
>
> D. Prevent practising of errors through careful developmental work and short sets of practice exercises.
>
> E. Introduce only one new development and more time for practice. (Underhill, 1977, pp. 39-40)

Neuropsychology. Early studies of adults with acalculia were carried out by neurologists and psychiatrists who hoped to obtain insights into the functioning of the brain. It was suggestive to assume, in line with other basic cognitive abilities (e.g., speech, motorics), a brain zone for calculating as well, a so-called "computational center" (Benton, 1987). Within mathematics education research as well, there is a branch regarding cerebral architecture as determining calculatory performance (Tarnopol & Tarnopol, 1979; Teyler, 1984). The final failure of the search for a computational center in the brain by examining brain injuries is ascribed to the high cognitive demand of calculating. This comprises linguistically coded forms of expression, perceptions, and representations. In all these fields, calculating performance seems to be impairable by cerebral lesions, a fact that causes diffuse injuries to lead to impaired calculating performance as well. Nevertheless, these studies established that an impairment of calculating performance was, in almost all cases, accompanied by defects in the optical area, a fact which caused the studies to emphasize visualization or visual mental representations within the calculating processes (Hartje, 1987). The construct of minimal cerebral dysfunction (MCD) brought some understanding for the subfactors underlying calculus development. Impairment in the following areas has proved detrimental to learning numbers and arithmetical operations: (a) in the tactile-kinesthetic area; (b) in auditory perception, storage, and seriality; (c) in visual perception, storage, and seriality; and (d) in intermodality.

Within this approach, disturbances in the development of very young children, in particular, in the tactile-kinesthetic area, with their ensuing impairment of the body schema and the disturbance of spatial orientation are made responsible for a reduced performance in calculus.

With this reduced view, however, which locates the causal factors for dyscalculia in the student's brain, no methodological-didactical measures can be derived. This is why it has proved impossible as yet to develop specific special methods for these children.

Curricular error analysis. The analysis of student errors in arithmetical operations shows that there seem to be certain universals in the child's reasoning that cause special error types in case of curricular demands. These errors have a fine structure (e.g., typical errors in written subtraction and in multiplication with 0 and 1; Cox, 1975), which remains constant even over a period of 50 years, independent of the instructional method used. On the one hand, certain error patterns and error configurations, which are typical for student populations, may serve to derive didactical measures for the entire class aimed at prevention or treatment. On the other hand, it can be shown that these error patterns are typical for the individual as well (Ginsburg, 1983). It has also been shown, however, that different (erroneous) solving strategies may lead to the same error pattern, and, hence, nothing definite can be said about the specific reasoning processes involved in solving this arithmetical problem. This can be achieved only by studying individual cases in the form of clinical interviews; the thinking aloud method not only revealing misconceptions of relevant concepts and misleading strategies of algorithms, but also clarifying more general (erroneous) strategies of receiving and processing information (Radatz, 1980).

Within the error analysis approach, treatment addresses the hierarchical sequencing of learning and analyzes the level of the structure of mental representation of arithmetical operations (Niegemann, 1988). This is based on the assumption in developmental psychology that learning consists of changing individual activity levels, and that the direction these changes will take can be predicted by a curricular description as well. In this sense, dyscalculia is conceived of differently to other approaches, as dyscalculia students are characterized, compared to their peers, by a form of work situated on a lower level of mental representation, the didactical measures extending to raising that level within the same problem.

Beyond that, error-sensitive diagnostic procedures can be developed that localize the cause of the error during the process of solving.

Cognitive psychology. Models of cognitive psychology arise, for the field of arithmetics, from merging the Piagetian tradition with approaches from psychological theories of information processing. As a rule, a student's erroneous solution is compared to a theoretical model of the adult expert. Unsuccessful attempts at solution are explained by misguided steps within the solving sequence. For correct expert solutions, flow diagrams can be established, which must also be absolved by the student. The cognitive-psychological effort lies in differentiating between the individual steps constituting the process of solving (Allardice & Ginsburg, 1983; Resnick & Neches, 1984). Here, the models developed for problem classes permit a more precise description of the errors the student has made, which, in contrast to curricular error analysis, is not focused on a quasi-objective hierarchy of difficulty, but rather on the cognitive steps that have to be absolved by the problem-solving subject.

From the perspective of cognitive psychology, dyscalculia is conceived of (a) as a quantitative problem, erroneous algorithms ("bugs" in the major programs guiding the individual steps) occurring in a multitude of content areas, thus leading to a massing of errors; and (b) as a qualitative defect in the sense of disturbance in an essential cognitive unit. In line with the computer metaphor, this is localized in memory, in the central processor, and so forth.

The treatment measures derived from that then consist of encouraging optimal solving strategies, in learning by insight, and in training metacognitions. This is significant, insofar as it does not favor, in contrast to other approaches, a drill-and-practice method, but rather the interplay between automatized procedural knowledge and processes of understanding. This is also evident from the students' attempts to correct errors by repair (Brown & van Lehn, 1980). Besides illustrating idiosyncratic reasoning processes in dyscalculia students, this also shows their desire to carry out algorithms in a syntactically perfect way while ignoring the semantic level.

> Inspection of the bugs . . . shows that they tend to "look right" and to obey most of the important syntactic rules for written calculation: the digit structure is respected, there is only a single digit per column, all the columns are filled, and so forth. In the sense of being an orderly and reasonable response to a problem situation, the buggy algorithms look quite sensible. But each of the bugs violates fundamental mathematical constraints. In this sense, they violate the conceptual meaning, or semantics . . . . (Resnick, 1984, p. 3)

For this reason, simply exercising algorithms does not seem sufficient for dyscalculia children, but rather the accompanying methods of recognizing and deciding must be learnt at the same time and thus thematized in the classroom.

1.3 The Cognitive Demands of Teaching

For the analysis of dyscalculia, it is increasingly evident that a multitude of factors concerning the subject matter to be learned, the social structure of the class and of the environment, and the student's personality may impede or prevent the processes of learning. For this reason, analysis of the individual case is mandatory for diagnostics. For this purpose, it has proved favorable to study the cognitive demand in teaching elementary arithmetics. Independent of the textbook used and the methodological approach, four stages can be distinguished in introducing and affirming new concepts:

1. The operation is built up by activities involving concrete material while respecting the quantitative structure. This requires the students to look back at the activity executed by remembering or visually anticipating the steps that still have to be carried out.

2. As compared to concrete activity, the focus here is on the iconic representation of the operation on worksheets and in textbooks. The iconic representation is accompanied by digital work. Here again, a visual representation of the operation or of the activity represented iconically in two dimensions is necessary. The static representation requires the student to represent the activity cognitively as a temporal-spatial one.

3. There is transition to a logical-unintuitive activity in the digit area, increasingly giving up visual meaning. Nevertheless, the student is asked to visually imagine the operations by means of intuitive correlates of activity und to have an auditory memory in learning. Disturbances occur in case of operative concretism and in case of impediments of long-term auditory memory.

4. Automatizations in the sign area are aimed for as a last item (number space up to 20, number facts like 1 x 1). This requires the student to have an associative memory, which is impeded in case of disturbances of short-term memory.

5. Situated across the previous four stages are the word problems. They require the students to have acquired not only reading performance but also the ability to transform verbal statements into iconic representations. Besides, the student is required to use everyday experience and knowledge about the world. Impediments occur in this stage through disturbances of verbal understanding, but also through reduced ability to generate visual images and to operate with them.

The impediments in learning arithmetical subject matter are caused by the area-specific effects of individual cognitive weaknesses. Diagnosis thus requires a holistic method in which the curricular error patterns are related to underlying basic cognitive demand. The treatment of dyscalculia accordingly runs on two tracks: The abilities (differentiated verbal understanding, memory, intuition, spatial orientation), which are necessary for learning mathematics, must be developed by suitable training methods; at the same time, working on the subject matter and tackling erroneous strategies in order to prevent the occurrence of long-term gaps and knowledge deficits with negative emotional effects.

2. GIFTED STUDENTS

The problem of mathematically highly gifted children has two parts: identifying extreme mathematical talent and finding appropriate support for these children. It has proved to be rather difficult to identify mathematically highly gifted children. To be able to solve mathematically demanding problems requires a rich knowledge about numbers and number relationships, which is normally not available to elementary school students. For this reason, a (probable) extreme gift can be predicted only by means of general personality factors in this age group. Higly gifted children become noticeable as preschoolers by learning to read very early, asking questions about complicated facts, developing curiosity for complex situations, having an excellent memory, and easily being able to generalize to new situations and problem formulations. They are wide awake, and their problem solutions are characterized by originality and creativity (Bhattacharya, 1982; Heller & Feldhusen, 1986). While the future highly gifted have high intelligence (Jellen & Verduin, 1986), the IQ span is larger among the highly gifted than it is between students with a learning disability and highly gifted ones (Snider, 1986). For this reason, simply establishing IQ is only a limited predictor of high gifts. This needs to be differentiated as to areas.

Identification via aptitude tests is made difficult by the fact that standardized tests for 1st graders (e.g., the frequently used SAT) differentiate insufficiently between mathematically good students and extremely gifted ones. The development of diagnostic methods for the second half of elementary school must at present be considered skeptically (Wilmot, 1983). In a way similar to that for mathematically creative adults (Michael, 1977), some characteristics of mathematically highly gifted children can be given, however. Already at the age of 7 or 8, they "mathematize" their environment, giving particular attention to the mathematical aspects of the phenomena they perceive. They realize spatial and quantitative relationships and functional dependencies in a variety of situations, that is, they see the world "by mathematical eyes" (Krutetskii, 1976, p. 302). Even in the first grades, it is observed that these children never tire to do mathematics and have an excellent memory for mathematical materials, relationships, proofs, and methods of solution.

Among the highly gifted children, three groups can be identified: the analytical type, the geometric type, and the harmonious type.

> Analytic thinkers possess a mathematically abstract cast of mind. In their thinking, a well-developed verbal-logical component predominates over a weak visual-pictorial one. They function easily with abstract patterns and show no need for visual supports when considering mathematical relationships. They will, in fact, employ complicated analytical methods to attack problems, even when vi-

sual approaches would yield much simpler solutions. They prefer abstract situations and will attempt to translate concrete problems into abstract terms whenever possible. They may have weakly developed spatial visualization abilities, especially for three-dimensional relationships. In school they are more likely to excel in arithmetic and algebra than in geometry.

Geometric thinkers exhibit a mathematically pictorial cast of mind. Their thinking is driven by a well-developed visual component that impels them to interpret visually expressions of abstract mathematical relationships, sometimes in very ingenious ways. Although their verbal-logical abilities may be quite well developed, they persist in trying to operate with visual schemes, even when a problem is readily solved by analytic means and the use of visual images is superfluous or difficult. Indeed, these students frequently find that functional relationships and analytical formulas become understandable and convincing only when given a visual interpretation.

Harmonic thinkers exhibit a relative equilibrium between the extremes of the other two types. They possess both well-developed verbal-logical and well-developed visual-pictorial abilities, and when given a problem, they are usually capable of producing solutions of both kinds. Krutetskii (1976) observed two subtypes among harmonic thinkers: those with an inclination for mental operations without the use of visual means, and those with an inclination for mental operations with the use of visual means. In other words, although harmonic thinkers are perfectly capable or representing relationships pictorially, some prefer to do so while others see no need for it.

In summary, we can identify from Krutetskii's work the following significant traits of the mathematically gifted (1976, pp. 350-351):

1. Formalized perception of mathematical material and grasp of the formal structure of problems.
2. Logical thought about quantitative and spatial relationships and the ability to think in mathematical symbols.
3. Rapid and broad generalization of mathematical objects, relations, and operations.
4. Curtailment of mathematical reasoning and the ability to think in curtailed structures.
5. Flexibility of mental processes.
6. Striving for clarity, simplicity, economy, and rationality of solutions.
7. Rapid and free reconstruction of a mental process as well as reversibility of mathematical reasoning.
8. Generalized memory for mathematical relationships, characteristics, arguments, proofs, methods of solution, and principles of problem-solving.
9. A mathematical cast of mind.
10. Energy and persistence in solving problems. (House, 1987, pp. 15-16)

For teaching mathematically highly gifted children, problems take two directions: social integration and emotional status, and their adequate promotion by teaching or by organizational measures. Their social integration into the class is often made difficult by frequent personality factors of highly gifted children. They tend to be introverted and are unable to understand, because of their quickness of mind, why other students are so slow, or are not understood themselves. Because of their idiosyncratic style of learning, they prefer learning independently of the others, like discovering in games and open problem situations, and submit at best to peer teaching (Brown, 1991).

The pedagogical concepts to promote highly gifted students consist essentially of two parts: (a) Differentiated curricula are developed (Stanley, 1977, 1979), and (b) they absolve basic curricular units in acceleration programs. Because of their individual styles of learning, mathematically highly gifted students require a teaching method of their own. They prefer a mixture of problem approach, discovery approach, and polytechnical approach, which enables them to mathematize different areas of knowledge like social studies, natural sciences, and so forth (Clendening & Davis, 1983).

Despite this enrichment, mathematically highly gifted students marshal the subject matter in school, college, and university considerably faster than their peers. For this reason, they overleap, in the school subject of mathematics or in other subjects, the subject matter by one or several grades, as far as this is possible for reasons of school organization. The results obtained with acceleration models in the past have been remarkable (Barkovich & George, 1980; Benbow, 1991). In particular, the Study of Mathematically Precocious Youth (SMPY) at Johns Hopkins University has followed, among the more than 10,000 young people identified as mathematically highly gifted and taken into the program, the development of more than 3,000 until adult age in a long-term study, confirming these results. In contrast to fears frequently stated in the Federal Repuplic of Germany concerning negative social and emotional effects in children who overleap classes and are thus transferred to a referential group inadequate for them, the American acceleration programs proved to be favorable in emotional aspects as well.

To conclude, it must be stated that mathematically highly gifted students profit most from teaching programs that stress higher reasoning strategies and general heuristics. These must not necessarily refer to mathematics, as these students are able to acquire the subject matter in independent learning rather rapidly.

3. RESEARCH METHODOLOGY

The methods and instruments for studying retarded and gifted students share a focus upon the individual and his or her specific thought processes. Thus the clinical case study method is used for both groups. Besides worksheets and erroneous (or highly creative) problem solutions, students are asked to verbalize their thoughts while working on a task ("thinking aloud method"). This research method may reveal "regularities of behavior – especially regularities that can be related to theories about how internal information processing proceeds" (Resnick & Ford, 1981). Methodological problems can arise when students are (partly) incapable of verbalizing their thought processes. Retarded students may lack the necessary verbal abilities, whereas gifted students' thoughts seem to be so fast and enriched with diverging associations that verbalization disturbs the problem solution. Thus a "post-thinking-aloud procedure" is often applied by interviewing students about the nature of their thoughts after they have completed their solving process.

For retarded and gifted students, research has a strong theoretical orientation. Possible shortcomings of the interview technique (i.e., conducting the interview in a specific way that leads the student to answer in accordance with a certain theoretical model) must be controlled by accepting only those hypotheses and interpretations that are shared by several observers.

REFERENCES

Allardice, B. S., & Ginsburg, H. P. (1983). Children's psychological difficulties in mathematics. In H. P. Ginsburg (Ed.), *The development of mathematical thinking* (pp. 319-350). New York: Academic Press.

Bartkovich, K. G., & George, W. C. (1980). *Teaching the gifted and talented in the mathematics classroom.* Washington, DC: National Education Association.

Benbow, C. P. (1991). Mathematically talented children: Can acceleration meet their educational needs? In N. Colangelo & G. A. Davis (Eds.), *Handbook of gifted education* (pp. 154-165). Boston: Allyn & Bacon.

Benton, A. L. (1987). Mathematical disabilities and the Gerstmann syndrome. In G. Deloche & X. Seron (Eds.), *Mathematical disabilities* (pp. 111-120). Hillsdale, NJ: Erlbaum.

Bhattacharya, D. N. (1982). *Gifted children in mathematics: Case studies.* Doctoral dissertation, State University of New York at Buffalo, New York.

Brown, M. D. (1991). *The relationship between traditional instructional methods, contract activity packages, and math achievement of fourth grade gifted students.* Doctoral dissertation, University of Southern Mississippi, Mississippi.

Brown, J. S., & Van Lehn, K. (1980). Repair theory: A generative theory of bugs in procedural skills. *Cognitive Science, 4,* 379-426.

Clendening, C. P., & Davies, R. A. (1983). *Challenging the gifted - Curriculum enrichment and acceleration models.* New York: Bowker.

Cox., L. S. (1975). Systematic errors in the four vertical algorithms in normal and handicapped population. *Journal for Research in Mathematics Education, 4,* 202-220.

Ginsburg, H. P. (1977). *Children's arithmetic: The learning process.* New York: Van Nostrand.

Ginsburg, H. P. (1983). *The development of mathematical thinking.* New York: Academic Press.

Grissemann, H., & Weber, A. (1982). *Spezielle Rechenstörungen - Ursachen und Therapie.* Bern: Huber.

Hartje, W. (1987). The effect of spatial disorders on arithmetical skills. In G. Deloche & X. Seron (Eds.), *Mathematical disabilities* (pp. 121-135). Hillsdale, NJ: Erlbaum.

Heller, K. A., & Feldhusen, J. F. (Eds.). (1986). *Identifying and nurturing the gifted: An international perspective.* Stuttgart: Huber.

House, P. A. (Ed.). (1987). *Providing opportunities for the mathematically gifted, K-12.* Reston: NCTM.

Jellen, H. G., & Verduin, J. R. (1986). *Handbook for differential education of the gifted.* Carbondale, IL: Southern Illinois University Press.

Johnson, D. J., & Myklebust, H. R. (1971). *Lernschwächen - Ihre Formen und ihre Behandungen.* Stuttgart: Hippokrates.

Klauer, K. J. (1992). In Mathematik mehr leistungsschwache Mädchen, im Lesen und Rechtschreiben mehr leistungsschwache Jungen? *Zeitschrift für Entwicklungspsychologie und Pädagogische Psychologie, 24*(1), 48-65.

Kosc, L. (1974). Developmental dyscalculia. *Journal of Learning Disabilities, 7,* 164-177.

Krutetskii, V. A. (1976). *The psychology of mathematical abilities in schoolchildren.* Chicago, IL: University of Chicago Press.

Lorenz, J. H. (1982). Lernschwierigkeiten im Mathematikunterricht der Grundschule und Orientierungsstufe. In H. Bauersfeld, H.-W. Heymann, G. Krummheuer, J. H. Lorenz, & V. Reiß (Eds.), *Analysen zum Unterrichtshandeln* (pp. 168-209). Köln: Aulis.

Michael, W. B. (1977). Cognitive and affective components of creativity in mathematics and the physical sciences. In J. C. Stanley, W. C. George, & C. H. Solano (Eds.), *The gifted and the creative: A fifty-year perspective* (pp. 141-172). Baltimore, MD: Johns Hopkins University Press.

Niegemann, H. M. (1988). Neue Wege in der pädagogischen -Diagnostik: Fehleranalyse und Fehlerdiagnostik im Mathematikunterricht. *Heilpädagogische Forschung, 14(2)*, 77-82.

Radatz, H. (1980). *Fehleranalysen im Mathematikunterricht.* Braunschweig: Vieweg.

Resnick, L. B. (1983). Toward a cognitive theory of instruction. In S. Paris, G. M. Olson, & H. W. Stevenson (Eds.), *Learning and motivation in the classroom* (pp. 5-38). Hillsdale, NJ: Erlbaum.

Resnick, L. B. (1984). Beyond error analysis: The role of understanding in elementary school arithmetic. In H. N. Cheek (Ed.), *Diagnostic and prescriptive mathematics issues, ideas, and insights* (pp. 2-14). Kent, OH: Research Council for Diagnostic and Prescriptive Mathematics

Resnick, L. B., & Neches, R. (1984). Factors affecting individual differences in learning ability. In R. J. Sternberg (Ed.), *Advances in the psychology of human intelligence* (pp. 275-323). Hillsdale, NJ: Erlbaum.

Resnick, L.B., & Ford, W.W. (1981). *The psychology of mathematics for instruction.* Hillsdale, NJ: Erlbaum.

Snider, J. H. (1986). *Designing a program for gifted mathematics students in junior high/middle schools.* Doctoral dissertation, George Peabody College for Teachers of Vanderbilt University, Tennessee.

Stanley, J. C. (1977). Rationale of the Study of Mathematically Precocious Youth (SMPY) during its first five years of promoting educational accelaration. In J. C. Stanley, W. C. George, & C. H. Solano (Eds.), *The gifted and the creative: A fifty-year perspective* (pp. 75-112). Baltimore, MD: Johns Hopkins University Press.

Stanley, J. C. (1979). The study and facilitation of talent for mathematics. In NSSE (Eds.), *The gifted and the talented: Their education and development* (pp. 169-185). Chicago, IL: University of Chicago Press.

Tarnopol, M., & Tarnopol, L. (1979). Brain function and arithmetic disability. *Focus on Learning Problems in Mathematics, 1*, 23-39.

Teyler, T. T. (1984). Brain functioning and mathematical abilities. In H. N. Cheek (Ed.), *Diagnostic and prescriptive mathematics – Issues, ideas and insights* (pp. 15-20). Kent, OH: Research Council for Diagnostic and Prescritive Mathematics.

Underhill, R. G. (1977). *Teaching elementary school mathematics.* Columbus, OH: Bell & Howell.

Wilmot, B. A. (1983). *The design, administration, and analysis of an instrument which identifies mathematically gifted students in grades four, five and six.* Doctoral dissertation, University of Illinois at Urbana-Champaign, Illinois.

SHOULD GIRLS AND BOYS BE TAUGHT DIFFERENTLY?

Gila Hanna
Toronto

1. INTRODUCTION

The current social and political climate in much of the world dictates that public education provide the same opportunity to females as it does to males. This is in sharp contrast to a traditional view that assigned entirely different spheres of activity and achievement to the two sexes and accordingly saw unequal access to education for males and females as part of the natural order.

In this context, the issue of women's participation in mathematics is a particularly thorny one. Though women are attending universities in ever-increasing numbers, and often in greater numbers than men, they are acutely underrepresented in the natural sciences, in mathematics, and in engineering. In Canada and the United States, the proportion of women among those receiving bachelor's degrees in mathematics is constantly rising, but has only recently reached the 45% mark, and women still account for no more than 20% of those obtaining master's and doctoral degrees in that subject.

One might well ask to what extent this underrepresentation is due to inequality of opportunity in the educational system. Clearly it is not rooted in the educational system alone, nor can it be ascribed entirely to the nature of mathematics. Some of the reasons for it are those that account for the differential representation of men and women across occupations in general, and are thus very varied and complex. This more general issue has been the subject of investigation from various perspectives, such as the biological, historical, and sociocultural, in addition to the educational. Nevertheless, it was not unreasonable to postulate the existence of gender-related differences specific to the learning of mathematics, and a failure of the educational system to accommodate females. Over the past three decades, efforts to understand the low participation of women in mathematics have indeed spawned a large body of research into gender differences in mathematics learning and mathematics achievement.

In the course of this research, various hypotheses have been advanced involving variables that are psychological (e.g., spatial skills), environmental (e.g., school practices), or instructional (e.g., teaching methods), among others. There is a great deal of controversy on this topic, as there is in most

R. Biehler, R. W. Scholz, R. Sträßer, B. Winkelmann (Eds.),
Didactics of Mathematics as a Scientific Discipline, 303-314.

educational research. There is disagreement over the relevance of variables, over the way research questions are framed, over the methods used to collect and analyze data, and over the interpretation of research results. There is no consensus among researchers into gender differences in mathematics education, and it would be naive to expect a high degree of agreement to emerge in the near future.

Some researchers in this area have nevertheless suggested that girls need to be taught differently: In other words (to use the term employed in the heading of this section of the present book), there is a need for differential didactics by gender. The implication is that girls will continue to be denied an equal opportunity to learn mathematics until schools introduce separate teaching methods for boys and girls.

Whatever other faults they may have, I do not think present teaching methods should be blamed for this inequality of opportunity. I believe there is only teaching and better teaching, rather than teaching specific to girls or boys, and, in this paper, I will cite recent research results in support of this position. Accordingly, I would suggest that mathematics educators focus their efforts on better ways of teaching all students, male or female, rather than on special methods for teaching girls.

Differential didactics might well be justified if one could show that girls and boys, regardless of social or cultural origin, by virtue of their sex alone, have markedly different ways of learning in general or of learning mathematics in particular. It is precisely this claim that a number of researchers have made.

In the next section of this paper, I will examine critically some of the hypotheses put forward to describe or explain claimed gender differences in cognitive development in general and in styles of mathematics learning in particular, and will attempt to show that empirical evidence in support of these hypotheses is totally lacking or indirect and unconvincing. I will argue, furthermore, that such claims may work against the increased participation of women in the mathematical sciences.

In a subsequent section, I will then discuss the empirical evidence for the existence of gender-related differences in mathematics achievement, with the overall conclusion that at the primary and secondary levels these differences are small, where they exist at all, and are getting smaller.

My thesis, then, in light of the evidence I examine, is that proposals for differential didactics by gender represent an unwarranted and potentially harmful initiative for the solution of a problem that in any case is disappearing at the school level.

2. HYPOTHESES ON DIFFERENTIAL LEARNING

I first discuss two hypotheses arising directly from research in mathematics education, and then examine some hypotheses about gender differences in knowledge acquisition in general that stem from the work of Gilligan (1982) and that of Belenky, Clinchy, Golderberg, and Tarule (1986).

2.1 Rote Versus Autonomous Learning

Kimball (1989) examined some 150 studies on gender-related differences in mathematics achievement and noted a marked contrast between differences in classroom grades and differences in the results of standardized tests. Girls were consistently reported to perform better on classroom tests than on standardized tests. Furthermore, girls, on average, were reported to outperform boys on classroom tests, but to underperform them on standardized tests.

It is very problematic to compare the size of such differences, because standardized achievement tests and classroom tests differ in their psychometric properties, but these results have nevertheless given rise to hypotheses relating either to learning styles or to the possible inherent bias against females in some standardized tests.

To explain why gender differences in performance on standardized tests are more pronounced than those on classroom tests, perhaps the most significant learning-style explanation is that of "rote versus autonomous learning" put forward by Fennema and Peterson (1985). Their hypothesis is that girls have an advantage in classroom examinations, because they tend to take a rule-following and rote-learning approach, whereas boys get higher grades on external standardized tests (and eventually outstrip girls in mathematical understanding), because they have a more autonomous approach to learning. Because the concepts of "autonomous" and "rote" learning have not been operationalized, it is not possible to observe such behaviors directly. Kimball (1989) evaluated this hypothesis, however, by examining over 30 published research studies that had investigated variables that could be considered to be related to an autonomous or a rote style of learning. If girls engage less often than boys in extracurricular activities related to mathematics, for example, they might tend to rely on rote learning and memorization. Similarly, if boys display more rebellious attitudes, they might well be more autonomous learners. Kimball also examined evidence on the existence of links between these two presumed learning styles and performance on both classroom and standardized tests.

In assessing this hypothesis, Kimball concluded that there would be a need for more evidence "before we can evaluate its potential to explain sex-related differences in classroom and standardized achievement measures" (p. 206). To date, there is still no convincing evidence that girls adhere to a style of learning that can be branded rote as opposed to autonomous, nor any evidence that either of these presumed learning styles might be directly linked to achievement in mathematics.

2.2 Novelty Versus Familiarity

Another hypothesis offered to explain gender differences in mathematics achievement is that of "novelty versus familiarity," which suggests that boys are confident and motivated to do well when confronted with new and challenging tasks such as those met in mathematical studies, whereas girls are less confident and often feel lost in such situations. This hypothesis, too, is prompted by the observation that girls tend to do well on classroom tests that cover material explicitly taught by the teacher and thus familiar to the students, but not as well on standardized tests, which may present unfamiliar content.

The extensive study mentioned earlier led Kimball (1989) to the additional conclusions that (a) girls are not disadvantaged when the tests reflect classroom content, (b) girls do display avoidance and lack of confidence in the presence of novel tasks, and (c) girls have been observed to be less active in classrooms that emphasize competition (which is likely to require answers to novel problems). She thus concluded that there is some empirical evidence to support the hypothesis of novelty versus familiarity. Kimball's conclusion is guarded, however: "Although there is some theoretical and empirical support for this hypothesis, more empirical evidence is necessary to evaluate it" (p. 208). This hypothesis is challenged at a very basic level, however, by the results of a more recent study. According to Felson and Trudeau (1991), the hypothesis of novelty versus familiarity is not a useful one, because gender differences appear to depend upon the mathematical topic covered by the tests. Girls have been shown to outperform boys on certain standardized mathematics tests, indicating that the novelty of the task in itself is not a barrier to their achievement.

2.3 Separate Versus Connected Knowing

The hypothesis of "separate versus connected knowing" takes its inspiration and its wording from the work of Carol Gilligan (1982) on moral choices. In her book, *In a Different Voice,* she presents a model of moral development in which women are contextually focused, relationship-oriented, responsible, and caregiving, whereas men are abstract, individualistic, and dominating. Gilligan suggests that women have a special edge on moral issues, in that they are prepared to make moral judgments on the basis of a concern for relatedness among people rather than on the basis of abstract principles of justice. From her analysis she also concludes that women have a "connected" way of grasping situations, whereas that of men is "separate."

Though it is not immediately clear what implications this distinction might have for knowledge acquisition, it has been embraced, seemingly rather uncritically, by several researchers and mathematics educators. Brown (1984), for example, found Gilligan's analysis relevant to mathematics education and turned "towards a consideration, in a rather global way, of how it is that a Gilliganish perspective on morality might impinge on the study of mathematics" (p. 12).

From the idea that women, with their "different voice," learn best when connections are made explicit, others have gone on to suggest a pedagogy for mathematics education specific to women. In their view, the successful teaching of females would necessitate, among other steps, injecting the mathematics curriculum with a high degree of context, emphasizing connectedness among concepts and situations, and de-emphasizing the teaching of general principles (Buerk, 1985; Lee, 1989).

Instructional strategies along these lines may well turn out to be effective for both men and women. But there would seem to be little basis for the thesis of "connected knowing" advanced in their support. Those educators who so quickly attempted to translate this thesis into a rationale for differential didactics appear to have ignored the controversy engendered by Gilligan's analysis as a whole. What has come to be called the "Gilligan debate" consists of over 100 articles published during the period 1982 to 1992 in books and learned journals. As reported by Davis (1992), this debate among psychologists, sociologists, and human rights theorists, most of whom are self-declared feminists, has subjected Gilligan's model of moral development to extensive and often devastating criticism.

Some of Gilligan's critics have pointed out serious flaws in her methodology, questioning, for example, the validity of findings based only upon a sample composed primarily of Harvard students. Others have questioned the very possibility of a specifically female morality based upon care. Yet other critics have found the interpretation of her results to be wanting. It has been pointed out, for example, that the "ethic of care," which Gilligan observed in her female subjects, and which she interpreted as a sign of moral maturity and a source of strength specific to women, could quite validly be interpreted as a normal reaction by any oppressed group, male or female, to their oppression.

Some critics have also contended that the indecisiveness in moral issues that Gilligan observed in her female subjects is not necessarily an indication of superior moral standards. It could well be seen as an sign of subordination, and one not necessarily associated with gender, but perhaps with social position. As Davis (1992) reports, many critics have made the further point that sociological perspectives are more useful than psychological ones as explanations for perceived gender differences in moral development.

Gilligan's work may be important as a critique of sexist bias in the literature on human development. Because her study was of very modest proportions and her data open to alternative interpretations, however, her female-male dichotomy remains far from convincing. In any case, it would seem reckless to make uncritical application to the teaching of mathematics, as some appear to have done, of her contention that women grasp situations in distinctive ways.

2.4 Women's Ways of Knowing

This hypothesis for gender differences in cognition is known by the title of the book *Women's Ways of Knowing* by Belenky et al. (1986), who contend that women have distinct ways of knowing that have arisen from their life experiences as women. The authors also criticize existing theories of cognitive development, mainly for their reliance on studies that observed male students only. In their own research, they do make use of the epistemological work of Perry (1970), however, claiming to have fleshed out those areas that they believe to be sketchy or missing in his model of intellectual development. The book reports on a thorough study of 135 women from various social groups. All the data, in the form of interview notes and responses to a series of open-ended questions, deal with the role of gender in the self-perception of these women, in their vision of their own future development, in their formal learning experiences, and in their ways of making moral judgments. A major conclusion of the authors is that the women they studied felt voiceless and alienated in the course of their education, an education that the authors believe to have been tailored to the needs of men, and that, by implication, this feeling is common to women in general.

The authors further conclude that women have a distinctive way of acquiring and constructing knowledge, and claim to have identified five "epistemological categories" proper to female cognition. They state that women follow a sequence from dependence to autonomy through five specific stages: silence knowing (the experience of being voiceless), received knowing (the capacity of receiving knowledge from external authorities), subjective knowing (the perception of knowledge as privately intuited), procedural knowing (the ability to apply knowledge), and constructed knowing (the ability to create knowledge).

Although this book is not about mathematics, it has had a great deal of influence in the field of gender and mathematics. Seeing in it convincing evidence for a conflict between the ways in which mathematics is presently taught and the ways in which they think women learn best, some researchers have concluded that mathematics education must take into account distinct masculine and feminine ways of learning.

Becker (1991) and Damarin (1990), for example, have attempted to spell out in some detail the implications of this book for mathematics teaching (referring to their proposals as a "feminist perspective" on the teaching of mathematics). Damarin suggests that "women learn abstractions (such as mathematics principles) best if statements of rules are preceded by quiet observation, by listening to others, and by personal experiences that women can relate to the abstractions" (p. 148): in other words, if teaching conforms to the five stages suggested by Belenky et al.

Becker states that the ideas in this book have major "implications for how to encourage girls and women to pursue mathematics" (p. 1). Although she recognizes that the case for a specifically feminine cognitive strategy is not proven, conceding that "research can help provide evidence to support or refute the hypothesis of women's ways of knowing in mathematics" (p. 3), she nevertheless proceeds to suggest some implications of this hypothesis for the teaching of mathematics. She believes, for example, that the third stage of knowing postulated by Belenky et al., subjective knowing, "is a very powerful one for the knower and brings in women's intuitive way of knowing" (p. 4).

The "women's ways of knowing" hypothesis elaborated by Belenky et al. has taken its place among the many schemes of cognitive development that compete for attention. There does not appear to be any evidence in their study or elsewhere, however, that this scheme has an advantage in any general sense over others, nor even that it is particularly useful in understanding how women learn. Belenky et al. have certainly not proved that there are cognitive differences between male and female learners, as they claim. (It is not at all clear that one can even design a study to prove or disprove such a contention.)

Those who argue for an intrinsically feminine way of understanding mathematics, most of them feminists and all of them well-intentioned, are actually doing a disservice to education and to other women. (In other contexts, their views would quickly be labeled as "sexist.") In reinforcing the traditional view of women as caregivers who are better at personal relations than abstract ideas, they run the risk of portraying women as fundamentally unsuited for science. In suggesting that the traditional male-female dichotomies (such as logic vs. intuition, aggression vs. submission, and rigor vs. creativity) are valid and ingrained, if not inherent, they run the risk of perpetuating existing stereotypes, legitimizing gender differences in mathematics achievement, and providing a rationale for the relatively low participation of women in scientific pursuits in general.

3. GENDER AND ACHIEVEMENT

I will now turn my attention to studies of gender differences in mathematics achievement, examining meta-analyses of research papers published in the last 20 years as well as some international achievement surveys and national studies. All indicate rather clearly that gender differences in mathematics achievement are rapidly disappearing.

A meta-analysis is a synthesis of several studies with more or less similar designs, in which the results of the studies are analyzed to yield summary measures about the overall statistical significance and the effect size of a given outcome. Meta-analyses thus combine information across multiple empirical studies that measured the same outcome, and can provide clear and concise effect-size measures of that outcome.

In these studies, the term "achievement" refers to the results of standardized tests or textbook tests. Such tests are designed to measure competence in a general sense, but, in point of fact, most of the test items measure only the ability to recall facts and the ability to apply concepts to the solution of relatively short problems. It should be pointed out that both classroom tests and standardized tests have recently come under severe criticism. The view of learning as mastery of factual knowledge implicit in the design of such tests is today considered incomplete, and is inconsistent with the view currently held by many cognitive and educational psychologists that learning is active and constructive. However, assessment instruments capable of measuring adequately what is now referred to as "authentic" mathematical competence have yet to be designed. In the meantime, we have to rely on studies that used testing instruments and psychometric methods pervasive throughout the 1970s and 1980s. Fortunately these studies do give us reliable information on significant aspects of mathematical mastery.

3.1 Meta-Analyses

Hyde, Fennema, and Lamon (1990) examined about 100 studies published in the years 1967 to 1987 that used standardized mathematics tests and reported on gender differences in achievement. Their meta-analysis indicated that, in elementary and middle school, there were no gender differences, that small gender differences favoring males emerged in high school and in college, and that the magnitude of these gender differences had declined over a 20-year period.

As a measure of the magnitude of gender differences in the general population, the authors derived an effect size (*d* metric), defined as the mean for males minus the mean for females, divided by the mean within-sexes standard deviation. Effect sizes were calculated as a function of the cognitive level (e.g., computation, concepts, or problem solving), as a function of the mathematical content (arithmetic, algebra, geometry, calculus, or mixed), as a function of age, as a function of ethnicity, and, finally, as a function of the selectivity of the sample (population selected according to level of performance).

When the data were analyzed by cognitive level, all effect sizes proved to be small; for computation and for concepts, the small effects were in the girls' favor; while for problem-solving and mixed levels, they were in the boys' favor. Looking at the data by mathematics content, effect sizes were again very small for all topics. When examined by age, the data showed a "slight female superiority in performance in the elementary and middle school years. A moderate male superiority emerged in the high school years . . . and continued in the college years . . . as well as in adulthood" (p. 149). The researchers concluded that their meta-analysis provided "little support for the global conclusions" of previous studies that boys outperform girls in mathematics achievement (p. 151).

Another meta-analysis by Friedman (1989) investigated 98 studies done between 1974 and 1987, comprising journal articles, doctoral dissertations, and large nationwide assessments carried out in the United States. The author concluded that "the mean random effects model ... is minute ... indicating that we cannot say with 95% confidence that a sex difference exists in the general United States population of school-age youth" (p. 204). The analysis also showed that "the sex difference in favor of males is decreasing over short periods of time" (p. 205).

Friedman comments that the finding that "the average sex difference is now very small ... should have considerable practical import" (p. 206), presumably in the expectation that it would help to dispel the widespread perception that boys outperform girls in school mathematics.

A third meta-analysis conducted by Feingold (1988) reviewed research done over the previous 27 years and concluded that the magnitude of gender differences in cognitive abilities had declined markedly over that period. Though the achievement gap at the upper levels of high-school mathematics had remained constant, gender differences in verbal reasoning, abstract reasoning, spatial relations, numerical ability, and other areas of cognitive ability had declined precipitously.

3.2 International Surveys

An analysis of data collected in 1981 to 1982 by the Second International Mathematics Study (SIMS), which compared mathematics achievement in 20 countries at age 13, has shown not only that gender differences vary widely from country to country but also that they are smaller than differences among countries (Hanna, 1989). In some countries, girls outperformed boys in one to three of the five subtests; while in others, it was boys who did better on some of the subtests. In 5 of the 20 countries studied, no gender-related differences were observed.

The more recent International Assessment of Educational Progress (IAEP) studies carried out in 1988 and 1991 also concluded that there are no marked gender differences in mathematics achievement among 13-year-old students. The first IAEP study encompassed 12 student populations from nine countries. The findings were that "boys and girls were performing at about the same level in 10 of the 12 populations assessed. Only in Korea and Spain do boys at this age achieve significantly higher in mathematics than do girls." (Lapointe, Mead, & Phillips, 1989, p. 18).

The second IAEP study surveyed the mathematical performance of 13-year-old students in 20 countries, as well as that of 9-year-old students in 14 countries. The results indicated that there were few statistically significant differences in performance between the genders. One of the findings was that "the patterns of performance for males and females at age 9 ... are not the same as those seen at age 13." More precisely, where small gender differences did exist in favor of boys, they were found in some countries at age 9 and in other countries at age 13 (Lapointe, Mead, & Askew, 1992, p. 86).

3.3 National Studies

In their 4-year "Girls and Mathematics" study of English schools, Walden and Walkerdine (1982, 1985) and their research team not only administered achievement tests but also carried out extensive classroom observation and interviews with students and teachers. Though Walden and Walkerdine have serious reservations about achievement tests, they nevertheless drew the following conclusions from their test data: (a) There are few statistically significant sex differences in the elementary and 1st-year secondary tests; (b) in the 4th-year secondary test, girls consistently outperform boys; (c) there is no support in the data for the commonly held view that girls perform better only on computation or low-level items; and (d) despite the high performance of girls, ". . . it turned out subsequently to be considerably more difficult for girls to be entered for O level mathematics than boys" (1985, p. 46).

The results of a study conducted in France over the period 1985 to 1989 indicate that gender differences in mathematics achievement among 12- and 15-year-old students are nonexistent or very small. Girls outperformed boys on some tasks and did as well as boys on others, while on yet other tasks, boys did better than girls. Baudelot and Establet (1992) conclude that, on the basis of their results, it is difficult to uphold the prevalent belief that girls are inferior to boys in mathematics achievement.

4. CONCLUSION

The recent studies discussed here show that girls are not underachievers in school mathematics. On average, they perform as well as boys on most of the mathematics tests; on some tests, they outperform boys, whereas, on others, boys have the edge. When one considers their level of achievement in light of the observations made in many studies that boys often get more attention and time from teachers, that girls tend to have less confidence in their ability to do mathematics, and that, when it comes to mathematics achievement, parents often have lower expectations of their daughters, one must conclude that girls have benefited from undifferentiated mathematics instruction at least as much as boys.

Girls have made enormous strides in mathematics achievement at the secondary level and are pursuing mathematics at the postsecondary level in increasing numbers. This is no reason, of course, for researchers to ignore those gender differences that persist. But, in the past decade, we have seen far-reaching proposals for a differential didactics at the school level resting upon alleged differences in cognition between boys and girls. Are not both the validity and the relevance of this radical solution clearly undermined by the achievements of girls in mathematics over this very decade in the face of well-recognized obstacles? In any case, we have not seen good evidence for differences in cognition. What case has been made that women have "a different voice" or a monopoly on "connected knowing?" And if differences in cognitive strategy between men and women could actually be observed, should they then be regarded as biologically determined? If not, if they are socially constructed and thus subject to change, should educational strategy be to accommodate these differences and thus perpetuate them, or should it rather be to force changes in their social determinants?

Even if there were a short-term advantage from an educational point of view in accommodating alleged socially conditioned differences through a pedagogy of mathematics specific to women (and there would not seem to be convincing arguments even for such a short-term advantage), would such a step be in the longer-term interest of women? Should women not fear their assignment, throughout the educational system and beyond, to a different place and a different role on the basis of characteristics they are alleged to share? Would this not be to resurrect a past we have all deplored?

REFERENCES

Baudelot, C., & Establet, R. (1992). *Allez les filles!* Paris: Seuil.

Becker, J. R. (1991). *Women's ways of knowing in mathematics.* Paper presented at the invited symposium of the IOWME, Assisi, Italy.

Belenky, M. F., Clinchy, B. M., Golderberg, N. R., & Tarule, J. M. (1986). *Women's ways of knowing: The development of self, voice and mind.* New York: Basic Books.

Brown, S. I. (1984). The logic of problem generation: From morality and solving to de-posing and rebellion. *For the Learning of Mathematics, 4*(1), 9 29.

Buerk, D. (1985). The voices of women making sense of mathematics. *Journal of Education, 167*(3), 59-70.

Damarin, S. K. (1990). Teaching mathematics: A feminist perspective. In T. J. Cooney & C. R. Hirsh (Eds.), *Teaching and learning mathematics in the 1990s: 1990 Yearbook.* Reston, VA: National Council of Teachers of Mathematics.

Davis, K. (1992). Toward a feminist rhetoric: The Gilligan debate revisited. *Women's Studies International Forum, 15*(2), 219-231.

Feingold, A. (1988). Cognitive gender differences are disappearing. *American Psychologist, 23*(2), 95-103.

Felson, R. B., & Trudeau, L. (1991). Gender differences in mathematical performance. *Social Psychology Quarterly, 54*(2), 113-126.

Fennema, E., & Peterson, P. (1985). Autonomous learning behavior: A possible explanation of gender-related differences in mathematics. In L. C. Wilkinson & C. B. Marrett (Eds), *Gender influences in classroom interaction.* Orlando, FL: Academic Press.

Friedman, L. (1989). Mathematics and the gender gap: A meta-analysis of recent studies on sex differences in mathematical tasks. *Review of Educational Research, 59*(2), 185-213.

Gilligan, C. (1982). *In a different voice.* Cambridge, MA: Harvard University Press.

Hanna, G. (1989). Mathematics achievement of girls and boys in grade eight: Results from twenty countries. *Educational Studies in Mathematics, 20,* 225-232.

Hyde, J. S., Fennema, E., & Lamon, S. J. (1990). Gender differences in mathematics performance: A meta-analysis. *Psychological Bulletin, 107*(2), 139-155.

Kimball, M. M. (1989). A new perspective on women's math achievement. *Psychological Bulletin, 105*(2), 198-214.

Lapointe, A. E., Mead, N. A., & Phillips, G. W. (1989). *A world of differences: An international assessment of mathematics and science.* Princeton, NJ: Educational Testing Service.

Lapointe, A. E., Mead, N. A., & Askew, J. M. (1992). *Learning mathematics.* Princeton, NJ: Educational Testing Service.

Lee, L. (1989). Vers un enseignement des mathématiques qui s'adresse aux femmes. In L. Lafortune (Ed.), *Quelles différences?* Montréal: Remue-ménage.

Perry, W. (1970). *Forms of intellectual development in the college years.* New York: Holt, Rinehart, and Winston.
Walden, R., & Walkerdine, V. (1982). *Girls and mathematics: The early years.* Bedford Way Papers 8. London: University of London Institute of Education.
Walden, R., & Walkerdine, V. (1985). *Girls and mathematics: From primary to secondary schooling.* Bedford Way Papers 24. London: University of London Institute of Education.

FROM "MATHEMATICS FOR SOME" TO "MATHEMATICS FOR ALL"

Zalman Usiskin
Chicago

1. INTRODUCTION

There have been, in this century, two major developments in mathematics education. The first of these, continuing a movement that is several centuries old, is the teaching of more and more mathematics to more and more people. For instance, the study of algebra and geometry, which, even a century ago, was reserved for a small percentage of the population even in the most technological of our societies, is now a part of the core curriculum for all students in many countries. The second development, only within the past 30 years or so, has been the emergence of computer technology, which enables much mathematics to be done more easily than ever before, and enables some mathematics to be done that could not be done at all previously. As a result, more people are encountering and doing far more mathematics than ever before, and there is great pressure nowadays to teach a great deal of mathematics to all people.

In this paper, these developments are placed in an even longer historical framework than this century, and that framework as well as some recent work is used to suggest directions in which mathematics in school and society may be moving and should be moving.

2. DEFINITIONS OF TERMS

The word *all* in the title of this paper refers to all of the population except the mentally disabled, which means at least 95% of any age cohort. The relationship between "all" and "all students" varies by country and age level of the student. For instance, in the United States, about 71% of 18-year-olds graduate high school with their age cohort, and about 15% more earn their high school diplomas later. So, for the United States, "all" constitutes a population larger than those who finish high school. In contrast, in Japan, 95% is just about the percentage of students who graduate high school.

On the other hand, here the phrase *mathematics* for all refers to *school* mathematics for all, and so these remarks are not meant to apply in those places where children do not attend school, or cannot attend school, or choose not to attend. Mathematics *for* all refers at different times in this pa-

R. Biehler, R. W. Scholz, R. Sträßer, B. Winkelmann (Eds.),
Didactics of Mathematics as a Scientific Discipline, 315-326.

per to the mathematics that *has been* learned by all, that *is* being learned by all, that *could* be learned by all, that *should be* learned by all, or that *will be* learned by all.

The content of school mathematics is broad, including: skills and algorithms; properties and proofs; uses and mathematical models; and representations of many kinds, what in the secondary materials of UCSMP are termed the SPUR (S = skills, P = properties, U = uses, R = representations) dimensions of mathematics (UCSMP, 1990, 1991, 1992).

3. THE CURRENT STATE OF MATHEMATICS FOR ALL

In most of the world, all students are expected to learn a considerable amount of arithmetic. Until recently, because one needed to know paper-and-pencil skills in order to use arithmetic, the Skills dimension of arithmetic was the most emphasized everywhere. However, because of the emergence of calculators, in some countries there has been a decrease in the attention given to the skill dimension, and a corresponding increase in attention to both the Uses and Representations dimensions. Yet it is probably safe to say that in most classrooms in the world, the teaching of paper-and-pencil skills still dominates class time.

Elementary school teachers are fearful of the calculator, for they know that a calculator can perform all of the arithmetic they have been teaching. They understand that arithmetic is important for every child to know, but given the presence of a calculator, these teachers do not know what to teach, and they may stop teaching arithmetic entirely. This view is reflected in recommendations by some science educators in the United States that much of the time spent on mathematics in the elementary school can now be spent on science, because the content that has been taught is no longer needed. Indeed, in one report there is no index listing for arithmetic, though there are listings for algebra, geometry, and many other aspects of mathematics (AAAS, 1989). Thus, though it would seem that "arithmetic for all" is so ingrained in schooling that it will not leave, there is a distinct possibility that without a well-formed replacement for the structure that the algorithms of arithmetic imposed on the curriculum, much of the arithmetic curriculum might disappear. It is already the case that in some countries some of the more complicated arithmetic algorithms, such as long division, are not being taught to all students and not being tested. It is a case of "arithmetic for all" becoming "arithmetic for some."

Despite the fact that some mathematics is becoming obsolete, more and more mathematics is entering the curriculum. As an example, in the United States only a generation ago, most students encountered not one day of probability, and the only statistics taught was how to calculate the average of a set of numbers. A national report in 1959 recommended merely that an *optional* course in probability and statistics be available to 12th-grade students (CEEB, 1959). By 1975, only 16 years later, there was quite a change: A report recommended that statistics be taught at all levels of the curriculum (NACOME, 1975), a recommendation that has been repeated many times (see, e.g., NCTM, 1989).

Similar increases in the mathematics all students are expected to learn have happened in all countries. For example, students in almost all countries today are expected to know a great deal more about measurement than they used to know. In some countries, all students are expected to know some algebra and some geometry, and this algebra is quickly becoming quite graphical with an earlier study of functions, and there are trends that indicate the geometry is becoming quite a bit more visually sophisticated, with the increasing use of coordinates, isometries, and other transformations, and continuous deformations.

4. FROM ARITHMETIC FOR SOME TO ARITHMETIC FOR ALL

To obtain guidance regarding what may happen or what should be our policies toward these changes, it is useful to ask if there has ever previously been a time like ours, when there was a revolution in the amount of mathematics that the average citizen was expected to know. From a Western perspective, a corresponding revolution began in the 15th century.

Compared with the situation today, in the 15th century very little mathematics – only counting and the simplest of addition – was known to all people even in the most advanced of countries. Nowhere near 95% of children went to school, and arithmetic was one of the liberal arts, taught in colleges that few attended. We might say that in the 15th century all mathematics was for some. Dantzig (cited in Swetz, 1987) tells a story, which supposedly took place in the first half of the 15th century, of a trader in Germany who wanted his son to get the best mathematics education he could. The trader consulted a professor at a German university, who advised him that his son could learn to add and subtract at his university, but if he wanted to learn to multiply and divide, then he should go to Italy, where they were more advanced in such matters.

Yet, 500 years later, by the end of the last century, whenever there was compulsory schooling, arithmetic was present, and the expectations for arithmetic were quite formidable, with the complexity of the problems being enough to challenge any of us today.

Three fundamental developments changed the situation. The first, as the quote from Dantzig reminds us, was the increased amount and sophistication of trade between peoples. This increased the need for accurate records that were understandable to traders and to those who benefited from the trade: manufacturers of goods, owners of land from whom farmstuffs and minerals were gotten, and all others in the marketplace. Great numbers of people were engaged in these professions, and so the increasing need for mathematical knowhow in the marketplace was no small influence on the amount of mathematics known to the average citizen.

The second development was mathematical: the invention of algorithms that made it easier to do arithmetic than had previously been the case. Roman numerals were not well-suited for computation beyond addition and subtraction, and algorithms for multiplication and division were in their infancy in the 15th century. At the end of the 16th century, in 1585 when Simon Stevin first considered decimal places to the right of the unit's place, one of the main arguments he put forth for using them was that there existed algorithms for multiplication and division that could be applied to what he called "decimal fractions," and thus computation would be simplified. Within 30 years of Stevin's invention of decimals, logarithms had been invented and decimals were established as the preeminent way to represent numbers. The third development that enabled the expectation for competence in arithmetic to become universal was the invention of printing. Arithmetic skills are not easily learned; certainly they are not usually learned merely from one or two books that might be community property. Thus in order for competence in arithmetic to become universal, there had to be enough books to enable all students to have their own books. Printing made it possible to have enough books. Printing also helped to standardize the language of arithmetic throughout the Western world. Today's differences in notation throughout the world are minor – numerals and other symbols are the same, enabling traders worldwide to use the same arithmetic language.

Thus between 1400 and 1900, "arithmetic for some" became "arithmetic for all," and necessary for this were three developments: a societal need for the competence; the mathematical language and tools that made this competence a reasonable expectation; and technology that made it possible for this competence to be realized.

5. FROM ARITHMETIC FOR ALL TO ARITHMETIC AS A PART OF LITERACY

At the same time that arithmetic changed from being for some to for all, so did reading, and for the same reasons. An enlightened citizenry and an intelligent work force came to require both the ability to read and the ability to compute and apply arithmetic.

One needs only examine a daily newspaper to get an idea of the extent to which arithmetic is ingrained in our cultures and has become a necessary part of communication, indeed, a part of literacy. In various countries, I have found invariably the *median* number of numbers on a newspaper page is somewhere between 120 and 150. The *mean* number of numbers is far higher – the last time I calculated it for a Chicago newspaper, the mean number of numbers on a page was over 500, due to the plethora of numbers on the sports, weather, and business pages, and in the want ads.

These numbers are used in many ways: as counts, with a wide variety of counting units, and often large; more often as measures than counts; in scales of various kinds; as ratios; both interval and single-number estimates and exact values. There are various kinds of graph, sometimes daily analyses of lotteries, results of polls, many stock averages, and sports statistics, all of which could be simplified at times if algebraic formulas were used. There are advertisements with discounts given as percents, annual percentage rates for investments, dimensions of the articles being offered, computer specifications, powers of zoom lenses, and other technical information.

An exhaustive listing of numbers in the newspaper is not needed to make the point that to read a newspaper today requires that the reader be able to process mathematical information to an extent far beyond that required even one generation ago. It is often said that we are in an information age; it is the case that much of that information is numerical or pictorial, and thus is mathematical.

Concomitant with the evolution of arithmetic as a part of literacy has come a major change in the views of society toward who can be competent in these things. No longer is arithmetic seen as the province of a few. In places where arithmetic is a part of literacy, no longer is it seen as a subject that is so abstract that only a few can learn. In these places, competence in arithmetic skills is no longer viewed as an indicator of intelligence.

6. THE CURRENT STATE OF ALGEBRA AS A PART OF LITERACY

Could we replace "arithmetic" in the previous sections by any mathematics other than arithmetic? A reasonable first candidate is algebra, since, in some countries, algebra is already taught to all. But algebra does not have nearly the status that arithmetic has in society. Many well-educated people ask why algebra was required for them in school; they would never ask that about arithmetic. Many people have been taught algebraic skills and perhaps its properties, and they may have even been taught some graphical representations, but they never were taught the uses, and they do not see the societal need for all to learn algebra. Algebra is viewed by many people as so abstract that it does not have uses of its own.

If we view the newspaper as signaling what mathematics is needed by society, then we see how far we have to go before algebra becomes viewed as a part of literacy. There may be thousands of numbers, and tables, and graphs, and charts in newspapers, but it is seldom that one finds any algebra. It is unusual to find one overt example of algebra in a newspaper, despite the fact that there are simple formulas underlying many of the sports statistics, discounts, and business data. So if algebra becomes a part of literacy, it is unlikely to be the algebra that is now being taught.

Indeed, whereas the level of political analysis one finds in newspapers is often quite deep and requires a thorough knowledge of a nation's governmental system, even the simplest algebra – even when studied by the vast majority of people in a nation – is taboo. (Stephen Hawking tells the story of how the publisher of *A Brief History of Time* did not want any formulas in the book at all, and only reluctantly agreed to the inclusion of $E = mc^2$.) When such mathematics is presented to a popular audience, it is often preceded by cautionary statements such as "For those who understand such things . . ." Unlike arithmetic, algebra is still viewed as the province of a few, and proficiency in algebra is often considered as a sign of intelligence by those who do not use it.

7. WILL ALGEBRA FOR SOME BECOME ALGEBRA FOR ALL?

It is appropriate to ask whether we can ever expect algebra to become as much a part of literacy for future generations as arithmetic is now.Will algebra ever be as universal as arithmetic?

Following the history of arithmetic, the first component in the question of algebra for all would have to be a perceived need by society for that algebra. I believe the general view of the nonmathematical public is that algebra is certainly required for future engineers or scientists or for work with computers, statistics, economics, or any field that seems dependent on numbers; or for any field that uses science, such as medicine. The general public might also realize that the building trades, such as carpentry or plumbing, use algebraic formulas. It may well be that this is enough to ensure that algebra should be and will be taught to all.

A second argument for major attention to algebra and higher mathematics for the entire populace has appeared recently in the policy arenas of the advanced industrialized countries. It goes somewhat as follows: The economic well-being of a country must be based on having jobs for its people. The new jobs in the 21st century will be based on achievements in sectors such as biotechnology, telecommunications, computers and software, robotics and machine tools, and microelectronics. Better products in these areas require statistical quality control. To have statistical quality control, workers need to understand it, which requires that they have studied statistics and operations research, and for these a person needs a considerable amount of mathematics (Thurow, 1991). Thus, whereas, in the past, knowledge of this mathematics has been seen as an *individual* need, now the knowledge is seen as a *societal* need. That makes it even more likely that algebra will be for all students.

For a couple of hundred years there has existed the mathematical language and tools that make universal competence in algebra a reasonable expectation. Worldwide we use the Latin alphabet in elementary algebra; we use coordinate graphs to picture functions.The big change – within the past 7 years – is that there now exists technology that makes the graphing of functions and data, and even curve-fitting and data analysis accessible to all, with the ability to be taken anywhere one has a pocket, and user-friendly enough so that one does not need to know huge amounts of mathematics in order to use the technology. Not only is algebra more accessible, but so is elementary analysis.

The widely available technology does not yet cover *all* of algebra. There does not yet exist a symbolic algebra calculator that is easy to use and cheap, that can solve literal equations as well as numerical ones, a simpler form of *Derive, Mathematica,* or *Maple*, for under $100. Yet this technology seems certain to come. For this reason, I believe that algebra will become a subject for all, but not the same algebra that we now teach, and with it will come many of the concepts of elementary analysis and calculus.

8. WILL ALGEBRA FOR ALL BECOME ALGEBRA FOR SOME?

As with arithmetic, the technology does not necessarily suggest an increased emphasis on algebra in schools. Because the purpose of technology is to avoid work, to make it possible for us to direct machines to do tasks even when we do not understand how the machine works, the same technological advances that have made it possible to do great amounts of algebra easily may also make it less necessary for people to learn certain algebra.

For example, suppose we wished to predict future population from recent data and an exponential model. The data can be plotted without knowing algebra. Transforming the variable P to $\log P$ can be done simply by writing a formula if one is using a spreadsheet or by pressing a button if one is using a calculator, and then the points on the second graph can be found. The line of best fit can be found without *any* algebra; simply press another few buttons. This line can be used for predicting the population from the graph. Thus a problem whose solution in the past might have required a considerable amount of algebraic skill now can be solved by someone who has none of the traditional skills. Instead, needed are the facility of graphing functions using an automatic grapher and knowledge of the inverse relationship between the exponential and logarithmic functions.

On the most recent graphing calculators, there exists a key that solves any type of a large number of equations arithmetically, by successive approximation methods hidden from the user. A student who has this calculator does not need to know the quadratic formula in order to obtain the solutions to a quadratic equation to the nearest thousandth; nor does the student need to know the inverse trigonometric functions in order to solve a trigonometric equation.

Mathematics educators often make the assumption that an increasingly technological world requires more and more mathematics for all (consider, e.g., NCTM's *Algebra for Everyone,* 1990). However, what may be the case is that such a technological world requires more and more mathematics for *some*, but less for *all* due to the advances that those few make. Just as we use algebra to solve problems that the ancient Greeks solved or attempted to solve geometrically, and even many of us in mathematics have never learned how they did that, it is possible that future generations will learn how to use the latest technology to solve our algebra problems, and never learn how we did it using algebra.

The situation in algebra is made more interesting by spreadsheets, which have their own algebra. Possibly, in the near future, the language of spreadsheets will become the most commonly used algebraic language. Thus it may be that the algebraic language everyone comes to learn is a different one than the one historically taught in schools. The difference between algebra in school and algebra in the real world would seem to be akin to the difference between arithmetic in school and arithmetic in the newspaper. In school, the tendency in almost all countries is to concentrate on the Skills and Properties of algebra, while in the world at large, the Uses predominate, with Representations also being quite important.

The same technology that enables algebra questions to be treated without algebra also enables calculus questions to be treated without calculus. The very same software programs and calculator technology that enable one to avoid symbolic algebra also make it possible to avoid the symbolic manipulations of calculus and statistics. It is possible today to answer max-min problems without having to resort to derivatives; to obtain areas under curves without integrals. In many places we have justified algebra not on its own merits but on its importance in the more advanced mathematics of calculus and differential equations. But, with technology, these subjects, too, are not so advanced. We must be careful that, despite its importance, we do not lose algebra in school because of the other means we now have for tackling problems that used to be algebra.

9. *CAN* ALGEBRA AND CALCULUS CONCEPTS BE LEARNED?

In many countries, the national curriculum includes a study of algebra for everyone, a trend that is gaining favor in the United States (NCTM, 1990). Yet, even in those countries, many of the algebra teachers believe algebra cannot be learned by all. The argument that I heard on a visit to schools in Shanghai was the same as the one I hear in my own community: Students can all learn to *do* algebra, but they do not all *understand* what they are doing. Many more teachers think that calculus cannot be learned by all, for the subject matter itself is beyond the students.

If these subjects do not change both in the classroom and in the society at large, I agree with this point of view. Courses taught as exercises in formal structure do not have a broad enough base of appeal to become more than mechanics. If we teach only mechanical skills, we should not be surprised that our students learn only mechanical skills! But all of the current developments suggest that "algebra for all" will be quite different from the traditional algebra that we have been teaching, and I believe that it will include calculus.

The reason for my optimism can be found in any country where our language is not the mother tongue. When students begin learning a foreign language in senior high school, so that they are studying that language at about the same time that they take algebra through calculus, as in the United

States, they usually have a great deal of trouble learning it. Their accents are atrocious, it seems as if the language is beyond them, and only a small percentage seem to do well in their language study. Yet where the language is spoken, even younger children understand, read, and speak it well.

Of course, students' proficiency in their mother tongue is not due to any special brilliance, but because they are immersed in it and so become fluent in it. With instruction, virtually all of them learn to decode the multitudinous combinations of letters and other symbols that constitute their own written language. It is difficult to believe that any person who can learn to read and write and comprehend his or her native language does not possess the ability to read and write and comprehend algebraic symbolism, part of the language of mathematics.

But the ability to learn does not guarantee the realization of that ability. What makes it possible for children to learn languages is an environment in which these languages appear in context. Good foreign language teachers try to imitate this reality. For example, throughout the world where French is not spoken, the effective teacher of French tries to make the classroom into a bit of Montreal or Paris. The movements within mathematics education to put context into the mathematics, to utilize applications of mathematics in everyday teaching, and to engage students in classroom discussions, can be seen as an attempt to speak the language of mathematics in the classroom. These are the Uses of the SPUR characterization of understanding of mathematics. Since mathematics beyond arithmetic is not yet commonplace outside the classroom, this is a necessary move within the classroom if we are to achieve higher levels of mathematics performance for all.

Because mathematics is so much a language, it seems reasonable to conclude that many aspects of it are better learned when the child is younger than when the child is older. Another reason for the difficulty of calculus is probably because its ideas are often first encountered at ages later than the optimal ages for learning a language.

10. FROM ALGEBRA/CALCULUS FOR SOME TO ALGEBRA/CALCULUS FOR ALL

In the future, the algebra-calculus sequence will give less attention to algebraic techniques when solving problems, because these will be able to be done by hand-held machines and preprogrammed software. But the sequence will need to have increased emphasis on two aspects of algebra: the uses to which algebra, functions, and calculus can be put; and the importance of algebra as a language for communicating generalizations and functional relationships. Both of these aspects increase in importance because of computers. In the parlance of the SPUR characterization, algebra of the future will undoubtedly contain less of the Skills dimension and more of the Uses and Representations dimensions. The Properties dimension, due to the importance of the language of algebra, is likely to maintain its role in the curriculum. In particular, the broad properties of functions, of matrices, and of vectors will probably enter the domain of mathematics for all.

Critical in all this is that the use of algebra as a language of communication be encouraged. In addition to the current emphasis on variables in formulas and variables as unknowns, greater emphasis is needed on the uses of variables to generalize patterns, the use of variables as indicating places in spreadsheets or computer storage, and the use of variables as arguments in functions. This could be done in many ways, for example: (a) emphasize how much easier it is in many circumstances to apply a formula rather than read a table; (b) demonstrate how the language of algebra and functions and matrices and vectors makes it easier to handle certain problems; (c) show how some patterns and trends can be described algebraically more compactly than with graphs; and (d) show the power of functions to predict, and how picking the wrong function can lead to errors.

11. GEOMETRY FOR ALL?

Although in school geometry, students are taught as if the only planar shapes are polygonal or circular, and the only 3-dimensional shapes are spherical, cylindrical, or conical, every object in the world, from the chair you may be sitting on as you read this paper to the leaves of a tree considered individually or as a set of leaves, has a shape and a size. Computer graphics have greatly increased our ability to draw pictures to represent this world and to examine those pictures. They have made the Skills and Uses of geometry more accessible, and, as mentioned earlier, they have increased the importance of geometrical Representations of functions. The world is geometric.

Indeed, one could argue that the world is more obviously geometric than arithmetic. Perhaps that is the reason why deductive reasoning came to geometry before arithmetic or algebra. But despite this, the geometry curricula of countries differ more than the curricula in any other area of elementary mathematics (Travers & Westbury, 1990, p. 207). As a result, it was very difficult to assemble the geometry items at the 8th-grade level for the Second International Mathematics Study (Robitaille, 1989) The theorems of Euclid that dominate some curricula have been replaced with transformation and vector approaches in others.

Consequently, though all students today learn some geometry, they do not learn the same geometry. The situation might be characterized as "some geometry, but not the same geometry, for all." Unless there becomes some sort of worldwide standard, it is difficult to believe that this situation will change (see, also, Usiskin, 1987).

The situation with respect to geometric representations is quite different. Coordinate graphs and displays of data (bar graphs, histograms, circle graphs, etc.) are reasonably universal, found in newspapers and popular magazines, and their status has become much like arithmetic despite their relatively short history (bar graphs and line graphs are barely 200 years old; see Tufte, 1983). They have become a part of literacy, found in social science curricula as often as in mathematics. Only the first of the three reasons given in section 4 above seems to apply here; the societal need to transmit information and the power of a visual display to do so.

It seems likely, then, that sets of points will play an ever increasing role in the curriculum, but these may not be the traditional sets of points of Euclid, but more ordered pairs and triples, graphs of functions and relations, and representations of graphs and networks. If this is the case, the importance of coordinates and transformations will increase, and the traditional work with polygons and circles is likely to decrease or to be encountered by students earlier in their mathematics experience.

12. MATHEMATICAL SYSTEMS

The traditional role of geometry as a vehicle for displaying a mathematical system is already gone from many countries, and there does not seem to be much call for its return where it has left. Moreover, it does not seem that other parts of mathematics have picked up this loss. Less and less formal deduction is being taught in school mathematics courses.

Computers present particular problems to those who favor more work with deduction. Because of their ability to display example after example, computers encourage induction as a valid method of argument. Picture a triangle with its medians drawn. A student who is able to continuously deform this triangle on a screen, and who sees that the medians are still concurrent, will surely be less likely to think that a written demonstration of the concurrency is needed. Similarly, a student who can zoom in on the graph of a function to determine its maximum value to virtually any desired accuracy is not likely to see calculus as being as powerful as previous generations saw it. For this reason, the current condition in most countries, in which formal deduction is taught only to some, is not likely to change. Formal deduction may even be taught to fewer students in the future.

These developments reflect a fundamental problem for mathematics education. The requirement that results be deduced in order to be valid is one of the fundamental characteristics of mathematical thought; it is too important not to be taught to all. Yet it is imperative to take advantage of the power of technology to experiment with mathematics and thus to conjecture from experiments. Nevertheless, to go through school mathematics and do many experiments but have little experience with proofs is like going through science education with little attention to experiments but much experience deducing from assumed principles. To avoid proof – even to de-emphasize it – would seem to do a disservice to the discipline of mathematics.

13. SUMMARY

We are in an extraordinary time for mathematics, a time unlike any that has been seen for perhaps 400 to 500 years. The accessibility of mathematics to the population at large has increased dramatically due to advances in technology. These advances make it likely that more mathematics than ever before will become part of the fabric of everyone's education and everyday literacy. But the mathematics will not be a superset of what is taught today, for those things that can be done quickly and easily by computers are very likely to disappear from the curriculum. What will remain will probably be a more conceptual and more applied and more visual mathematics.

REFERENCES

American Association for the Advancement of Science (1989). *Science for all Americans.* Washington: AAAS.

College Entrance Examination Board Commission on Mathematics (1959). *Program for college preparatory mathematics.* New York: CEEB.

National Advisory Committee on Mathematical Education (NACOME) (1975). *Overview and analysis of school mathematics: Grades K-12.* Reston, VA: National Council of Teachers of Mathematics.

National Council of Teachers of Mathematics (1989). *Curriculum and evaluation standards for school mathematics.* Reston, VA: NCTM.

National Council of Teachers of Mathematics (1990). *Algebra for everyone* (edited by Edgar Edwards, Jr.). Reston, VA: NCTM.

Robitaillc, D. F. (1989). Students' achievements: Population A. In D. F. Robitaille & R. A. Garden (Eds.), *The IEA study of mathematics II: Contexts and outcomes of school mathematics.* Oxford: Pergamon.

Swetz, F. (1987). *Capitalism and school arithmetic: The New Math of the 15th Century.* LaSalle, IL: Open Court Publishing Co..

Thurow, L. (1991, October). *Public Investment.* Paper presented at the Economic Policy Institute Conference on Public Investment. Washington, DC.

Travers, K. J., & Westbury, I. (1989). *The IEA study of mathematics I: Analysis of mathematics curricula.* Oxford: Pergamon.

Tufte, E. (1983). *The visual display of quantitative information.* Cheshire, CT: Graphics Press.

University of Chicago School Mathematics Project (1990, 1991, 1992). *Transition mathematics. Algebra. Geometry. Advmnced algebra. Functions, statistics, and trigonometry. Precalculus and discrete mathematics.* Glenview, IL: Scott, Foresman.

Usiskin, Z. (1987). Resolving the continuing dilemmas in geometry. In *Learning and Teaching Geometry: The 1987 Yearbook of the National Council of Teachers of Mathematics.* Reston, VA: NCTM.

Acknowledgements

This paper is adapted from a talk given at a subplenary session of the 7th International Congress on Mathematical Education, (ICME-7) in Quebec City, August, 1992. I would like to thank my wife Karen for her help in organizing the talk.

CHAPTER 7

HISTORY AND EPISTEMOLOGY OF MATHEMATICS AND MATHEMATICS EDUCATION

edited and introduced
by
Rolf Biehler
Bielefeld

A theory of mathematical knowledge and its relation to individuals and social systems, a theory relating the mathematical learning processes in history within scientific communities to the learning processes and the knowledge development in individuals under conditions of schooling, would be quite helpful for the didactics of mathematics. This chapter is concerned with some aspects of this problem, and its papers refer to various referential sciences, for instance, to philosophy and history of mathematics and of science in general, sociology of knowledge and of education, or epistemology of mathematics.

The papers have more or less a common concern underlying their epistemological and historical analyses, namely, to overcome the isolation of mathematics and regard and teach it as a subject with broad relations to many other domains of human knowledge and activity. The mathematical problem and puzzle solver is not the model of the student aimed at; rather students should be encouraged to develop their personal relationship to

R. Biehler, R. W. Scholz, R. Sträßer, B. Winkelmann (Eds.),
Didactics of Mathematics as a Scientific Discipline, 327-333.

mathematics as part of culture and society. It has frequently been suggested that epistemology and history of mathematics should become a topic of mathematics education to foster mathematical metacognition and meta-knowledge on mathematics and its learning. For instance, Papert (1980) is a prominent advocate of the idea of children as epistemologists. NCTM (1969, 1989) are examples of classroom uses of history of mathematics. Up to now, actual classroom practice has seldom taken up these topics explicitly, although epistemological problems are everywhere in the everyday classroom. However, direct classroom use would be only one possible practical application of the research presented in this chapter. Its indirect significance through teacher education (see chapter 2) and curriculum design (see chapter 1) and its relevance for other domains of didactics such as psychological research (chapter 5) is even more important, as has been convincingly argued by, for instance, Steiner (1987).

The "modern" concern of philosophy and epistemology of mathematics with the didactics of mathematics can be traced back at least as far as René Thom's critique of the *new math* reform. His statement, "In fact, whether one wishes it or not, all mathematical pedagogy, even if scarcely coherent, rests on a philosophy of mathematics" (1973, p. 204), is one of the sources cited most often by mathematics educators who are arguing for an epistemological and historical reflection on mathematics as part of the research domains of didactics of mathematics. However, Thom's belief is itself a reaction to the new math reform that was based fairly explicitly on a philosophy of mathematics rooted in the Bourbaki interpretation of mathematics as well as in set theory, logic, and work on the so-called foundations of mathematics (Steiner, 1965a, 1965b). The reflection and conscious change or choice of the implicit assumptions about the process of didactical transposition to which philosophical aspects belong can be seen as part of the rationalization and theoretization of practical activity of preparing mathematics for teachers and students within didactics of mathematics.

An early sketch of a research program in this area was formulated by Otte, Jahnke, Mies, and Schubring:

> The didactics of mathematics requires a "philosophy of mathematics" in the sense of Thom for a series of fundamental issues: (a) questions regarding the relationship between mathematical abstraction and experience; (b) the difficulties involved in grasping the inherent regularity of mathematical research processes, which are directly relevant to the problems involved in a productive acquisition of mathematical concepts; (c) the complex relationships between mathematics and its applications in social practice, which play a multifaceted role within the discussion on the content and construction of a mathematics curriculum and its integration into general education; and finally (d) the problem of interrelationship between the theoretical system of mathematics and the contents and methods of mathematics instruction. (Otte, Janke, Mies, & Schubring, 1974, p. 6, translated, R.B.)

For the interests of didactics of mathematics, it is particularly important that the history of mathematics is blind without epistemology and that the epistemology of mathematics is empty without history. This is a famous dictum of Imre Lakatos that modifies a famous dictum of Kant. Internationally, the *International Study Group on the Relations Between History and Pedagogy of Mathematics* focuses on history as it relates to didactics. In some countries, for instance, in Germany, a series of conferences has been initiated on this topic, see Steiner and Winter (1985), Steiner (1990), and Jahnke, Knoche, and Otte (1991), who also provide overviews on activities in other countries (UK, France, Italy, USA). More than the history of mathematics, epistemological aspects of mathematics seem to be an integrated aspect of didactical research. Vergnaud (1990), for instance, gives an informative summary of the role of epistemology in the psychology of mathematics education (see also chapter 5, this volume). The didactical research on epistemological obstacles met in history and, in a transformed way, in the learning process of students particularly illustrates how the history of mathematics can be used and is relevant for the psychology of mathematics education.

Paul Ernest's article on *the philosophy of mathematics and the didactics of mathematics* aims at a comprehensive picture of the relation between the two. He describes the change in the philosophy of mathematics itself from a prescriptive, absolutist account to a broad spectrum of social views of mathematics. These developments are intimately related to widespread currents in transdisciplinary thought. Ernest formulates criteria for a philosophy of mathematics that are adequate for the didactics of mathematics. He provides empirical evidence on René Thom's thesis of the relevance of philosophy by analyzing various educational movements in mathematics education, tracing back the influence of different philosophies of mathematics and their combination with pedagogical philosophies. Social philosophies of mathematics that acknowledge culture-embeddedness resonate with the aims of critical mathematics education. This is a topic that *Mogens Niss* (this chapter) discusses in his paper on the basis of his analysis of the role of mathematics in society and in connection with educating for democracy. However, Ernest is right to emphasize that the same philosophy of mathematics is compatible with different styles of education. Bourbaki mathematics was associated with an activity-based discovery style of pedagogy as well as with a transmission style of pedagogy. Similarly, empirical research on teachers' cognitions and behavior has shown that there may be quite a mismatch between teachers' verbally subscribed philosophies of mathematics and their teaching practice (see, also, Cooney, this volume; Hoyles, 1992). Further on, he describes how a social constructivist view of mathematics and mathematics education may provide a theoretical framework for developing pedagogical principles and a new theory of teaching and learning mathematics that links together the social framework of mathematics education, classroom interaction, and individual work by students (see chapter 3, this volume).

In their paper on *the human subject in mathematics education and in the history of mathematics, Michael Otte and Falk Seeger* start by reviewing different reasons for using history in mathematics education. A major reason is that historical studies can counterbalance a mere technical treatment of mathematics and can reveal the involvement of subjects and their intentions and difficulties in mathematical thinking as well as the fact that there is not just one mathematics but many different forms of mathematics. Revealing the historicity of contemporary mathematics and appreciating the multiplicity of perspectives on mathematics may provide new self-awareness in developing one's own personal relationship to mathematics. Otte and Seeger's approach resonates with Ernest in the sense of overcoming the positivist-formalist doctrine of mathematics. However, it is still a problem to understand that, nevertheless, mathematics presents itself as a highly formalized and depersonalized body of knowledge, and it is far from clear why this is the case and how this can be related to personal development. The authors interpret mathematics as *theoretical* knowledge whose specificity is a form of generality that is a result of a division of labor in the sciences. Its formal character is closely related to the historical rise of "relational or functionalist thinking" in contrast to substantialist thinking. This distinction is elaborated with regard to two different identity principles in mathematics and principles of individualization in society.

Scientific knowledge as a product of division of labor enters into conflict with common (everyday) knowledge. This conflict cannot be resolved without scientific knowledge, because it is pervasive in contemporary society (Niss, this chapter). However, the self-image of science is not appropriate for being introduced in its reasoning, and the philosophical foundations and its historical genesis and roots become relevant for the resolution of the conflict between scientific and everyday knowledge. Compulsory general education cannot do without theoretical knowledge that opens up a universe of experience that is rich enough to allow a very great variety of members of society to participate. However, the theoretical character of knowledge causes the problem of meaning of mathematics (see, also, Steinbring, this volume). Development of meaning is regarded in a *maximal loop approach*, as a *journey* that brings the subject into contact with as many different perspectives on mathematics as possible. From the perspective of general education, the domain-specificity of knowledge cannot be the last word: The historicity of one's own perspective has to become conscious in the context of experiencing the multiplicity of perspectives in the classroom.

In his article on *mathematics in society, Mogens Niss* differentiates the social and cultural view of mathematics through an analysis of *mathematics in society* as a pure and applied science, as a system of instruments, as a field of aesthetics, and as a teaching subject. Being very complex issues, mathematical modeling and the problem of validating mathematical models are central for understanding the role of mathematics in the development and shaping of society. The eminent role of a country's mathematical qualification structure for its development becomes clear, especially from an international comparative perspective. However, discerning its concrete role is still a topic that requires more research. Mathematics' role is changing in history; recently, its influence and significance have been enhanced through the development and use of computers, although, in common discourse, mathematics is not visible behind computers and its applications. This is an example of the social *invisibility of mathematics,* which constitutes a relevance paradox for mathematics education that is central to Niss' analysis. The generality of mathematics, which is also analyzed by Otte and Seeger, is a specific characteristic that is responsible for mathematics unexpectedly being disguised in many diverse contexts in which people who actually apply mathematics are not regarded as mathematicians but as bankers, engineers, traffic planners, and so forth. Niss points to the need for further research on what one may also call a social epistemology of mathematics. An attempt in this direction is Fischer (1992), whose paper can be read as relating the perspective of Niss and Ernest in this chapter.

From his analysis, Niss reconceptualizes the task of preparing mathematics for students as working on the *justification problem* (of a certain mathematics education for a population), the *possibility problem* (what can be taught), and the *implementation problem.* This relates to the discussion in chapter 1 of preparing mathematics for students, however, now, from the (normative and critical) perspective of society. The didactical transposition should not only be concerned with making scholarly knowledge in mathematics teachable, it also has to relate and question its content in relation to the role of mathematics in society. A particular normative interpretation consists in viewing mathematics education as part of a general education (Allgemeinbildung) for a democratic society. Niss elaborates the consequences of such a conception that go far beyond educating for "intelligent citizenship." It aims at counterbalancing expert rule in society by using mathematics to provide *insight into the general,* that is, the acquisition of an overview and an understanding of main development patterns. However, it may be the case that students refuse to receive such an education, and the relevance paradox may reproduce itself on a psychological level in that students think that "mathematics is useless to me, but at the same time I know that I am useless without mathematics." This indicates that coordinating the goals for society with those of individual learners may be a difficult problem in practice.

Jim Kaput analyzes *the representational roles of technology in connecting mathematics with authentic experience.* His major concern is also related to overcoming the isolation of formal mathematics. The role of representations is regarded as crucial here, since the isolation of mathematics and the difficulties of students are partly due to its specific representations. Complementary to the analysis by Mogens Niss, which emphasizes social aspects, simulation and mathematical modeling are reanalyzed by Kaput from a representational and cognitive point of view with regard to the aim of connecting mathematics to students' authentic experience. This strategy of extracting mathematical forms from authentic experience is confronted with the strategy of casting mathematics in more "natural forms" as exemplified by using traditional manipulatives. A major thesis of the paper is that *computer technology* offers new representational freedom and flexibility that may support a new attack on the "island problem" of formal mathematics. In this sense, Kaput's paper is related to chapter 4 (technology) of this volume. Kaput develops a theoretical background for his representational perspective that combines insights from the history and epistemology of mathematics concerning the role of representations with cognitive theories on the role of representations for thinking processes. A first major historical change related to technology is the transition from inert to interactive media. Second, computer-assisted representation systems can become real *action systems,* that is, systems with an operative function similar to traditional symbol systems but that support different representational strategies than those underlying the traditional formal notations of mathematics. Third, a physical linkage of representational systems becomes possible. Examples of new software are briefly reviewed from this representational perspective (see, also, Tall, this volume).

A major example in Kaput's paper is the analysis of an extended scenario of a *lived-in* simulation, in which students can "drive" *MathCars,* an interactive simulation environment that is designed to provide qualitatively new experiences from which students might develop qualitatively new understandings of algebra, calculus, and physical motion and the role of modeling in this context. This may once more bring closer together that which became separated in history due to the division of labor between mathematics and physics. The author finishes by emphasizing the need for empirical research and for intervention by teachers so that students may really benefit from the new representational opportuinities.

This concrete part of Kaput's paper is related to a research-based preparation of mathematics for students (chapter 1). In this, it exemplifies the long path back from philosophical and epistemological reflection to constructive work with its own problems, regularities, and need for empirical evaluation.

REFERENCES

Fischer, R. (1992). The "human factor" in pure and applied mathematics. Systems everywhere: Their impact on mathematics education. *For the Learning of Mathematics, 12*(3), 9-18.

Hoyles, C. (1992). Mathematics teaching and mathematics teacher: A meta-case study. *For the Learning of Mathematics, 12*(3), 32-45.

Jahnke, H.-N., Knoche, N., & Otte, M. (1991). *Das Verhältnis von Geschichte und Didaktik der Mathematik.* Antrag für ein Symposium, Institut für Didaktik der Mathematik, Universität Bielefeld. [Proceedings in preparation]

NCTM (1969). *Historical topics for the mathematics classroom. 31st NCTM yearbook.* Wahington, DC: The National Council of Teachers of Mathematics.
NCTM (1989). Historical topics for the mathematics classroom (2nd ed.). Reston, VA: The National Council of Teachers of Mathematics.,
Otte, M., Jahnke, H. N., Mies, T., & Schubring, G. (1974). Vorwort. In M. Otte (Ed.), *Mathematiker über die Mathematik* (pp. 5-23). Berlin: Springer.
Papert, S. (1980). *Mindstorms: Children, computer and powerful ideas.* New York: Basic Books.
Steiner, H.-G. (1965a). Mathematische Grundlagenstandpunkte und die Reform des Mathematikunterrichts. *Mathematisch-Physikalische Semesterberichte, XII*(1), 1-22.
Steiner, H.-G. (1965b). Menge, Struktur, Abbildung als Leitbegriffe für den modernen mathematischen Unterricht. *Der Mathematikunterricht, 11*(1), 5-19.
Steiner, H.-G. (1987). Philosophical and epistemological aspects of mathematics and their interaction with theory and practice in mathematics education. *For the Learning of Mathematics, 7*(1), 7-13.
Steiner, H.-G. (Ed.). (1990). *Mathematikdidaktik-Bildungsgeschichte-Wissenschaftsgeschichte II. IDM-Reihe Untersuchungen zum Mathematikunterricht 15.* Köln: Aulis.
Steiner, H.-G., & Winter, H. (Eds.). (1985). *Mathematikdidaktik-Bildungsgeschichte-Wissenschaftsgeschichte. IDM-Reihe Untersuchungen zum Mathematikunterricht 12.* Köln: Aulis.
Thom, R. (1973). Modern mathematics: Does it exist? In A. G. Howson (Ed.), *Developments in mathematical education* (pp. 194-209). Cambridge: Cambridge University Press.
Vergnaud, G. (1990). Epistemology and psychology of mathematics education. In P. Nesher & J. Kilpatrick (Eds.), *Mathematics and cognition: A research synthesis by the International Group for the Psychology of Mathematics Education* (pp. 14-30). Cambridge: Cambridge University Press.

THE PHILOSOPHY OF MATHEMATICS AND THE DIDACTICS OF MATHEMATICS

Paul Ernest
Exeter

1. DEVELOPMENTS IN THE PHILOSOPHY OF MATHEMATICS

The 20th century has seen the flowering of the philosophy of mathematics as a field of professional research. There have been a number of developments of the utmost importance for the didactics of mathematics. The first has been a shift from a concern to give a *prescriptive* (or normative) account to a *descriptive* (or naturalistic) account of mathematics (Ernest, 1991, 1992). Two traditional assumptions concerning the nature of mathematics are that (a) mathematical knowledge is absolutely secure objective knowledge, the cornerstone of all human knowledge and rationality (the assumption of absolutism), and (b) that mathematical objects such as numbers, sets and geometric objects all exist in some objective superhuman realm (the assumption of Platonism). The prescriptive tradition has sought to reformulate mathematical knowledge in order to validate these assumptions. Recently, there has been a shift in academic philosophy of mathematics from attempts to erect absolutist epistemological systems (the projects of Logicism and Formalism) to ontological concerns, but the two assumptions of prescriptive philosophy of mathematics still dominate the field. This traditional approach is represented by Benecerraf and Putnam (1983).

In contrast, a *descriptive* or naturalistic turn in the philosophy of mathematics has been emerging more recently. This is still an ill-defined movement, which Aspray and Kitcher (1988) term a "maverick" tradition. What binds this movement together is a shared rejection of the epistemological and ontological assumptions of prescriptive philosophy of mathematics, and a positive concern to broaden the scope of the philosophy of mathematics to that of giving an account of mathematics acknowledging the centrality of mathematical practice and social processes. Thus the concern is to describe the naturally occurring epistemological and more generally philosophical practices of the discipline, rather than to legislate normatively.

Much of the rejection of the prescriptive task of the philosophy of mathematics comes from a view, which is spreading amongst the communities of mathematicians, educationists and, to a lesser extent, philosophers, that the foundations of mathematics are not as secure as was supposed. Gödel's

R. Biehler, R. W. Scholz, R. Sträßer, B. Winkelmann (Eds.),
Didactics of Mathematics as a Scientific Discipline, 335-349.

(1931) first incompleteness theorem has shown that formal axiomatics and proofs must fail to capture all the truths of most interesting mathematical systems (those at least as strong as the theory of Peano arithmetic). His second incompleteness theorem shows that, in such systems, consistency is indemonstrable without adopting more assumptions than in the system itself. Together, these results severely weakened Hilbert's Formalism and Frege, Russell and Whitehead's Logicism. This has forced a concession from even the most computationally minded that human creativity cannot be replaced by mechanized deduction (Wang, 1974). More generally, it is increasingly accepted that any body of knowledge rests on assumptions that cannot themselves be given a secure foundation, on pain of infinite regress (Lakatos, 1976; Popper, 1979). There is also a growing dissatisfaction amongst mathematicians, philosophers and other scholars with the traditional narrow focus of the philosophy of mathematics, limited to foundational epistemology and ontology (Tiles, 1991; Tymoczko, 1986).

A number of authors have proposed that the task of the philosophy of mathematics is to account for mathematics more fully, including the "human face" of mathematics. Publications by Davis and Hersh (1980), Ernest (1991), Kitcher (1984), Lakatos (1976, 1978), Putnam (1975), Tymoczko (1976), Wang (1974) and Wittgenstein (1953, 1956), for example, have suggested new fallibilist, quasi-empirical or social constructivist views of mathematics. This descriptive or naturalistic turn in the philosophy of mathematics is represented by Tymoczko (1986).

The shift from prescriptive to descriptive accounts parallels a second shift from objectivist accounts of mathematics and mathematical knowledge to social accounts (possibly with subjective accounts seen as intermediary position). Although this seems to be an immediate corollary of the descriptive turn, there is still tremendous resistance from many philosophers and mathematicians to the notion that social processes and practices might be constitutively central to mathematics. Putnam (1975) and Machover (1983), for example, acknowledge that absolute foundations for mathematical knowledge are lacking, but are far from agreeing that mathematics is at base social. Karl Popper has been very influential in promoting the view that all scientific knowledge is fallible (his philosophy of science is termed "critical fallibilism"). But he resists any notion that scientific knowledge is constitutively social (Popper, 1979). Even his protégé Imre Lakatos, who perhaps made the most decisive contributions to the maverick tradition in philosophy of mathematics, in his later years argued for the primacy of logic and objectivity over the social, at least in his accounts of scientific knowledge (Lakatos, 1978).

The various different descriptive social philosophies of mathematics making up the "maverick" tradition share a number of assumptions and implications. They view mathematics as the outcome of social processes and understand mathematics to be fallible and eternally open to revision, both in terms of its proofs and its concepts. They reject the notion that there is a unique, rigid and permanently enduring hierarchical structure and accept instead the view that mathematics is made up of many overlapping structures. These, like a forest, dissolve and re-form. Since mathematical knowledge is always open to revision, the processes of creating mathematics gain in philosophical significance, for there is no ultimate product to focus on exclusively. Consequently, both the history and practice of mathematicians acquire a major epistemological significance (as well as needing to be accounted for naturalistically for descriptive purposes). This significance makes mathematics quasi-empirical, and not wholly disjoint from empirical science, as traditional philosophies of mathematics assert (Lakatos, 1978; Quine, 1960). The boundaries between the different areas of knowledge and human activity are not absolute, which means that mathematics is context-bound and value-laden, and not pure, remote and untouched by social issues such as gender, race and culture.

These concerns herald a third shift: a broadening of the concerns of the philosophy of mathematics (Körner, 1960; Tymoczko, 1986). A set of adequacy criteria for the accommodation of the shift towards a naturalistic and social orientation is as follows:

> A proposed philosophy of mathematics should . . . account for:
> (i) Mathematical knowledge: its nature, justification and genesis.
> (ii) The objects of mathematics: their nature and origins.
> (iii) The applications of mathematics: its effectiveness in science, technology, and other realms.
> (iv) Mathematical practice: the activities of mathematicians, both in the present and the past. (Ernest, 1991, p. 27)

To this should be added the need for an outline account of the learning of mathematics, because the transmission of mathematical knowledge from generation to generation is central to the social practice of mathematics; also, the learning of mathematics cannot be separated from the parallel practices of mathematicians in creating and communicating new mathematical knowledge (Ernest, in press). As well as being central to the *didactics* of mathematics, a theory of learning is also an aspect of the human-mathematics interaction that the *philosophy* of mathematics should also accommodate.

Developments in descriptive social philosophies of mathematics have parallels in widespread currents in transdisciplinary thought. Thus developments in the history of mathematics (Kline, Joseph, Høyrup, Szabo), cultural studies of mathematics (Bishop, Wilder, Mackenzie), anthropology of mathematics and ethnomathematics (Ascher, Crump, D'Ambrosio, Gerdes, Zaslavsky), the sociology of science, knowledge and mathematics (Bloor, Fisher, Restivo, Fuller), the rhetoric of science (Billig, Knorr-Cetina), interdisciplinary post-structuralist and post-modernist thought (Foucault, Walkerdine, Lyotard), semiotics (Rotman, Eco), social constructionist psychology (Gergen, Harré, Shotter), evolutionary epistemology (Campbell, Rav), the philosophy of science (Feyerabend, Hacking, Kuhn, Laudan) and philosophy in general (Rorty, Toulmin) are all looking towards social constructivist accounts of knowledge. These developments therefore also support the didactical consequences of social philosophies of mathematics discussed below.

2. THE PHILOSOPHY OF MATHEMATICS EDUCATION

The central claim of this chapter is that different positions in the philosophy of mathematics have significantly different implications for the didactics of mathematics as Thom, Hersh and Steiner claim:

> In fact, whether one wishes it or not, all mathematical pedagogy, even if scarcely coherent, rests on a philosophy of mathematics. (Thom, 1973, p. 204)

> The issue, then, is not, What is the best way to teach? but, What is mathematics really all about? . . . Controversies about . . . teaching cannot be resolved without confronting problems about the nature of mathematics. (Hersh, 1979, p. 34)

> *Thesis 1*
> Generally speaking, all more or less elaborated conceptions, epistemologies, methodologies, philosophies of mathematics (in the large or in part) contain – often in an implicit way – ideas, orientations or germs for theories on the teaching and learning of mathematics . . . .

> *Thesis 2*
> Concepts for the teaching and learning of mathematics – more specifically: goals and objectives (taxonomies), syllabi, textbooks, curricula, teaching methodologies, didactical principles, learning theories, mathematics education research designs (models, paradigms, theories, etc.), but likewise teachers' conceptions of mathematics and mathematics teaching as well as students' perceptions of mathematics – carry with them or even rest upon (often in an implicit way) particular philosophical and epistemological views of mathematics. (Steiner, 1987, p. 8)

Any philosophy of mathematics has powerful implications for social and educational issues and many didactic consequences. However, these are not all strictly logical deductions from the position, and a number of aims, values and additional conceptions must be assumed in addition to the philosophy of mathematics *per se* (Ernest, 1991). Because the link is not one of logical implication, it is theoretically possible to consistently associate a philosophy of mathematics with almost any educational practice or didactic approach. Both a neo-behaviourist (such as Ausubel) and a radical constructivist are concerned to ascertain what a child knows before commencing teaching, despite having diametrically opposite epistemologies. Likewise, a traditional purist mathematician and a social constructivist may both favour a multicultural approach to mathematics, but for different reasons (the former to humanize mathematics, the latter to show it is the social construction of humanity).

Although there is no logical necessity for, for example, a transmission-style pedagogy to be associated with an absolutist, objectivist epistemology and philosophy of mathematics, such associations often are the case (Ernest, 1988, 1989, 1991). This is due to the resonances and sympathies between different aspects of philosophies, ideologies and belief systems, which form links and associations in moves towards maximum consistency and coherence.

3. DIDACTICAL CONSEQUENCES OF PRESCRIPTIVE, OBJECTIVIST PHILOSOPHIES OF MATHEMATICS

Many didactical consequences of prescriptive philosophies, such as Logicism and Formalism, follow from their identification of mathematics with rigid and logically structured mathematical theories following the Euclidean/Cartesian paradigm of mathematics as an objective, absolute, incorrigible body of knowledge. According to such views, mathematics rests on certain foundations, such as logic, and rises from its base to heights of abstraction and generality. The structure that supports the edifice is that of deductive logic, which locks it into a fixed and rigid hierarchy. Consequently, mathematical knowledge is viewed as timeless, although new theories and truths may be added; it is superhuman and ahistorical, for the history of mathematics is irrelevant to the nature and justification of mathematical knowledge; it is pure isolated knowledge, which happens to be useful because of its universal validity; it is value-free and culture-free, for the same reason.

Such a view of mathematics may be related to current developments in British mathematics education. An absolutist conception of mathematics (and knowledge in general) underpins the British National Curriculum in mathematics. For this identifies the mathematics curriculum as a rigid hierarchical structure of five Attainment Targets, comprising items of knowledge and skill at 10 discrete levels. The hierarchical structure of the National Curriculum may be viewed as a "fractional distillation device," because it serves to separate off different fractions of the school population by class/gender/race and future occupation (Dowling & Noss, 1990; Ernest, 1991).

An important didactic consequence of absolutist philosophies of mathematics is that they support a transmissive teaching approach based on the broadcast metaphor. If mathematics is a pre-existing and superhuman body of knowledge, then its teaching is a matter of efficient transmission. The emphasis is on the content, and any obstacles in coming to terms with it would be due to the learner's poor grasp (or the teacher's unclear exposition) of the ready-made knowledge being transmitted. Such views of mathematics may be associated with humanistic approaches to mathematics teaching, but these may merely seek to ameliorate the problem arising from the intrinsic nature of mathematics (i.e., its objective purity, abstractness and difficulty).

4. PROGRESSIVE ABSOLUTISM AND ITS DIDACTICAL CONSEQUENCES

The intuitionism developed by Brouwer (1913) can be viewed as both prescriptive and descriptive. It is prescriptive in attempting to secure the foundations of a part of mathematics on a constructive basis. But there is also an attempt to construct a system that is faithful to the subjective experience of the mathematician: Construction, algorithm and agency are all important both in the construction of mathematics and in its foundations. This fidelity to subjective experience (i.e., descriptiveness) may account for the popularity of intuitionism, despite its weaknesses (Machover, 1983). This perspective has been termed "progressive absolutism" (Confrey, 1981; Ernest, 1991). It views knowledge as both progressive and open ended, but presupposes some underlying shared and absolute truth to which subjective constructions tend as to a limit.

Some contrast with the educational outcomes of prescriptive philosophies of mathematics is provided by the progressive education movement. This provided the background to progressive elementary education in Britain, with its emphasis on exploration, activity and child-centredness. In mathematics education, this movement has encouraged problem-solving, discussion, investigational approaches and a respect for the creations of the learner. On the negative side, this perspective can be over-protective of the learner, wishing to shield her or him from the stigma and hurt of getting answers wrong (ticks but not crosses are used). It often fails to engage with real-life social and political issues; not only the importance of examinations but also confidence in critical social arithmetic, essential to gaining power over one's adult life.

Many of these weaknesses arise from the fact that progressive mathematics education is based on a humanized absolutist conception of mathematics that regards mathematics as pure and absolute. Although a progressive pedagogy expects learners to build meaning actively on the basis of exploration, conjectures, and other constructive processes, there is an underlying assumption that there is a correct body of mathematical knowledge that will emerge from construction. Thus, although the pedagogy is not based on the broadcast metaphor, it assumes that there are truths to be discovered, which are self-evidently correct once found. The focus is therefore on the deep constructive activity of the learner, assumed to produce the required mathematical knowledge and truth. The teacher's role is restricted to that of midwife, facilitator and corrector when the learner goes astray; not the leader in the negotiation of meaning and knowledge. These views underpin certain versions of constructivism (not the radical constructivism of von Glasersfeld, 1983) and reveal their underlying epistemological weakness.

5. DIDACTICAL CONSEQUENCES OF SOCIAL PHILOSOPHIES OF MATHEMATICS

A social view of mathematics has important implications for the didactics of mathematics and educational issues including those of mathematics and gender, race and multiculture, since it recognizes the social import and value-laden nature of mathematics; for pedagogy, by supporting fully investigational and problem-solving approaches as paralleling the means by which mathematical knowledge is generated; as well as enabling challenges to hierarchical views of mathematics, learning and ability due to a rejection of fixed and objective epistemological hierarchies. Since social views of mathematics acknowledge its fallibilism and culture-embeddedness, and thus look critically at received knowledge structures and their relations with society, they resonate with the aims of critical mathematics education to educate confident problem posers and solvers able to critically evaluate the social uses of mathematics. Some of the didactical consequences of the main social philosophies of mathematics are as follows.

Wittgenstein (1952, 1956) offers a powerful social vision of mathematics. His key contribution is to recognize the subjective and social basis of certainty that following a rule in mathematics or logic does not involve logical compulsion. Instead, it is based on the tacit or conscious decision to accept the rules of a "language game," which are grounded in pre-existing social "forms of life." For didactics, Wittgenstein's importance is to show that the "certainty" and "necessity" of mathematics are the result of social processes of development, and that all knowledge, including that in education, presupposes the acquisition of language in meaningful, already existing social contexts and interactions. Through his concepts of language games and forms of life, Wittgenstein acknowledges not only the primacy of social context but also its multifaceted nature. Thus, he anticipates the notion that human activities fall into a set of different practices with different purposes, associated language games, resources and participants. Thus he anticipates much foundational thought in modern philosophy and educational theory that affords primacy to culture, context or discursive practice. However Wittgenstein's approach is synchronic rather than diachronic, that is, he emphasizes existing social structures and linguistic use-patterns but not their historical development (Foucault's "archaeology of knowledge").

Lakatos (1976, 1978) goes beyond Wittgenstein's insight to show more fully the historical and conceptual change-basis of all of the concepts, terms, symbolism, theorems, proofs and theories of mathematics. The historical dimension can show why particular concepts and results were developed in mathematics, based on particular problems and difficulties encountered historically. For didactics, his importance is due to bringing in the historical dimension, and for showing that the methodology of mathematics as used by practising mathematicians does not differ in kind from the heuristics of problem-solving in the classroom. He also shows the import of conventions, agreement and power in the warranting of mathematical productions (mathematical knowledge), which so strikingly parallels classroom developments.

Kitcher (1984) extends more systematically the historico-cultural basis of mathematics, and, in particular, shows the role of mathematical authorities (e.g., teachers) in communicating mathematical knowledge at both the disciplinary and the didactic levels. He ascribes an epistemological role to authorities as providing warrants for mathematical assertions, which then later become warranted for an individual by reason or some similar process. Thus he admits three types of warrant: sensory experience, authority and reason, which apply both to the cultural and individual development of mathematical knowledge.

Davis and Hersh (1980, 1988) elaborate on and extend the insights of Lakatos. Their unique contribution is to demonstrate the cultural nature of mathematics, how it has both an inner and outer aspect. Whereas previous scholars have emphasized the internal history of mathematics, these authors demonstrate that mathematics permeates and shapes all aspects of social and cultural life, and is, in turn, shaped by social forces. Didactically, their position is important, because it transcends the pure-applied and academic-folk mathematics boundaries and shows that mathematical activity is universal, multicultural, and cannot be divorced entirely from its social context of use. Furthermore, having power over the social forms, manifestations and uses of mathematics is a vital goal of mathematics education.

The social constructivist view (Ernest, 1991, 1992, in press) offers a synthesis of the philosophies indicated here, and makes a number of their features more explicit. Beyond this, it is novel in several features, including the proposal that individual and disciplinary knowledge of mathematics are mutually interdependent, and that they recreate each other through interpersonal interaction, mediated by texts or other linguistic, symbolic or iconic representations (possibly at a distance but modelled on conversation). It suggests that the development of new mathematical knowledge, as well as new subjective understandings of mathematics, are derived from interpersonal negotiations and dialogue; that is, that learning and making mathematics emerge from similar processes. There is also a particular stress on the tacit and linguistic knowledge shared by members of a culture, which provides a basis for their acquisition of mathematical knowledge. Finally, a suggestion is made that mathematicians and others, through extended work with symbols, construct such convincing imagined "math-worlds" that the objects of mathematics seem to have an independent existence. Didactically, this parallels an inverted problem, that symbolic manipulations often do not lead to the construction of subjective math-worlds, leading to problems of incomprehension, alienation and failure. These problems lead to one of the contradictions of mathematics education. Mathematics is, on the face of it, the most rational of all subjects, since its conclusions are legitimated by reason alone. Yet when the reasoning behind mathematics is not understood, because of the strict rigour and abstract symbolism needed for precision and power, it becomes the most irrational and authoritarian of subjects.

Many didactic consequences flow from social constructivism. One is the importance of the linguistic basis of the understanding of mathematics. Children begin schooling with a rich vocabulary (half that of an adult) and a set of mathematical terms and notions. They can already sort, count, locate, play, make, design, plan, explain, argue, and maybe measure: all the activities Bishop (1988) identifies as the cultural basis of mathematics. According to social constructivism, ontogeny, if not recapitulating, at least parallels phylogeny. The developing child's "culture" includes all the proto-mathematical ideas, actions and terms needed for the meaningful foundation of formal school mathematics, and social constructivism supports the view that formal instruction should build on this foundation.

6. CONSTRUCTIVE EDUCATIONAL OUTCOMES OF A SOCIAL VIEW OF MATHEMATICS

The various social views of mathematics, social constructivism in particular, when combined with parallel social theories described above, give rise to a number of features of significance for the mathematics classroom. In outline, these include:

1. The social and cultural context within which all mathematics occurs, including interpersonal relationships, social institutions and power relations.
2. The social processes involved in the determination, construction and negotiation of mathematical concepts, methods, symbolism, arguments and results.
3. The historico-cultural context of mathematics, the sources and uses of the artifacts, tools and concepts involved.
4. The linguistic basis of mathematical knowledge, and, in particular, the role of the special symbolism in mathematics.
5. Education is an intentional activity, and so there are the values, purposes and goals underpinning the processes of mathematics education.
6. Mathematics depends crucially on the subjective construction of meaning, and the ability to construct, call up, and enter the resultant personally imagined math-worlds, since there is no "real-world" described by mathematics. The paradigm for these is the social worlds of meaning every child learns to construct through participation in communicative social practices.
7. Mathematics (including mathematical knowledge) is a discursive social practice that is not wholly disjoint from other social practices or areas of knowledge; the separateness of mathematics from other school subjects (and out-of-school practices) is a construction.

6.1 Principles of a Social Constructivist Pedagogy of Mathematics

Respect for learner's meanings and prior knowledge. Knowledge and understanding of mathematics depends on the learner's prior knowledge of language and sense-making. Instead of discounting prior learning and imposing a completely new discursive practice of formal school mathematics, which disregards and discounts the value of children's out-of-school knowledge (but still implicitly depends on it), a social constructivist pedagogy should consciously build on that knowledge and meaning. Thus the oral dimension is vitally important, as is getting children to describe their ideas, interpretations, methods, strategies and out-of-school contexts and meaning worlds. Attention should be paid to developing and extending their vocabulary and the associated meanings for terms like large, small, next to, between, above, angle, how many, different, alike, less, more, number, shape, and so forth.

Building on child-methods through the negotiation of knowledge. Much of the symbolism, conventions and knowledge within school mathematics (and mathematics in general) is arbitrary and depends on the decision of the appropriate community and its utility in the pursuit of certain goals. These processes should be made explicit in a number of ways, including the following: (a) The setting up of didactical situations in which learners develop their own algorithms for solving problems, then, through discussion, compare and streamline them, then compare them with standard algorithms. Walther (1984) provides an example in which the multiplication algorithm is so developed. (b) Offering young learners the opportunity to develop their own representations and symbolism, as in Hughes (1986), where pre-school children developed their own numeral notation. Through processes of social negotiation they can then be presented with standard notation. (c) The explicit recognition of the arbitrary definitions of mathematics and the underlying rationale for them is needed: for example how $2^0, \ldots, n^0$ are defined and the pattern they support; and why 0^0 is not. (d) In the solution of mathematical problems, there may be no unique "right" answer or method, just as in an English language essay. Correctness usually only applies to matching a convention or its consequences and is less relevant to higher-level or creative work in mathematics as in any other school subject.

The inseparability of mathematics and applications (and the centrality of motivation and relevance). Mathematics teaching, especially at the higher levels, needs to result not only in learner knowledge of symbolism, algorithms and formal methods and systems. Following Sneed (1971), scientific knowledge can be understood as comprising a repertoire of interpretations and applications in a variety of domains of human practice as well as formal theory. Steiner (1987) recounts Jahnke's extension of this notion to mathematics, and Dowling (1991) likewise has theorized a range of contexts of application and practice as central to mathematical knowledge. A social constructivist pedagogy would not separate the intellectual tools of mathematics from their uses. Thus the curriculum would treat concepts, methods and other tools in the light of (a) their historical and cultural origins and the problems they serve; (b) current uses and applications, including the mastery of a chosen central selection; and (c) contexts of use of direct meaning to the lives and interests of the learners. Mellin-Olsen (1987) provides examples of such projects from Norway.

6.2 A Social Constructivist Theory of Learning Mathematics

A sketch of a social constructivist theory of mathematics learning and school mathematical activity related to current philosophical work (Ernest, in press) is offered here as a final didactical consequence. This has three levels: the *social context* (including classroom, teacher, learners, etc.), the *frame* surrounding any task or activity, and the linguistically presented *task or activity* around which school mathematics pivots.

Social context. The context of the mathematics classroom is a complex, organized social form of life that includes:

1. persons, interpersonal relationships, patterns of authority, student-teacher roles, modes of interaction, and so forth;
2. material resources, including writing media, calculators, microcomputers, texts representing school mathematical knowledge, furniture, an institutionalized location and routinized times;
3. the language of school mathematics (and its social regulation), including: (a) the content of school mathematics: the symbols, concepts, conventions, definitions, symbolic procedures and linguistic presentations of mathematical knowledge; and (b) modes of communication: written, iconic and oral modes, modes of representation and rhetorical forms, including rhetorical styles for written and spoken mathematics.

For example, teacher-student dialogue (typically asymmetric in classroom forms) takes place at two levels: spoken and written. In written "dialogue," students submit texts (written work on set tasks) to the teacher, who responds in a stylized way to their content and form (ticks and crosses, marks awarded represented as fractions, crossings out, brief written comments, etc.).

This theorization draws on a number of sources that regard language and the social context as inextricably fused: Wittgenstein's philosophy, Foucault's theory of *discursive practices,* Vygotsky and Activity Theory, Halliday and sociolinguistics. For applications to the learning of mathematics, see Walkerdine (1988), Pimm (1986) and Ernest (1991).

Frame. This concept is elaborated in a number of different ways by Marvin Minsky, Erving Goffman and others, and applied to mathematical activity by Davis (1984) and Ernest (1987), albeit in an information-processing orientation. It resembles Papert and Lawler's concept of microworld, and that of "solution space" in problem-solving research. Frames concern a specific (but growing) range of tasks and activities, and each is associated with a particular set of representations, linguistic and otherwise, a set of intellectual tools, both symbolic and conceptual (and possibly a set of manipulable tools, such as rulers or calculators).

Frames have a dual existence, both public and private. The public aspect of a frame corresponds to a mathematical topic or problem type and the associated language and intellectual tools. It constitutes what is taken as shared by a number of persons, although different instantiations of a frame will vary, for example, across time and social location.

In its private aspect, a frame is constructed individually by each person (learner or teacher) as a sense-making and activity-performing device (resembling a "schema"). The meanings, conceptual tools and goal types make up a math-world, which is a subjective construction associated with the frame, at least in outline (specific details may be filled in during particular tasks). Each individual's personal construction of a frame is associated with a body of cases of previous uses of the frame, sets of symbolic and conceptual tools, and stereotypical goals. Social interaction allows some meshing of the individually constructed frames, and a crucial feature of frames is that they are genetic, continually developing and growing as a result of interaction and use (the varieties of frame use and growth correspond to Donald Norman's categories of schema use: tuning, routine use or assimilation, application, restructuring or accommodation).

The process of frame utilization and growth requires the learner internalizing and pursuing an activity-related goal (as in Leont'ev's version of Activity Theory). Particularly in the engagement with and performance of non-routine tasks, the learner will be making effort and success-likelihood estimations, and may disengage from the goals and give up the task or seek assistance from others. The learner may lack confidence and need reassurance; or may not be able to make the transformations unaided (i.e., lack a tool, or not know which to apply) in order to achieve the goal. Then the task lies within the learner's Zone of Proximal Development, and assistance enables the learner to make the symbolic transformations, hence to extend the appropriate frame so that ultimately she or he can undertake this challenging type of task unaided.

Task or activity. Typically, a task is a text presented by someone in authority (the teacher), specifying a starting point, intended to elicit a frame (a task in a sequence may assume a frame is in use), and indicating a goal state: where the transformation of signs is meant to lead. The theorization of tasks draws on Activity Theory and semiotic analyses of mathematics (e.g., Rotman, 1988) as well as cognitive science approaches. Mathematics education sources include Christiansen, Howson and Otte (1986), Cobb (1986), Mellin-Olsen (1987), Davis (1984), Skemp (1982), and Ernest (1987). From a semiotic perspective, a completed mathematical task is a sequential transformation of, say, n signs inscribed by the learner, implicitly derived by n-1 transformations. The first sign is a representation of the task as initially construed (the text as originally given, curtailed, or some other mode of representation, such as a figure); the last is a representation of the final symbolic state, intended to satisfy the goal requirements as interpreted by the learner. The rhetorical requirements of the social context determine which sign representations and which steps are acceptable. Indeed, the rhetorical mode of representation of these transformations with the final goal representation is the major focus for negotiation between learner and teacher, both during production and after the completion of the transformational sequence.

Following Saussure's analysis of a sign into signifier and signified, it can be said that transformations take place on either or both of these levels of signification. Signifieds vary with interpreter and context, and are far from uniquely given. The level of signifieds is a private math-world constructed individually, although, in a degenerate activity, it may be minimal, corresponding to Skemp and Mellin-Olsen's notion of "instrumental understanding." Signifiers are represented publicly, but to signify for the learner (or teacher), they have to be attended to, perceived, and construed as symbols. The structure of a successfully completed task can be represented linearly as a text, but it does not show the complex non-linear process of its genesis. Finally, the levels of signifier and signified are relative; they are all the time in mutual interaction, shifting, reconstructing themselves. What constitutes a sign itself varies: Any teacher-set task is itself a sign, with the text as signifier, and its teacher goal (and possibly frame) as signified.

This theory suggests some of the multi-levelled complexity involved in a learner carrying out a mathematical activity. This includes the construction of a math-world, one or more thought experiments or "journeys" in it, and the construction of a text addressing the rhetorical demands of written mathematics in the particular social (school) context. Any such activity needs to be situated in a student's learning history in the social context of the mathematics classroom in order to situate their learning activities. Ernest (1993) provides a fuller account and an example of this theory applied to a case study of a learner.

7. CONCLUSION

This theory sketch offers a synthesis combining learners' constructions of meaning with their public symbolic activities situated in the social context of school mathematics. One of the strengths of the approach is that it is able to take account of the demands of the rhetoric of school mathematics, something largely missing in research on learning, but necessitated by a social constructivist view of mathematics.

This concludes a brief review of the philosophy of mathematics and the didactics of mathematics. The treatment of the former is a balanced account of developments in philosophy, albeit from one perspective. However, in reviewing didactical implications, arbitrary choices have been made and personal preferences compressed into a short account. So I claim neither to offer a comprehensive review nor an adequate justification, but indicate instead part of a research agenda.

REFERENCES

Aspray, W., & Kitcher, P. (Eds.). (1988). *History and philosophy of modern mathematics.* Minneapolis, MN: University of Minnesota Press.

Benecerraf, P., & Putnam, H. (Eds.). (1983). *Philosophy of mathematics: Selected readings* (rev. ed.). Cambridge: Cambridge University Press.

Bishop, A. J. (1988). *Mathematical enculturation.* Dordrecht, Netherlands: Kluwer.

Brouwer, L. E. J. (1913). Intuitionism and formalism. *Bulletin of the American Mathematical Society. 20,* 81-96.

Christiansen, B., Howson, A. G., & Otte, M. (Eds.). (1986). *Perspectives on mathematics education.* Dordrecht, Netherlands: Reidel.

Cobb, P. (1986). Contexts, goals, beliefs, and learning mathematics. *For the Learning of Mathematics, 6*(2), 2-9.

Confrey, J. (1981). Conceptual change analysis: Implications for mathematics and curriculum, *Curriculum Inquiry, 11*(5), 243-257.

Davis, P. J., & Hersh, R. (1980). *The mathematical experience.* Boston, MA: Birkhauser.

Davis, P. J., & Hersh, R. (1988). *Descartes' dream.* London: Penguin.

Davis, R. B. (1984). *Learning mathematics.* Beckenham, Kent: Croom Helm.

Dowling, P. (1991). The contextualising of mathematics: Towards a theoretical map. In M. Harris (Ed.), *Schools, mathematics and work* (pp. 93-120). London: Falmer.

Dowling, P., & Noss, R. (1991). *Mathematics versus the National Curriculum.* London: Falmer.

Ernest, P. (1987). A model of the cognitive meaning of mathematical expressions. *British Journal of Educational Psychology, 57,* 343-370.

Ernest, P. (1989). The impact of beliefs on the teaching of mathematics. In C. Keitel, P. Damerow, A. Bishop, & P. Gerdes (Eds.), *Mathematics, education and society* (pp. 99-101). Paris: UNESCO.

Ernest, P. (1989, July). *Mathematics-related belief systems.* Poster presented at the 13th Psychology of Mathematics Education Conference, Paris.

Ernest, P. (1991). *The philosophy of mathematics education.* London: Falmer.

Ernest, P. (1992). The nature of mathematics: Towards a social constructivist account. *Science and Education, 1*(1), 89-100.

Ernest, P. (1993). *Mathematical activity and rhetoric: Towards a social constructivist account.* Paper submitted to the 17th International Conference on the Psychology of Mathematics Eduction, July 1993, Tokyo.

Ernest, P. (in press) *Social constructivism as a philosophy of mathematics.* Albany, NY: SUNY Press.

Glasersfeld, E. von (1983). Learning as a constructive activity. In Janvier, C. (Ed.), *Problems of representation in the teaching and learning of mathematics* (pp. 3-17). Hillsdale, NJ: Erlbaum.

Gödel, K. (1931). Über formal unentscheidbare Satze der Principia Mathematica und Verwandter Systeme I. *Monatshefte für Mathematik und Physik, 38,* 173-198.

Hersh, R. (1979). Some proposals for reviving the philosophy of mathematics. *Advances in Mathematics, 31,* 31-50.

Hughes, M. (1986). *Children and number.* Oxford: Blackwell.

Kitcher, P. (1984). *The nature of mathematical knowledge.* New York: Oxford University Press.

Körner, S. (1960). *The philosophy of mathematics.* London: Hutchinson.

Lakatos, I. (1976). *Proofs and refutations.* Cambridge: Cambridge University Press.

Lakatos, I. (1978). *Philosophical papers* (Vols. 1 - 2). Cambridge: Cambridge University Press.

Machover, M. (1983). Towards a new philosophy of mathematics. *British Journal for the Philosophy of Science, 34,* 1- 11.

Mellin-Olsen, S. (1987). *The politics of mathematics education.* Dordrecht, Netherlands: Reidel.
Pimm, D. (1986). *Speaking mathematically.* Oxford: Blackwell.
Popper, K. (1979). *Objective knowledge.* Oxford: Oxford University Press.
Putnam, H. (1975). *Mathematics, matter and method.* Cambridge: Cambridge University Press.
Quine, W. V. O. (1960). *Word and object.* Cambridge, MA: MIT Press.
Rotman, B. (1988). Towards a semiotics of mathematics. *Semiotica 72*(1/2), 1-35.
Skemp, R. R. (1982). Communicating mathematics: Surface structures and deep structures. *Visible Language, 16*(3), 281-288.
Sneed, J. (1971). *The logical structure of mathematical physics.* Dordrecht, Netherlands: Reidel.
Steiner, H. G. (1987). Philosophical and epistemological aspects of mathematics and their interaction with theory and practice in mathematics education. *For the Learning of Mathematics, 7*(1), 7-13.
Thom, R. (1973). Modern mathematics: Does it exist? In A. G. Howson (Ed.), *Developments in mathematical education* (pp. 194-209). Cambridge: Cambridge University Press.
Tiles, M. (1991). *Mathematics and the image of reason.* London: Routledge.
Tymoczko, T. (Ed.). (1986). *New directions in the philosophy of mathematics.* Boston, MA: Birkhauser.
Walkerdine, V. (1988). *The mastery of reason.* London: Routledge.
Walther, G. (1984). Mathematical activity in an educational context. In R. Morris (Ed.), *Studies in mathematics education 3* (pp. 69-88). Paris: UNESCO.
Wang, H. (1974). *From mathematics to philosophy.* London: Routledge.
Wittgenstein, L. (1953). *Philosophical investigation.* Oxford: Blackwell.
Wittgenstein, L. (1979). *Remarks on the foundations of mathematics* (rev. ed.). Cambridge, MA: MIT Press.

THE HUMAN SUBJECT IN MATHEMATICS EDUCATION AND IN THE HISTORY OF MATHEMATICS

Michael Otte and Falk Seeger
Bielefeld

1. INTRODUCTION

Problems of the theory of mathematics education are fundamentally philosophical problems. Since Kant, the philosophical as well as the scientific debate on knowing has been divided between thorough-going relativism, all knowledge held to be just a representation of the subject's particular perspective on reality on the one side, and the claim that there are self-authenticating experiences or methods that guarantee direct knowledge of reality. In order to transform this dichotomy into a productive "paradox" (knowledge is relative and objective at the same time), we have to explore the "objectivity of the subjective." This exploration will essentially have to take an evolutionary or historical view. The present paper tries to break ground for the undertaking of such an exploration of the historical objectivity of the subject. It can be understood as an attempt to sketch some very general outlines of the relation between the history of mathematics and mathematics education. We take it to be a highly important goal of mathematics education that the knowledge it helps students develop is not only of a factual kind, being distant from the subject, but that it is personal in the sense that it is also knowledge about the subject's self. It is a truism that not only is mathematics a historical phenomenon but also that what we understand as the subject is the result of history as reflected in the self-image of the scientific disciplines.

What we would like to do is the following: We start with a brief review of the reasons to employ history in mathematics teaching. The conclusion of this review is that the benefit of historical understanding originates in the perspectives of metaknowledge and metacognition it necessitates. We argue that metaknowledge and metacognition are part and parcel of a relational conception of knowledge – as opposed to a substantialist conception. If knowledge is seen to reside in the relation between things, it follows that the relation to the human subject – metaknowledge – is involved. We then discuss how metaknowledge under the influence of literacy and print can be understood as a variation in perspective. With the spread of print, we find a growing focus on the individual as the source of knowledge. We then try to

R. Biehler, R. W. Scholz, R. Sträßer, B. Winkelmann (Eds.),
Didactics of Mathematics as a Scientific Discipline, 351-365.

give a picture of the role of individualism as a historical type of reasoning in mathematics as well as in school. Looking for reasons for this individualist version of epistemology, we step back to gain a view of the influence of society at large on certain notions of science, knowledge, and school learning. We find that the positivist perspective cannot take into account the clash between substantialist and functionalist perspectives on knowing and meaning. Mathematical formalism cannot be treated in a formal way, as did positivism. We conclude that the complementarity of substantialist and functionalist aspects of knowing and meaning requires a historical perspective – giving a picture of the subject and the self that goes beyond individualism as it highlights its historical origins and becoming.

2. REASONS TO EMPLOY A HISTORICAL PERSPECTIVE IN THE MATHEMATICS CLASSROOM

History has traditionally been used as a source to stimulate students' motivation for doing mathematics, and we would like to shortly review some of the arguments in the following part. It seems rather obvious that such an employment of history is unsatisfactory, as the student very quickly learns that the real stuff comes only after the "storytelling" is over.

The idea of linking the studies of mathematics with those of the history of that subject was quite popular in Germany during the 19th century (Gebhardt, 1912). This idea has been revitalized by, above all, Otto Toeplitz whose posthumously published book, *Die Entwicklung der Infinitesimalrechnung. Eine Einführung nach der genetischen Methode* (1949), has been widely appreciated although it was not successful in the stricter sense. Toeplitz's endeavors are to be understood in relation to the many activities of German mathematicians, like Weyl, Speiser, Dehn, Siegel, and others, who tried to place mathematical production into a broader cultural context. Since 1945, Bourbakism has overthrown these attempts, and only during the last 15 years has a certain change of attitude taken place. Nevertheless, the only text specifically devoted to the introduction of historical ideas into the mathematical classroom at school level is the collection of sources compiled by Popp (Popp, 1968). At the university level, the situation is a little more favorable. But still there is only one text devoted to the introduction into a mathematical discipline via historical argumentation (Scharlau & Opolka, 1980). Our own activities started in 1979 when the Volkswagen Foundation financed our proposal (Otte, 1977) enabling us to organize an international as well as interdisciplinary meeting on *Epistemological and social problems of the sciences in the early 19th century* (Jahnke & Otte, 1981).

Often historical themes are introduced into the classroom in order to counterbalance the technical treatment of mathematical ideas. Mathematics is approached then by asking for its connections to other areas of human cultural activity. Positivist-formalistic conceptions of mathematics, on the contrary, start from the specificity of mathematics in comparison to the other sciences and in contrast to other fields of human experience. It seems obvious that this notion of positivism may well function as a stumbling block in the process of making mathematics meaningful in the classroom just because it emphasizes the unconnectedness of the mathematical experience to other fields of experience. This notion of the specificity of mathematics in the spirit of positivism is still rather typical for the use of history in the mathematics classroom quoted above. Accordingly, the critique of the positivist-formalist doctrine will form one major intention of the present paper.

The specificity of mathematics is not to be seen in any particularity of method or content but rather in the fact that science in general is a child of the social division of labor. From this results the formal character of mathematics and the abstract image of science. Mathematics according to such an understanding is hypothetico-deductive reasoning.

The formal character of mathematics, historically, is connected closely to an epistemological insight that is essential for modernity, that is, the idea of "relational thinking." According to this notion, the content of theoretical concepts does not refer to things but to relations between things (see Cassirer, 1953). The essence of scientific thinking in general consists, as Max Born once said, in the discovery that relations can be controlled as well as communicated, whereas phenomena or things cannot. This holds also with respect to data that seem to speak for themselves by communicating their meaning. For example, the information that 7,000 people have been killed through traffic accidents in a certain country during a certain period of time takes on a fundamentally different meaning if it is supplemented with the additional information that the respective figures have ranged between 10,000 and 15,000 during preceding periods compared to the information that these figures had always been smaller than 2,000.

There is another motivation to include the history of mathematics into mathematics teaching. The claim of an absolute objectivity of knowledge cannot be justified, because there is not only one correct interpretation or just one possible meaning of a piece of knowledge. Theoretical terms are valuable as means of cognitive activity just because they represent idealizations that cannot be dissolved exhaustively into one particular possible interpretation or application. Formal mathematical knowledge as well as everyday procedural knowledge is silent beyond the representation given in the sense of Wittgenstein's dictum: "What can be shown cannot be said" (1974, 4.1212). A representation, accordingly, hides that which it does not express. In contrast to a representation, a theoretical concept per se expresses nothing and says very many things. Each successful application of a concept shows it in a different light; each of its representations leads to a different conclusion or to a different activity. The power of a concept in the "relational" meaning thus cannot be seen in its mapping of reality but in the potential relations it opens up. The "potentiality" of theoretical concepts is also gained in the process of historically reconstructing the development of a mathematical concept or a mathematical idea. History provides us with the insight that there is not one mathematics, and this insight might encourage and strengthen the learner with respect to her or his own personality and approach to knowledge. Thus, the meaning of "relational" also applies to the relation of knowledge to the human subject, as is very well put in Max Born's statement quoted above.

Mathematics education has to take into account that there is no knowledge without metaknowledge, that one cannot learn a theoretical concept without learning something about concepts, in order to understand what kind of entities those are. This metaknowledge can, however, be developed by means of historical studies.

3. MEDIATEDNESS, METAKNOWLEDGE, THE INDIVIDUAL, AND LITERACY

Education aims at organizing processes of learning. Such processes always undergo a dual determination, as learning is always at the same time metalearning. Even the mechanical learning of rules, like learning to execute an algorithm, is accompanied by a second-order learning, by metalearning (Bateson, 1983).

The increase in mechanical learning, which represents a phenomenon of metalearning, and the individual variation show that learning and knowledge are always reflexive, or that the genesis of knowledge is to be understood first with regard to its relationship to the world of objects, and second, with regard to the subject's inner world, the relationship to the self. Difference and connection between learning and metalearning thus have to be discussed against the background of the distinction between subject and object. This distinction is absolutely dependent on communication. We communicate with the subject about the object, and not vice versa. Cognition and communication very much depend on the means and media. Both changed deeply with the invention of the phonetic alphabet in antiquity and with the invention of the printing press during the 15th century. J. Goody, among others, has investigated the impact of literacy on human thought. He writes:

> . . . it is not accidental that major steps in the development of what we now call "science" followed the introduction of major changes in the channels of communication in Babylonia (writing), Ancient Greece (the alphabet), and in Western Europe (printing). (1977, p. 51)
>
> The specific proposition is that writing, and more specific alphabetic literacy, made it possible to scrutinise discourse in a different kind of way by giving oral communication a semi-permanent form; this scrutinity increased the potentiality for cumulative knowledge, especially knowledge of an abstract kind, because it changed the nature of communication beyond face-to-face contact as well as the system for the storage of information . . . . No longer did the problem of memory storage dominate man's intellectual life; the human mind was freed to study static

> "text" (rather then be limited by participation in the dynamic "utterance"). (1977, p. 37)

In historical terms, the role of printing was predominantly that it fundamentally changed the relation between people and knowledge and thereby the concept of knowledge in society as well as the individual's position (cf. Glück, 1987). Mathematics played a greater role than other kinds of knowledge in those processes, because it helped to develop new technologies as well as to organize and systematize the knowledge and experience of the practitioner. Descartes' algebraization of geometry, for instance, was primarily intended to bring new order into the geometrical knowledge of Greek antiquity as well as that of the artisans and mechanics of his time.

Cognitively, the distinction between subject and object and communication between subjects has the advantage of permitting a change of perspective on the object. All representation of objectivity is based on a variation of perspective. It yields the advantage of a double check of reasoning, and the still greater advantage of developing logic and methodology. This double check or this possibility of an alternative perspective is greatly enhanced by literacy, in particular, since the invention of the printing press.

Printing made it possible to compare statements exactly. Different readers could discuss a specific argument that was located precisely within identical copies. Text became autonomous from interpretation by an established authority. Contradictions and connections between arguments became clearly visible. Before the printing press, to study medicine meant to study Galen, to engage in physics or geography was to read Ptolemy, and to learn mathematics meant to study Euclid's *Elements*. Texts were only considered truthful and trustworthy during the Middle Ages if the name of the author was indicated as well as those of the compilator and the commentator. Statements on the order of "Hippocrates said . . ." or "Pliny tells us . . ." were markers of a proven discourse. Only afterwards did it become possible to surpass ancient authority and to check conflicting or incomplete verdicts rendered by their teaching against the great book of Nature or against own experience. Discourse was no longer able to justify its claims by referring to the supporting authority of another, and it was constrained increasingly to become self-authorized. Enlightenment assumptions and revolutionary experience coalesced with printing technology. With the availability of identical texts, not only the content of an argument but its style and particular expression became relevant too. And this fostered individualism (see, also, Havelock, 1986; Ong, 1982, for other accounts of the rise of individualism in relation to print and literacy).

4. IDENTITY IN MATHEMATICS

The formation of any theory begins with certain principles of individuation that serve to establish the ontology of the theory, that is, the claims for the existence of the objects about which the theory speaks or wants to speak. In more or less close analogy, any society may be characterized by specifying the principles and social mechanisms of personal individualization.

In mathematics, we are familiar with two types of identity principle, which we should like to characterize with the names of Leibniz and Grassmann respectively. The first type goes back to Leibniz' principle of indiscernibility, or his *principium identitatis indiscernibilium*, which actually dates back to Aristotle. It consists in the thesis that there are no two substances that resemble each other entirely but only differ quantitatively, because then their "complete concepts," that means, the concept that characterizes the substance completely, would coincide. The ultimate goal of classical science, which is, in general, to be accomplished only by God through an infinite analysis, lies in the determination of the individual substances. The goal of modern science, on the contrary, is to be seen in appropriate generalizations that orient technical action and prediction. Modern science assumes that reality is general rather than specific, and, therefore, that knowledge is related toward an abstract formal order rather than being knowledge of a specific, substantive order (a formal order allows more freedom to the individual than a substantive order).

According to Leibniz's principle, there is no distinction without a motive, without a reason, as Leibniz stresses in his correspondence with Clarke (Alexander, 1956, p. 36). The second constructivist principle of individuation in mathematics is based on precisely the opposite view, that is, on a distinction without motive and reason. "It is," writes Grassmann, "irrelevant in what respect one element differs from another, for it is specified simply as being different, without assigning a real content to the difference. Our science shares this notion of element with combinatorics" (Grassmann, 1844/1969, p. 47; our translation, M.O./F.S.). This kind of individuation is, first of all, a process by which a certain perspective and style of reasoning is introduced that guides our cognitive activities.

As distinct from the substantialist principle of identity, the problem here is to be understood as a functionalist principle of identity. Two objects are equal if they are functionally equivalent in a certain way specified by theory. This principle has been much emphasized in AI research (cf., e.g., Bundy, 1983, p. 42). In order to establish equality in the context of an axiomatic theory, we would have to single out those functions and predicates that make up the substitution axioms that distinguish equality from other equivalence relations, those n-ary functions f or predicates p that are compatible with the equality relation.

Leibniz's theory of identity derives from the traditional Aristotelian theory of substances. Substances are the subjects of predication. They are in this sense the prerequisite of properties and relations, and they cannot, as modern analytical philosophy believes, be reduced to bundles of qualities. Otherweise all knowledge would be analytic and loose its connection with reality. A equals B or $A = B$ means that A and B are appearances of the same substance. In Fregian terminology, this is stated by saying that *A* and *B* represent different intensions of the same extension or that they are representations with a shared referent but different meanings. The other, Grassmanian view interprets "*A* is *B*" in terms of the idea of shared qualities of different things. In modern mathematics, as in modern analytical philosophy, it is not the substances that matter but the relations. The objects of a theory are equivalence classes of unidentified elements, which are constituted according to a functional principle of operativity.

As the objects of a theory become identical with their descriptions, we get applications of knowledge as a problem in their own right. Theories become realities sui generis.

A good example, which illustrates the two identity principles, is the equation $A = B$ provided by economic exchange. Every commodity contains use value and exchange value. Empirical abstraction starts with use value as the notion that constitutes the individual goods and finds out about the exchange value only a posteriori on the market place. Theoretical abstraction considers exchange value as the essentialy quantitative representation of an independent substance, namely, economic value as such. Economic activity in a capitalist society makes exchange value its real end and the use value a means to this.

In school, too, we are familiar with these two types of individuation: the substantialist and the functionalist. They characterize the transition from arithmetic to algebra. For children, numbers at first have a shape and a life of their own. 1/2 is a privileged fraction, and it is easier to calculate with than the other rational numbers. Substantialist reasoning inquires into the properties, the essence, the meaning. Functionalist reasoning treats all those things as identical that function in the same way within a certain context. Thus, if we assume that numbers are determined above all by the fact that they lend themselves to calculations according to certain axiomatically preestablished principles, then these numbers can be designated by general symbols $x, y, \ldots$, and so forth. In algebra, calculating is thus done with indeterminate or "general" numbers, that is, with variables that designate numbers only with regard to the fact that they can be treated arithmetically.

A second example is the following: Students, as a rule, have difficulties with equations, because they have interpreted and learned the equality sign in the sense of "yields." This "input-output" interpretation represents a "direct" understanding of the equation. The concept of equation has not yet been transformed into an object of mathematical reflection; a relational or functional understanding has not yet been achieved.

5. IDENTITY AND THE SOCIETAL SUBJECT

Societies also are based on principles of individuation and are determined according to which principles of identity or individualization they encounter. In a way parallel to the above distinctions, an organic and a functionalist principle of identity exist in society as well. On the one hand, people are determined by their individual personalities, and, on the other hand, by the functions they assume in the larger society characterized by division of labor. Every social individual is a contradiction in itself insofar as it has both an organic-biological and personal existence and, on the other side, is integrated into society by the roles it fulfills. The exemplification of the two conceptions of an equation $A = B$ in terms of economic value of commodities is a direct expression of different conceptions of society.

Aristotle regarded society as a substance, and this view persisted up to the 15th or 16th century. But society is a rather unusual substance, in that humans have a capacity to think and choose the ends they pursue. There is undeniably a tension between the view that society is a substance and the view that humans are free agents. A single metaphor for society, which prevailed from antiquity to the beginning of capitalism, was that of an organism, whereas, for modern capitalist society, another analogy came to seem more appropriate: the analogy with a set or an aggregate. The analogy of the set has been pervasive in the thought produced in capitalist society as the analogy of the organism was in precapitalist society. In traditional precapitalist society, there did not exist a contradiction or tension with respect to the definition of the individual. In precapitalist formations, the forms of social relations that correspond to these are personal dependence. In capitalist society, there is personal independence based on objective dependence. We may, in summary, note that the complexity of our reasoning and of our personality in general increases with the complexity and formality of our social relations. Individualism is a product of social history, not of nature. It is also a product of social division of labor that leads to conflicts between the world of science and the everyday world.

This problem has been investigated with reference to the problems of science education (see, e.g., DiSessa, 1982) and it has been described in a rather general setting by the British philosopher Gilbert Ryle in his book *Dilemmas*. He writes:

> . . . we seem to find clashes between the things that scientists tell us about our furniture, clothes and limbs and the things that we tell about them. We are apt to express these felt rivalries by saying that the world whose parts and members are described by scientists is different from the world whose parts and members we describe ourselves, and yet, since there can be only one world, one of these seeming worlds must be a dummy-world. (Ryle, 1964, p. 68)

In fact, theorists do not describe chairs, clothes, or limbs at all, as we wanted to say by describing the different principles of individuation. And Ryle therefore concludes that if the feuds between science and common knowledge are to be dissolved, their dissolution can come only "from drawing uncompromising contrasts between their businesses" (p. 81). It is a better policy to remind people "how different and independent their trades actually are" than to pretend that all "are really fellow-workers in some joint but unobvious missionary enterprise" (p. 81).

The conlusion Ryle proposes is difficult to maintain in a society that as a whole has been transformed into a "laboratory" for complex technologies. To point to the fact of a radical division of labor that prevails in our societies is no more sufficient when the society as a whole irrevocably and completely depends on science and technology, and that demands that everybody be educated scientifically to a certain degree. Application of knowledge is a sociohistorical process that is more strongly influenced by knowledge about humans than by knowledge about objects. Nowhere is the technologically or scientifically manufacturable taken as a guideline for action. Political or social considerations always interfere.

Our picture of science can now be sketched more completely. As science is a social system too, it inherits the dichotomies that beset society. For instance, it is not as purely objective as might appear so far. It must seem almost obvious that much of the dynamics and orientation of theoretical knowledge is governed by the self-image and the desires or wishes of the cognitive subject, by that which it considers as relevant. Otherwise, discontinuities and revolutions in the history of science could not be explained and would even remain unthinkable. In this manner, normative and objective aspects of science become inseparably entangled, and human interactions with objective reality take different forms in analogy to different forms of social interactions.

Positivist science in general tends to ignore such involvements and bases its activities on a strict separation between subject and object as well as on the assumption of an independent but knowable reality. It thereby excludes the problems of knowledge application from its proper concern, too.

The sciences begin with the distinction between subject and object, or their activity is based on it, but they are not aware of this fact. They do not see what they assume operatively. They operate with existing things, but do not concern themselves with the essence or with the reality of this existence. Being is, as Kant said, no real predicate of logic (Kant, 1787, B 626). Essence or existence, however, are important categories for the dynamics of the learning process, as this process is at the same time a process of developing the subject or the personality. From this, it can be concluded that the self-image of science may not be appropriate for being introduced into its reasoning.

Epistemologically, recent centuries were under the sign of nominalism. The evolution of industrial capitalism was accompanied by a state of mind that understood the mental process as overcoming a limiting philosophy having medieval roots. This has led to the idea that there is complete freedom in forming concepts. Only after humanity, as it is said, took the liberty of creating its own concepts according to its own goals did reasoning become, on the one hand, a means toward any purpose, and, on the other hand, was divested of objective meaning. It became disengaged and instrumental at the same time. Shedding scholastics as a transition to nominalism! Nominalism prevails with Augustine's "discovery of subjectivism," which shifts the aporias of history and of mechanical motion into the subject (cf. von Weizsäcker, 1971, p. 435).

With regard to mathematics, this philosophy seems to come to a head with Georg Cantor's repeatedly quoted, world-famous statement of 1883: "The essence of mathematics lies precisely in its freedom;" the other half of the quote which deals with the metaphysical constraints on applied mathematics always being suppressed and forgotten.

Again, this double process that liberated knowledge from its situational context and the individual from its metaphysical religious and social bounds depends very much on the development of literacy and printed media. In the limit of idealization, all of mathematics can be regarded as a collection of grammatically correct potential texts. And this kind of knowledge requires a new kind of mind, a mind that could achieve knowledge without social involvement, the mind of the rational, detached individual.

6. THE PROBLEM OF MEANING

On the one hand, general compulsory education, as institutionalized in our schools, has always been dependent of this kind of knowledge. It has always relied on the theoretization of knowledge, despite the fact that this theoretization permanently causes practical pedagogical troubles and difficulties. Only the complexity of theoretical generalizations, which are determined objectively as well as socially, makes it possible to link individual development to the complex possibilities of the "real world:" Theoretization opens up a universe of experience that is rich enough to allow a very great variety of members of society to participate and to develop and simultaneously keeps all the conflicting immediate demands and wishes at a distance leaving a certain necessary autonomy to the school. On the other hand, this theoretization and abstractness of knowledge causes a fundamental problem in education, namely, the problem of meaning.

> The real problem which confronts mathematics teaching is not that of rigor, but the problem of the development of "meaning," of the "existence" of mathematical objects. (Thom, 1972, p. 202)

This fundamental problem of mathematics teaching in the framework of compulsory general education comes from the antiempiricist nature of theoretical knowledge in particular. Theoretical terms are neither concrete objects nor properties of such objects and are not names of empirical objects either. The content of a theoretical term consists of relationships between things and not of things (their properties, etc.) themselves. J. Bruner has stated this problem to be the problem of the indirect nature of scientific knowledge.

> [Scientific concepts] are inferences we draw from certain regularities in our observations. This is all very familiar to us .... To a young student, who is used to thinking of things that either exist or do not exist, it is hard to tell the truth in answer to his question whether pressure *really* exists. (Bruner, 1960, p. 69)

Stating the fundamental problem of mathematical education in such a way by no means expresses a danger of forgetting the pedagogical, organizational, social, or psychological implications of education and schooling. On the contrary, formulating the central issue of didactics in this way implies, for instance, that a direct teaching of scientific concepts is impossible. There is no simple way of relaying a theoretical concept from teacher to student. Concepts cannot be handed over like concrete objects, because they are not such things nor their names.

According to how we have explained this problem of meaning, it seems obvious that this problem is not a technical problem but a problem of the development of the individual. And this question of subjective development is not to be restricted to the cognitive aspects either but implies a development of personality. This links mathematics education to philosophy and history. The reference to history adds, or better, makes aware of, a philosophical element to scientific reasoning, which refers precisely to the role the subject's self-understanding has.

In a certain sense, the difficulty is contained in Leibniz' principle of the identity of indiscernibles, which was mentioned already. Now Bertrand Russell has pointed out that it follows from the analysis of this principle that a subject or a substance is either nothing else but the sum of its properties, thus losing its subject character, or that it cannot be defined at all. From this, Russell concludes that Leibniz' principle makes no sense. As opposed to that, we believe that Russell's analysis expresses a profound philosophical difficulty. This problem has become salient in our time, in particular, in the subject-machine problem, but it has more or less explicitly played a key role in the development of fundamental concepts of mathematics since the 17th and the 18th centuries. Among these, the concept of (mechanical) motion would have to be named first, and, correspondingly, the concept of function (cf. Bibler, 1967) . But also in the philosophy of psychology, it has played a major role, as reflected in Gestalt psychology's statement that the whole is more than the sum of its parts (cf., e.g., Wertheimer, 1925/1967).

In the present debate on the question "Can computers think?" the result is that the (human) subject can neither be identified with the totality of his or her presently accessible properties, for, otherwise, she or he could be simulated on a suitably programmed computer, nor that the subject can be conceived of as a substance beyond all his or her properties, because, otherwise she or he would be inaccessible to his or her own self-reflection. In this case, the human subject would lose his or her subject character, as human cognition differs from mere information by the fact that the subject not only knows but knows that she or he knows. Knowledge and metaknowledge are inseparably connected in human cognition. Thus, it may not be a solution to argue, that the "whole is one its parts" (Minsky, 1988; Minsky & Papert, 1973).

It may thus be said that certain current problems with science and technology (the problem of meaning, the man-machine problem, the problem of interpretation of quantum physics, etc.) have got nominalism into difficulties, difficulties that have to do with the subject's self-image, and that all the aporias and dualisms, the polarity of the finite and the infinite, the paradoxes of motion, of time, and of the present, and so forth are manifested mainly in the subject, indeed in the cognizing subject's self-image as a (potentially) universal and actually always limited being. As a consequence, we conclude that thinking has an existential import and that theory can be understood as a "mode of life."

Now mathematics education has to do with the relationship between mathematics and the human subject, and requires that these problems are dealt with in the sense that an effort at historical reconstruction is made, which must also make use of philosophical insights that were present before the rise of individualism and so-called subjective turn. In this sense, von Weizsäcker writes:

> I would not have turned to studying classical philosophy if I had not encountered inconsistencies in the conceptual traditions of modern physics and humanities and in modern philosophy which I could only hope to understand by going back to their historical sources. To me, the great steps of progress in modern times like the emergence of the exact natural sciences, the shaping of subjectivity and the growth of a historical consciousness seem to have been paid for by certain constrictions to the questions raised and the concepts formed. Whoever inquires at one spot into the problems of modern times, say, into the foundations of physics, will rediscover the very same structures which have already been discovered by the Greek philosophers, although from another perspective. (von Weizsäcker, 1992, p. 440; our translation, M.O./F.S.)

This quotation may not be understood as a plea to recur in mathematics education to Greek philosophy as a general didactical strategy, even though this might prove worthwhile at varying occasions. It should be read as a plea to teach mathematics as a historically grown subject in the sense in which Foucault put it that "real science recognizes its own history without feeling attacked" (Martin, 1988, p. 12).

If we try to condense what has been said so far about the parallelism between the principle of identity in mathematics and the process of identity formation of the human subject, it becomes quite clear that the contribution of mathematics education to the self cannot be seen as residing in changes of the self as a substance. Focusing instead on the relational aspects and on the processes of becoming, we find the relation of the self to time to be of utmost importance (cf. Brockmeier, 1991). The self has to be understood as intimately connected to processes that develop in time, identity being what remains constant in the flux of time. The development of the self, then, is better conceived of as a "way," a "journey along the road," very much in analogy to what happens to the protagonist in the classical *Bildungsroman* (cf. Bakhtin, 1981). Using the metaphor of the equivalence or identity of the two sides of an equation, the aim of mathematics education cannot be to arrive at the shortest possible way to the conclusion that *A* equals *A*. We argue that this "minimal loop" approach (cf. Churchman, 1968) is closely connected to a formalist view in mathematics education that largely excludes a historical perspective on the becoming of mathematical concepts and theories. A "maximal loop" approach to furthering the development of mathematics as part of the development of the self instead seeks to bring the subject into contact with as many different perspectives on mathematics as possible along the "way." We have given some reasons above why this should necessarily also be a "journey into the past," as we have to learn from history how the subject itself is involved in the processes of constructing meaning.

7. CONCLUDING REMARKS

We have tried to underline above that the perspectivity of knowledge is a necessary by-product of literacy and literal culture. If one takes a look now at the conditions that frame a realization of the above deliberations in the mathematics classroom, it becomes clear that it has to be taken into account that learning in classrooms is mostly an outcome of an oral discourse as part of an oral culture. The development of meaning, thus, cannot be seen only in the decontextualization, in the liberation from the concrete situation that was made possible through literacy. Constructing meaning in classroom learning is a result of contextualization and situatedness that is typical for the discourse in schools. The linearization and individualization of literal thinking (Havelock, 1986) has to be complemented by the orality of the classroom, by conversation and discussion, which all put the subject in relation to other subjects and make her or him experience that their own perspective is only one among different possible ones.

The importance of a historical perspective extends well beyond the students' discovery that similar problems existed a long time ago and that their obsolescence seems unwarranted. In the course of a historical study, the process of constructing meaning comes into focus. In this way, it is conceivable that substantial and functional thinking are not only steps in a process of evolution that culminates in functional thinking as having the most general claim to truth and objectivity. The different modes of thinking will rather be understood as resulting from a certain worldview, and it becomes clear that the universal claim of our own worldview is only a relative one.

The relation between universality and particularity is a key to an understanding of the role of the human subject. The epistemological situation of the subject has been styled above by a potential universality and an actual particularity or limitation. At present, many models of the human subject in mathematics education start from the basic assumption that the subject organizes knowledge in different domains that are not necessarily connected by highly general structures forming a coherent system. This model sharply differs from models of the past that focused on a general ability or a general structure as an outcome of learning. In any case, to underline the domain-specificity of knowledge or the subjectivity of domains of experience seems to be important. In view of what has been said about the historicity of the subject, domain-specificity cannot be the last word. By no means can it be a goal of mathematics education to teach the students, starting from their domains of subjective experience, a range of domain-specific knowledge and techniques turning them into experts in selected fields. The goal of mathematics education, as it were, is general education. And how could the core of a general education be better styled than as being the experience of the multiplicity of perspectives that rests on being conscious of the historicity of the own personal perspective?

Subjective domains of experience are the outcome of social and collective processes of learning and the outcome of an interiorization of relations and processes between humans. These processes are characterized by a transition from the interpsychological to the intrapsychological plane (Vygotsky, 1987). The subjective experience of "multi-voicedness," which makes it possible to put the general in relation to the particular, needs collective processes in the mathematics classroom that have to be cultivated by mathematics education as a discipline.

REFERENCES

Alexander, H. G. (Ed.). (1956). *The Leibniz-Clarke correspondence.* Manchester: Manchester University Press.

Bakhtin, M. M. (1981). *The dialogic imagination - Four essays* [Edited by M. Holquist]. Austin, TX: University of Texas Press.

Bateson, G. (1973). *Steps to an ecology of mind.* St. Albans, Hertfordshire: Paladin.

Bibler, V. S. (1967). Die Genese des Begriffs der Bewegung [I. Maschke-Luschberger, Trans.]. In A. S. Arsen'ev, V. S. Bibler, & B. M. Kedrov (Eds.), *Analyse des sich entwickelnden Begriffs* (pp. 99-196). Moskau: Nauka.

Brockmeier, J. (1991). The construction of time, language, and self. *Quarterly Newsletter of the Laboratory of Comparative Human Cognition, 13*(2), 42-52.

Bruner, J (1960). *The process of education.* New York: Vintage Books.

Bundy, A. (1983). *The computer modelling of mathematical reasoning.* London: Academic Press.

Cassirer, E. (1953). *Substance and function.* New York: Dover.

Churchman, C. W. (1968). *Challenge to reason.* New York: McGraw-Hill.

DiSessa, A. A. (1982). Unlearning Aristotelian physics. *Cognitive Science, 6,* 37,-76.

Gebhardt, M. (1912). *Die Geschichte der Mathematik im mathematischen Unterricht.* IMUK-Abhandlung IV, 6. Leipzig: Teubner.

Glück, H. (1987). *Schrift und Schriftlichkeit - Eine sprach- und kulturwissenschaftliche Studie.* Stuttgart: J. B. Metzler.

Goody, J. (1977). *The domestication of the savage mind.* Cambridge: Cambridge University Press.

Grassmann, H. (1969). *Die lineale Ausdehnungslehre.* New York: Chelsea Publ. Co. [Original work published 1844]

Havelock, E. A. (1986). *The muse learns to write. Reflections on orality and literacy from antiquity to the present.* New Haven, CT: Yale University Press.
Jahnke N., & Otte M. (Eds). (1981). *Epistemological and social problems of the sciences in the early 19th century.* Dordrecht, Netherlands: Reidel.
Kant, I. (1787). *Kritik der reinen Vernunft* (2nd ed.). Riga: J. F. Hartknoch.
Martin, R. (1988). Truth, power, self: An interview with Michel Foucault, October 25, 1982. In L. H. Martin, H. Gutman, & P. H. Hutton (Eds.), *Technologies of the self: A seminar with Michel Foucault* (pp. 9-15). London: Tavistock.
Minsky, M. L. (1988). *The society of mind.* London: Heinemann.
Minsky, M. L., & Papert, S. (1973). *Artificial intelligence.* Eugene, OR: Oregon State System of Higher Education.
Ong, W. J. (1982). *Orality and literacy. The technologizing of the word.* London: Methuen.
Otte, M. (1977). Zum Verhältnis von Wissenschafts- und Bildungsprozess. *Zentralblatt für Didaktik der Mathematik, 9,* 205-209.
Popp, W. (1968). *Wege des exakten Denkens - Vier Jahrtausende Mathematik.* München: Ehrenwirth.
Ryle, G. (1964). *Dilemmas.* Cambridge: Cambridge University Press.
Scharlau, W., & Opolka, H. (1980). *Von Fermat bis Minkowski.* Heidelberg: Springer.
Taylor, Ch. (1989). *Sources of the self.* Cambridge, MA: Harvard University Press.
Thom, R. (1973). Modern mathematics, does it exist? In A. G. Howson (Ed.), *Developments in mathematical education.* Cambridge: Cambridge University Press.
Toeplitz, O. (1949). *Die Entwicklung der Infinitesimalrechnung.* Berlin: Springer.
Vygotsky, L. S. (1981). The genesis of higher mental functions. In J. V. Wertsch (Ed.), *The concept of activity in Soviet psychology* (pp. 144-188). Armonk, NY: M. E. Sharpe.
Vygotsky, L. S. (1987). Thinking and speech. In R. W. Rieber & A. S. Carton (Eds.), *The collected works of L. S. Vygotsky.* (Vol. 1, pp. 38-285). New York: Plenum Press.
Weizsäcker, C. F. von (1971). *Die Einheit der Natur.* München: Hanser.
Weizsäcker, C. F. von (1992). *Zeit und Wissen.* München: Hanser.
Wertheimer, M. (1967). *Drei Abhandlungen zur Gestalttheorie.* Darmstadt: Wissenschaftliche Buchgesellschaft. [Original work published 1925]
Wittgenstein, L.(1974) *Tractatus logico-philosophicus* [D. F. Pears & B. F. McGuiness, Trans.] (2nd ed.). London: Routledge & Kegan Paul.

MATHEMATICS IN SOCIETY

Mogens Niss
Roskilde

1. INTRODUCTION: MATHEMATICS AS A DISCIPLINE

Mathematics is a discipline in several different respects. It is a *science* in an epistemological sense, oriented towards developing, describing and *understanding* objects, phenomena, relationships, mechanisms, and so forth belonging to some domain. When this domain consists of what we usually think of as mathematical entities, mathematics acts as a *pure* science. In this capacity, mathematics aims at internal self-development and self-understanding, independent of the world outside except for the fact that mathematics is exercised by human beings interacting with each other and working in societal institutions in accordance with social norms and habits. If, on the other hand, the domain under consideration lies outside of mathematics, typically within some other scientific field, mathematics serves as an *applied* science. In this capacity, mathematics is activated to help to understand and develop aspects of various extra-mathematical areas. Needless to say, mathematics as a pure science provides crucial contributions to mathematics as an applied science, although often with a great delay. The difference between these two aspects of mathematics is a question of the focus of attention rather than of mathematical content matter. Whether pure or applied, mathematics as a science serves to generate *knowledge* and *insight.*

Mathematics is also a system of *instruments*, products as well as processes, that can assist *decisions* and *actions* related to the mastering of extra-mathematical practice areas. (That such decisions and actions will often be based on scientific knowledge and insight, whether mathematical or extra-mathematical, is quite true but not essential in the present context.) Thus mathematics provides *tools* for the exercise of a very wide range of *social practices* and techniques.

Mathematics is a field of *aesthetics* capable of giving experiences of beauty, joy and excitement to many of those who indulge in it. In this respect, mathematics resembles an art form such as sculpture, painting, architecture and music – all of which also have certain content aspects in common with mathematics.

The transmission, the dissemination and the furtherance of mathematics as a discipline require mathematics to be learnt by new generations. As the

R. Biehler, R. W. Scholz, R. Sträßer, B. Winkelmann (Eds.),
Didactics of Mathematics as a Scientific Discipline, 367-378.

learning of mathematics does not take place spontaneously and automatically, mathematics needs to be taught. So, mathematics is also a *teaching subject* in the educational systems of our societies.

If we agree that mathematics is constituted as a discipline by its *five-fold nature* as a pure science, an applied science, a system of instruments, a field of aesthetics and as a teaching subject, we are prepared to undertake an analysis of the social rôle and significance of mathematics as a discipline.

2. THE RÔLES OF MATHEMATICS IN SOCIETY

Every society maintains, supports and finances mathematical activity in all the above respects in such a way and to such an extent that it is clear that society attributes prime importance to mathematics. However, there are many other essential sciences, crucial systems of instruments for social practice, marvellous fields of aesthetics and fundamental teaching subjects, each of which are much less favoured by society than is mathematics. Internationally considered, mathematics is apparently rather unique in the position it occupies in almost every country. How come?

The answer seems to be (a) that mathematics, probably more than any other discipline, has all the above-mentioned five properties *at the same time*; (b) that the most important of these properties is "the unreasonable effectiveness of mathematics" (Wigner, 1960) as an applied science and as a system of instruments for social practice, both of which are very *general* and pertain to an incredibly broad range of extra-mathematical subjects and practice areas; and (c) that mathematics – due to (b) – is intimately linked to the functioning and development of society at large. So, the social rôle and significance of mathematics as a discipline is not characterized alone by its status as a pure science but is related to the way in which mathematics is activated to deal with matters and issues beyond mathematics itself. Of course, this should not be taken to imply that mathematics in its capacity as a pure science is of no significance to society. Not only is the pure mathematics of today often the applied mathematics of tomorrow, but – as is the case with any science – mathematics exerts an impact on mind and culture as well.

More specifically (see, e.g., Booss & Krickeberg, 1976; Friedman 1988-1990; Khoury & Parsons, 1981; Rosen, 1972-1973; Steen, 1978; Wan, 1989), mathematics is connected to the functioning and development of society in the following ways:

1. As a science applied to, and in, *other scientific subjects,* mathematics plays an increasingly important part in the formulation and foundation of many scientific disciplines as well as in the methods and techniques they employ. This is true for the entire ranges of the physical, the engineering and the biological sciences, for information science, economics, sociology, linguistics and for dozens of other disciplines as well, although the way in which mathematics is involved in them varies considerably with the disci-

pline (Howson, Kahane, Lauginie, & de Turckheim, 1988, pp. 1-4). These disciplines all have important and well-known social applications and implications, and since mathematics is instrumental for their rôles, it inherits an indirect, yet central, significance for society and its functioning.

2. Mathematics is involved more directly in a number of *specialized practice areas* (some of which are also supported by separate scientific disciplines). To mention just a few: prediction, decision-making and control in the *social sphere*; description and forecasting of phenomena and events in *segments of nature*, perhaps modified by man and society; utilization and allocation of *natural resources*, renewable or extinguishable; and design, operation and regulation of *industrial and socio-technical systems*. Mathematical tools of varying degrees of sophistication are involved in every one of these sectors of social practice.

3. Thirdly, mathematics is an essential but, ironically enough, often ignored element in a broad variety of *general*, that is, non-specialist, *areas of practice* in everyday life in society: representation of numbers; elementary business and money transactions; calendars; geographical coordinates; measurement of time, space, weight, currency; all sorts of graphical representations and tables; work and art drawings; shapes of objects; codes. All of this penetrates innumerable aspects of modern life. The unproblematic mastering of these elements for private and social life – the possession of basic numeracy – is a simple necessity in the same way as literacy is.

The crux of the linking of mathematics to the functioning and development of society as indicated in Points 1 to 3 is the application of mathematics to a variety of extra-mathematical areas. This is brought about by *mathematical modelling*, that is, the construction and utilization of mathematical models. I shall confine myself to emphasizing two aspects of mathematical models and modelling. First, in contrast to a commonly held assumption, the foundation, place and rôle of mathematical models in extra-mathematical areas vary tremendously with the area and cannot be understood or judged on mathematical grounds alone. This fact underlies the second point: The single most important point related to mathematical modelling is the *validation* of models. Implying all sorts of scientific, philosophical, technical and practical issues, the validation of models is a matter of abundant complexity and controversy (cf. Booss-Bavnbek, 1991). Many extra-mathematical fields (e.g., weather forecasting, actuary science, insurance practice) are based on mathematical models and modelling to an extent that make model validity the key criterion of quality.

4. Finally, because mathematics is socially important in all the respects outlined in Points 1 to 3, individuals' acquisition of mathematical qualifications constitutes a marked feature of society. All experience shows that the obtaining and maintenance of mathematical qualifications is far from being a straightforward and unproblematic affair. In fact, mathematical qualifications at appropriate levels and in sufficient amounts form a scarce resource

in most places in the world. Therefore, the presence, distribution and fostering of this resource in the population is itself a matter of social significance and consequence. Not only do societies invest efforts and resources in establishing systems to generate mathematical competence in their citizens. (Differences in qualification levels across countries are a dynamic factor that generates development – or the opposite.) The material, cultural, social and job conditions of an individual are strongly influenced by the level of mathematical competence possessed by that individual (see, e.g., Damerow, Dunkley, Nebres, & Werry, 1984; Department of Education and Science, 1982: *The Cockcroft Report;* Keitel, Damerow, Bishop, & Gerdes, 1989; Morris, 1981), as is the status and prestige he or she enjoys. Thus, a country's mathematical qualification structure has an impact on the whole of society as well as on each of those who live and work in it.

What we have seen above is that mathematics has a crucial rôle in providing a basis for the functioning and development of society. This is true both from a technological and from a sociological perspective. Concerning technology, we should include not only *material technology* (i.e., physical objects and systems) but also what we may call *immaterial technology* and *cultural techniques*, terms that may compress what was outlined in Points 2 and 3 above. Altogether, if we add up the influence mathematics exerts on the cultural and mental circumstances in society, we cannot but conclude that mathematics is embedded in the material and immaterial *infrastructure* of society. Thus, mathematics contributes in a thorough way to the *shaping of society*, for better and for worse. (Further aspects of this are dealt with in Niss, 1985, but a lot of research ought to be done to identify and analyse the impact of mathematics on society in depth and detail.)

From a historical perspective, the rôle of mathematics in society has always been subject to change over time. At first sight, this change simply consists in growth. Mathematics continues to become involved in still new areas of activity in society. In so doing, it is often the case that mathematics tends to penetrate and qualitatively transform the areas of activity in which it occurs. The emergence and dissemination of computers constitutes another kind of (recent) change in the rôle of mathematics in society. The relationship between mathematics and computers is a dual one: They are vehicles for one another. Computers would hardly exist, and would definitely not be so socially important, without mathematics as a fundamental prerequisite for their design and functioning at all hard- and software levels. (This is not to say, of course, that mathematics is the *only* fundamental prerequisite. We only need refer to microelectronics.) Conversely, computers offer new opportunities for dealing with mathematical problems and tasks that previously could not be handled properly. They also open avenues for simulation, exploration and experimentation in and with mathematics that were not at our disposal in former times. Thus, computers serve as extremely efficient, and sometimes even indispensable, tools and amplifiers for various sorts of mathematical activity, but – and this is important – they do not change the nature of mathematical work *in principle*. Computers amplify the signal, they do not create it. Therefore, computers enhance the significance of mathematics in society, but, contrary to widely held beliefs, they do not replace it.

3. THE INVISIBILITY OF MATHEMATICS

Against the background outlined in the previous sections, it is a striking fact that although the social significance of mathematics seems to be ever *increasing* in scope and density, the place, rôle and function of mathematics are largely *invisible* to – and unrecognized by – the general public, decision makers and politicians. Moreover, they are even invisible to many of those who work in extra-mathematical fields that make extensive use of mathematical models and modelling. Even quite a few mathematicians and mathematics educators seem to have fairly unclear pictures of the rôle of mathematics in society as well.

This discrepancy between the objective social significance of mathematics and its subjective invisibility constitutes one form of what the author often calls the *relevance paradox* (Niss, 1979) formed by the simultaneous objective relevance and subjective irrelevance of mathematics. The inherent irony is stressed by the fact that the widespread subjective irrelevance of mathematics does not prevent most societies from investing relatively visible amounts of resources in maintaining research, education and other activities in and with mathematics. This is undoubtedly based on the – subconscious? – conviction that, ultimately, mathematics *is* essential to (at least) the scientific, technological and material welfare of society.

Separate research is needed to investigate the causes of the relative invisibility of mathematics in society. Probably the key to an explanation lies in the fact that mathematics can never be found on the surface of the matters to which attention is paid. It is always embedded in, is a direct or indirect prerequisite for, or is disguised by the matter "proper." For instance, insurance premiums are given in terms of sums of money. The calculations behind are difficult (and debatable) and not accessible to the lay person. Weather forecasts are presented to us as phenomenological statements, accompanied by numerical indications of, say, expected temperature, wind direction and speed. The enormous amount of expertise, mathematical modelling work and computations on which the forecasts are based do not form part of the presentation. Pin codes, strip codes, magnetic cards and so forth may have a mathematical appearance (at least when we have forgotten our credit card pin code), but convey no impression of the amount of sophisticated mathematics that was involved in designing the systems, or of the coding and cryptography problems that had to be solved. The same is true with the mathematics hidden in the representation, condensation and transmission of computerized pictures. Furthermore, we have often accustomed ourselves so much to using items with an explicit mathematical content that we do not think of them as having anything to do with mathematics. This is the case when we buy wall paint to cover a certain area, when we estimate the time it takes to perform an ordered sequence of actions (involving transportation), when we cut and sew a skirt, when we decide whether or not to engage in an investment venture such as buying a new car or a new house, or when we prepare a seven-person meal following a cookery book recipe meant for four persons.

With a whole lot of more complex matters that are on the agenda of public political debates – for instance, unemployment, economic (in)equity, immigration, planning of traffic systems, accident risks in industrial or power plants, environmental changes and so forth and so on – it is characteristic, on the one hand, that mathematical models form part of the basis of the conclusions drawn by the specialists working in the area. On the other hand, this is not at all taken into account in the public discourse, mainly because it is largely unknown to the public, and the extent to which it is not, is because the mathematics involved is then considered to be a technical thing that does not interfere with the substance.

We could go on giving examples at all levels. Common to all of them is that the mathematics involved lies in the background, belongs to their interior or foundation, not to their appearance. In one word, mathematics is invisible because it is *hidden*, not because it is absent. This is amplified – again a bit of a paradox – by the fact that mathematics is *general*. It is present – and often very unexpectedly so – in a remarkable variety of different and in other ways unrelated contexts. Therefore mathematics hardly has a territory of its own, well-defined in social terms, where we can go and find it. Rather than being clearly located in the world, mathematics is more like an all-permeating ether (though more real than the ether of 19th century physics!). An additional aspect in the same vein is that most of those who exercise mathematical activity in society, globally speaking, are not considered, neither by themselves nor by others, as mathematicians. They are scientists in other fields, or engineers, architects, traffic planners, meteorologists, economists, insurance or banking people, forestry specialists, chemical plant designers and so forth. They appreciate the mathematics they make use of but simply think of it as a necessary or convenient tool in the service of purposes to which mathematics is of no independent interest. In this respect, mathematics is invisible like the wood that we cannot see because of all the trees. Another factor that tends to further disguise and hide the presence and function of mathematics in society is that information technology is, in contrast, very visible indeed and often "steals the picture" from the substance it dresses.

It is in its generality that the rôle of mathematics in the world differs from that of almost any other discipline. This generality has two sides, the heterogeneity and width of the *range* of areas in which mathematics is activated, and the *nature* of this activation. Despite the diversity of areas, the involvement of mathematics in them is founded on a relatively limited set of general questions, approaches, theories, methods, results and techniques that are basically the same in all contexts even if they are dressed in a continuum of appearances. (This should not be taken to imply that mathematics as *an edifice* is of limited size.) Of course other scientific disciplines – such as physics, chemistry, biology, economics, philosophy, linguistics and so forth – possess and display kinds of crucial generality as well, but within more constrained (not to be mistaken for small) ranges.

4. MATHEMATICS EDUCATION IN SOCIETY: THREE PROBLEMS

If, for the occasion, we accept the sketch presented in the previous sections as a fair description of the rôles of mathematics in society, an obvious issue for further examination is the position of mathematics education in society.

Irrespective of its specific raison d'être, place and organization, education is always embedded in a social context. Hence it is not disjoint from the spheres of values and interests, or from ideological, political, economic and cultural circumstances. It is necessary, therefore, to invoke the classical distinction between *analytical* considerations, which attempt to be neutral, objective and disinterested, and *normative* considerations, which involve or presuppose values and standpoints, keeping in mind that the presence of values and standpoints does not imply the absence of reason and argument. In what follows, I shall begin by presenting elements of an analysis and conclude by remarks of a normative nature.

If mathematics education is considered in a social context, whether from an analytical or from constructive/normative (e.g., curricular) perspectives, three interrelated problems emerge.

The first one, which I could call the *justification problem*, deals with the reasons, motives and arguments for providing mathematics education to a given category of students. In order words, it focuses on the question "*why* mathematics education for this category of students?" Answers to this question express the overall purposes and goals of mathematics education and have to rely on and reflect perceptions of the rôle of mathematics in society, of the philosophy of mathematics, the socioeconomic and cultural structure, conditions and environment in society, ideological and political ideals, and thus vary with place and time.

On the supposition that the justification problem has been settled, the *possibility problem* appears. It is concentrated on the issue of whether or not it is in fact possible to give mathematics education to the students of the categories considered, while satisfying the purposes and goals expressed in the answers to the "why" question. So, the possibility problem contains mathematical components such as the aspects of mathematics that are actualized by the arguments put forward to justify mathematics education, including the specific aims and objectives of mathematics teaching and learning entailed by these reasons. On this basis, psychological components are evidently particularly crucial – "*who* can learn what? On which conditions, and under what circumstances?" The same is true with the boundary conditions and prerequisites necessary for pursuing (and achieving) the overall purposes and goals of mathematics education, as well as the specific aims and objectives of mathematics teaching and learning.

Last, but not least, enters the *implementation problem*. Assuming that the justification problem and the possibility problem have been tackled, the implementation problem deals with establishing the structural and organizational framework within which mathematics education is to take place. It further deals with providing the immaterial resources (e.g., content, curricula, pedagogy, teaching methods, teacher education, working forms), the human resources (teachers, consultants, mathematics educators) and the material resources (classrooms, textbooks, technology) for the realization of mathematics education. The implementation problem also includes issues related to the philosophy and modes of assessment. In other words, the implementation problem focuses on the questions of "*how?*" and "*what?*" As there is a continuum of answers to these questions, varying with, and depending strongly on, the concrete circumstances, this problem is of a less universal nature than the other two.

It is important to note that these three problems represent an *analytical reconstruction*. Society does not normally see, articulate or tackle them as they are stated here. The point is that no educational system that provides mathematics education can avoid dealing with these problems directly or indirectly, and that explicit or implicit versions of them constitute the main driving forces of reform in mathematics education.

When one considers how these three problems have been tackled as a function of place and time, it appears that conflicting sets of answers exist, not only as regards the implementation problem, where differences would be expected, but also with respect to the more fundamental justification and possibility problems. For instance, it is an often observed phenomenon that the political and administrative authorities in a society give answers that differ considerably from those suggested by the majority of mathematics teachers and educators in that society, who, in turn, may well be in disagreement with the dominant views of research mathematicians, while many people in the arts, humanities and "soft" social sciences share a fourth set of views of mathematics education.

The dominant interest of society at large in relation to mathematics education is to provide for the *utilization*, *maintainance* and *development* of mathematics as an *applied science* and as an *instrument for practice* as means for technological and socioeconomic development, with the ultimate purpose of increasing the material wealth in society. Herein lies, in most countries, the general answer to the problem of justifying mathematics education for the general population. However, because it is recognized that all this presupposes that mathematics thrives as a pure science, society has a derived interest in providing as advanced a mathematics education as can be afforded to a smaller number of students.

So, society aims at supplying itself with mathematical *expertise*. On the other hand, society does not consist only of system components and anonymous power centres but also of individual human beings on behalf of whom society holds ideals, visions and beliefs. Therefore, societies further take an interest – of a varying degree – in providing mathematical prerequisites to the population at large to master their private and social lives as individuals and citizens. This is often called *numeracy*, *matheracy* (D'Ambrosio, 1985), *mathemacy* or *mathematical culture*, and constitutes another kind of justification of mathematics education.

It seems to be a tendency in many countries for mathematics educators, while agreeing with society on the importance of fostering mathematical expertise for the handling of extra-mathematical matters of socioeconomic significance, to place much more emphasis on the fostering of mathematical competence for private and social citizenship. Often we will encounter research mathematicians who focus on the educational requirements needed for the development of mathematics as a (pure) science, viewed as an element of human culture and belonging to the great accomplishments of mankind. In contradistinction, many professionals in the arts, humanities and non-mathematical social sciences (e.g., general educationalists) tend to think that society's legitimate interest in generating mathematical expertise should not lead us to infer that a substantial mathematics education of major segments of the population would be justified beyond the limit of functional numeracy. (This may be considered as an instance of the gap between "the two cultures," the scientific and the humanistic culture, identified by C. P. Snow already in 1959 but in vivid existence still today, although in new forms.)

The latter point of view is intimately connected with a certain position on the possibility problem. It is a widely held assumption in many quarters of society that not everybody can learn mathematics, or to be more precise: Perhaps it is possible to educate any normal person in mathematics, but the costs in terms of resources, time, effort and mental commitment and pain it takes to do so exceed what we ought to pay. The political and administrative authorities in a country not infrequently agree with people in the humanistic culture on this conclusion, but disagree as far as the premises are concerned, in that society thinks of the money and resource costs, whereas the humanists have the human costs of effort, commitment, and even suffering in mind. Some research mathematicians would agree that only relatively few individuals can *really* learn mathematics, and that it is a waste of costly and scarce resources to insist on bringing mathematics education to less able students, most of whom would rather prefer to be without it.

5. MATHEMATICS EDUCATION FOR DEMOCRACY

So far, I have attempted a neutral analysis of the overall position of mathematics education in society. Time has now come to look at mathematics education from a critical, and hence a normative, perspective. The considerations that follow are based on the assumption that no society is one of stable harmony, consensus and uniformity of interests. Differences and conflicts in life situations and conditions, in values and interests, exist at a multitude of levels and in many different forms. The most fundamental (potential) conflict is probably the one between the ruling segments of society, that is, the groups in economic, political, administrative and ideological power, and the individual citizen. This conflict is present irrespective of political system and is not so much to do with the individuals in or out of power, but rather with the asymmetry in the relationship between the rulers and those being ruled, no matter the basis on which the rulers have been recruited.

This pertains to mathematics education as provided by the educational system of society, simply because that system is under social rule (cf. Niss, 1981). So, there is likely to be a lack of harmony between what society thinks and wants as expressed by its representatives and authorities put in charge of mathematics education, what mathematics teachers, mathematicians, mathematics educators and users of mathematics, respectively, think and want, and what corporations, employers, parents and – last but definitely not least – students think and want.

We can now ask this question: Should the task to be solved by mathematics education for the population at large be confined to generating mathematical expertise and the numeracy needed for everyday private and social life? My answer is "no," for the following reasons.

In previous sections, we have seen that mathematics is instrumental in the shaping of society. Combined with the fact that mathematical competence is a much-demanded resource, which is scarce because it is difficult and demanding for the individual to acquire, this contributes to the creation of *expert rule* in society. Those not in possession of mathematical competence beyond what is used in everyday life will be excluded from influencing important processes in society that have a considerable impact on their lives as individuals and citizens. If we want to stimulate a democratic development of society – as distinct from both an authoritarian and a populist one – the fostering of intelligent and concerned citizenship is of crucial importance. (Reflections along these lines form part of a strong tradition established in Denmark during the last 20 years, Niss, 1979; for recent contributions, see, e.g., Nissen, 1993; Skovsmose, 1992.) I believe that I have argued, in this paper, for the key rôle of post-elementary mathematical qualifications for the exercise of such citizenship.

Of course, it is neither realistic nor desirable to educate everybody to become mathematical professionals, "experts." It is, however, possible to provide the common citizen with an insight into the nature of the experts' expertise, but at a cost. It should be a central purpose of mathematics education for all to be what in German – in the absence of an appropriate English word – is called *allgemeinbildende*, that is, having an educational value that is general and goes beyond the immediate and specific context in which it was established.

What does it mean for mathematics education to be of a general educational value? In contrast to the classical claims that mathematics has a special formative capacity of fostering and transferring context-free mental faculties such as logical thinking, precision in expression and language, a systematic attitude to problems, a sense of order and so forth (which is probably true to some extent), my emphasis is different: Mathematics education should be provided to everybody in order to give *insight into "the general,"* by which I mean: the constitutive features of and the essential driving forces behind the development of nature, society and the lives of human beings. Insight into the general does not consist in facts and skills alone and for their own sake, but serves the acquisition of overview, knowledge and judgment of main patterns, connections and mechanisms in the world; the ultimate end being to create prerequisites for taking positions on and acting towards processes of significance to society and the individual. For instance, it belongs to the realm of *Allgemeinbildung*, in a macro-economic mathematical model, to be able to distinguish between and identify the economic assumptions, the mathematical components and their technical representation in a computer program. Generally speaking, it is, however, yet another non-trivial research task to determine the specific mathematical elements in general education.

Even if we suppose that mathematics education should provide general education to the common citizen, the pursuit of this purpose presents us with severe problems. Perhaps the most important one is that many students will tend to refuse to receive it. This is due to the combined obstacle of the invisibility of mathematics in society and the fact that mathematics is a difficult subject to learn, regardless of the approaches applied. So, the relevance paradox that exists on the social level is reflected on the individual and psychological level: If the importance of mathematics in society and in ordinary people's everyday lives is invisible to the individual, why should the individual bother to take the (often non-neglible) pain involved in learning mathematics? On the other hand, most youngsters know, as an empiricial and sociological fact, that mathematical competence – even if for unclear reasons – is a key to attractive education and job opportunities. Therefore, large groups of students suffer from a paradoxical dilemma on the personal level that could be put as follows: "Mathematics is useless to me, but at the same time I know that I am useless without mathematics" (this version of the relevance paradox was provided by my colleague Jens Højgaard Jensen).

This is where the implementation problem and the art of mathematical pedagogy manifest themselves and lead us to other places and other stories. No miraculous cure has been (or will be) found that can dissolve all the difficulties inherent in the learning and teaching of mathematics. There is, however, still much new land for us to explore and to reclaim if we want to provide a better and richer set of multi-dimensional and reflective mathematical qualifications to everybody in society, with the ultimate end of serving its democratical development.

REFERENCES

Booss, B., & Krickeberg, K. (Eds.). (1976). *Mathematisierung der Einzelwissenschaften.* Interdisciplinary Systems Research 24. Basel: Birkhäuser

Booss-Bavnbek, B. (1991). Against ill-founded irresponsible modelling. In M. Niss, W. Blum, & I. Huntly (Eds.), *Teaching of mathematical modelling and applications* (pp. 70-82). Chichester: Ellis Horwood.

D'Ambrosio, U. (1985). Mathematics education in a cultural setting. *International Journal of Mathematical Education in Science and Technology, 16*(4), 469-477.

Damerow, P., Dunkley, M. E., Nebres, B. F., & Werry, B. (Eds.). (1984). *Mathematics for all.* Science and Technology Document Series 20 (pp. 1-109). Paris: UNESCO.

Department of Education and Science (Welsh Office, Committee of Inquiry into the Teaching of Mathematics in Schools) (1982). *Mathematics counts (The Cockcroft Report).* London: Her Majesty's Stationary Office.

Friedman, A. (1988-1990). *Mathematics in industrial problems* (Parts 1-3). The IMA volumes in mathematics and its applications 16, 24, 31. New York: Springer.

Howson, A. G., Kahane, J.-P., Lauginie, P., & de Turckheim, E. (Eds.). (1988). *Mathematics as a service subject* (ICMI Study Series). Cambridge: Cambridge University Press.

Keitel, C., Damerow, P., Bishop, A., & Gerdes, P. (Eds.). (1989). *Mathematics, education, and society.* Science and Technology Document Series 35. Paris: UNESCO.

Khoury, S. J., & Parsons, T. D. (1981). *Mathematical methods in finance and economics.* New York: Elsevier North Holland.

Morris, R. (Ed.). (1981) *Studies in mathematics education 2.* Paris: UNESCO.

Niss, M. (1979). Om folkeskolelæreruddannelsen i det vigtige fag matematik. In P. Bollerslev (Ed.), *Den ny matematik i Danmark. En essaysamling.* Copenhagen: Gyldendal.

Niss, M. (1981). Goals as a reflection of the needs of society. In R. Morris (Ed.), *Studies in mathematics education 2* (pp. 1-21). Paris: UNESCO.

Niss, M. (1985). Mathematical education for the "automatical society". In R. Schaper (Ed.), *Hochschuldidaktik der Mathematik.* Alsbach-Bergstrasse: Leuchtturm-Verlag.

Nissen, G. (1993). Der Mathematik aus ihrer Isolation heraushelfen - Bericht über das dänische Projekt 'Mathematikunterricht und Demokratie'. In H. Schumann (Ed.), *Beiträge zum Mathematikunterricht, 1992* (pp. 35-41). Hildesheim: Verlag Franzbecker.

Rosen, R. (Ed.). (1972-1973). *Foundations of mathematical biology* (Vols. 1-3). New York: Academic Press

Skovsmose, O. (1992). Democratic competence and reflective knowing in mathematics. *For the Learning of Mathematics, 12*(2), 2-11.

Snow. C. P. (1959). *The two cultures and the scientific revolution.* Cambridge: Cambridge University Press. [An expanded version is (1964) *The two cultures: a second look.* Cambridge: Cambridge University Press].

Steen, L. A. (Ed.). (1978). *Mathematics today. Twelve informal essays.* New York: Springer.

Wan, F. Y. M. (1989). *Mathematical models and their analysis.* New York: Harper & Row.

Wigner, E. P. (1960). The unreasonable effectiveness of mathematics in the natural sciences. *Communications on Pure & Applied Mathematics, 13,* 1-14.

THE REPRESENTATIONAL ROLES OF TECHNOLOGY IN CONNECTING MATHEMATICS WITH AUTHENTIC EXPERIENCE

James J. Kaput
North Dartmouth

1. INTRODUCTION

This paper is an account of issues and opportunities associated with new, technologically based attempts to attack a central didactic problem of mathematics education: creating viable, functional connections between the world of authentic human experience and the formal systems of mathematics. These new attempts take the form of changes to the historically received representation systems, the introduction of new systems, and the dynamic linking of different systems. A companion problem, the elevation of levels of thinking involved in the doing of mathematics from low-level computation to higher level planning, strategic and structural thinking, is treated in another paper (Kaput, in press b). At the end of the paper, I will try to point out some of the unmet challenges in exploiting electronic technologies in the representational realm – a realm distinct from, for example, the use of artificial intelligence or the execution of massive computations.

But, in order to expose what is new, I will first examine features of the inherited systems and the traditional didactic approaches to these problems. I hope to bring to consciousness, to render explicit, certain of the grand, but tacit, strategies that have been, or can be used. And in order to do this, I shall briefly establish a framework for the discussion, some background laying out a strongly interactivist perspective on the relations between mental and physical structures and operations on these. More extended versions can be found in Kaput (1989, 1991, 1992), which also include references to the wider literature relating to these topics.

1.1 Background: The Interactive Perspective

I draw a distinction between two sources of structure in mathematical experience: mental and physical. Neither can be formed or be utilized without interaction with the other – although, with the use of mental structures, this interaction might well be delayed considerably in the sense that extensive mental operations can take place apart from direct physical activity. The

R. Biehler, R. W. Scholz, R. Sträßer, B. Winkelmann (Eds.),
Didactics of Mathematics as a Scientific Discipline, 379-397.

mental structures, however, are ultimately the result of physical activity on physical material. In the Piagetian tradition and following from Saussure, this distinction is sometimes characterized as a distinction between the signified (internal) and the signifier (external). These two sources of structure differ in another important aspect – the physical is observable while the mental is hypothetical, to be inferred from observations and accompanying theory (in an interactional style not unlike that being posited here).

The interactions are usually cyclical and sometimes very rapid. What one "sees" triggers, or, perhaps more accurately (Grossberg, 1980; Kosslyn & Koenig, 1992), resonates with, existing mental structures (I mean to include both perceptual and cognitive structures), which alters what one can see in the physical material with which one is dealing. One may see more detail, or perhaps a larger pattern, which, in turn, resonates with additional mental structures, and the process continues. This interaction may or may not involve the actual production of additional physical structure (writing, drawing, etc.). Different mental structures can be called into play by, what appears to an independent observer, exactly the same physical material. For example, the following array of disks in Figure 1 might be thought of as a representation of $3 \cdot 5 + 4$ (3 rows of 5 per row, etc.) or as a representation of $4 \cdot 4 + 3$ (4 columns of 4 per column, etc.), or any of a large set of possibilities, including nonmathematical ones, of course.

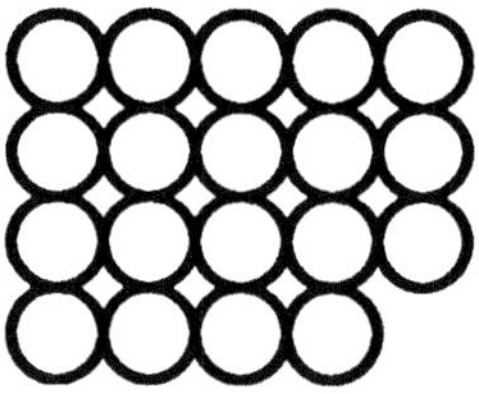

Figure 1. A number representation.

1.2 Background: Inert Versus Interactive Media

In the physical medium, there are two possibilities when one acts: (a) the "marks" one produces merely remain, maintaining their physical and hence perceptual permanence in the medium, as when one writes on paper with a pencil; or (b) the physical material responds in some way with representational elements that the "writer" did not directly produce, as when one hits the "=" sign on a calculator and some computational result appears. A physical medium that does not produce a reaction will be referred to as an *inert medium*, and one that has the capacity for physical response to inputs will be referred to as an *interactive medium*. (This distinction is discussed more fully in Kaput, 1992, in press b.) If the above in Figure 1 were an array of counting chips on a table, one could rearrange it, but, upon the rearrangement action, the chips would remain unchanged – they constitute a manipulable, but, nonetheless, an inert medium. In an interactive computer environment, the rearrangement of the screen array might effect a change in a linked numerical representation, or one might effect a change in the array by changing the numerical representation. The innovation inherent in interactive media is potentially profound in its impact on mathematics and mathematics education, as I shall attempt to explore below.

1.3 Background: Horizontal and Vertical Dimensions of Representation

Referential relationships between the mental and the physical usually take place in yet another referential context, when one set of physical materials is assumed to stand in some referential relationship to another. For example, suppose one has a situation from which one has isolated some particular aspect, say the motion of an oscillating spring, and one is attempting to describe that motion mathematically, that is, to create a mathematical model of the motion, for example, a function expressed in one or more representation systems. This situation might alternatively only be available to the person through a text description. These representational contexts are shown in Figure 2 in the horizontal dimension. Note that, for an individual, the referential relationship exists at the mental level in the form of integrated knowledge structures. From a social perspective, it might be the case that a group may take as shared or given that a referential relationship exists, although it would nonetheless involve mental operations on the part of the members of the group.

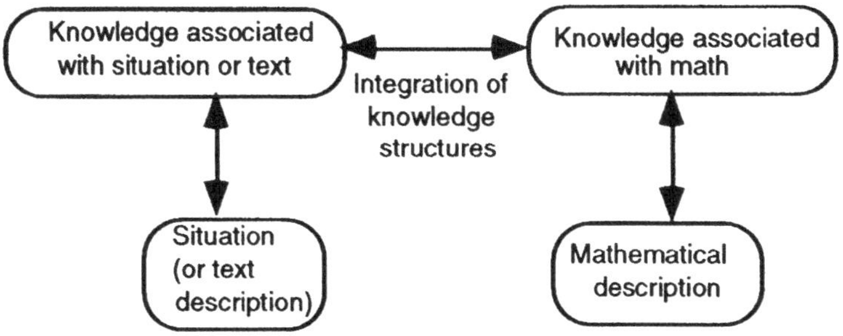

Figure 2. Horizontal and vertical dimensions of representation.

Other types of referential relationships might exist, as between two different representations of what mathematically mature adults take as a shared concept, say "the" concept of number and its constituent units. For example, suppose one has been given a formal decomposition of the number 19 written as 3(5) + 4, and one is asked to represent that decomposition as an array of counting chips. In this case, two different kinds of knowledge are involved and need to be integrated, and, usually, one is less complete or stable than the other. One is knowledge of physical groupings, arrays, and how to produce and recognize them. The other is knowledge of such formal decompositions, their syntax, and so on. The usual short-term aim of this sort of activity is to use the stable and elaborated physical knowledge to build knowledge of such formal decompositions. The longer-term aim is to build a rich, flexible, and elaborated understanding of numbers as collections of units of various sizes, in which different unit structures of particular numbers can be produced or recognized across many different contexts. This might be thought of as building fluent traversal of such diagrams as in Figure 2. Below I shall examine the horizontal dimension as it can be realized in interactive media, in which a physical link exists between the physical elements appearing in Figure 2.

1.4 Background: Action Systems Versus Display Systems

A diagram may be used to represent some aspect of some entity or situation, a skeletal anatomy, for example. It is intended to be observed. The same can be said of rhetorical algebra, which acted as a shorthand for natural language. It is entirely another matter to act physically in a system of symbols according to a prescribed or prescribable syntax, as in algebra as we now know it. These differences highlight an important distinction: between static display systems and operative active systems. As already noted, there are mental operations involved in *any* representational relationship, so the key difference with action systems is in the physical nature of the actions of the user on the representation, changing the physical state of the representation – which might take virtually any form, any state change initiated by the user. The active/display distinction reflects a slow and deep historical evolution. The earliest operative algebra did not appear until algebra began to emerge as an operative system in the 16th and 17th centuries. This development enabled new ways of thinking "characterized by the use of an operant symbolism, that is, a symbolism that not only abbreviates words, but represents the workings of the combinatory operations, or, in other words, a symbolism with which one operates" (Mahoney, 1980, p. 142). Bochner emphasizes the enormous historical import of this development:

> It brought about momentous innovations which gradually wrought a changeover from the inertness of traditional syllogistic schemata of Greek mathematics to the mobility of symbols and functions and mathematical relations. These symbols and functions and relations have penetrated into many areas of systematic thinking, much more than sometimes realized, and they have become the syllables of the language of science. (Bochner, 1966, pp. 185-186)

Later Bochner goes on to say

> . . . that various types of "equalities," "equivalences," "congruences," "homeomorphisms," etc. between objects of mathematics must be discerned, and strictly adhered to. However this is not enough. In mathematics there is the second requirement that one must know how to "operate" with mathematical objects, that is, to produce new objects out of given ones. Plato knew philosophically about the first requirement for mathematics, but not at all about the second. And Greek mathematics itself never developed the technical side of the second requirement, and this was its undoing. (Bochner, 1966, p. 313)

Leibniz' great achievement in the evolution of calculus can be characterized in the same terms – as the development of a system of symbols and operations on them that represented the transformations relating tangents and areas, rates and accumulations, and so on that constitute the conceptual heart of calculus. One can perform operations on symbols that outrun our ability to execute the conceptual operations that they represent; a major leap forward, not merely in mathematics, but for western civilization.

The new availability of interactive and representationally plastic media makes possible a wide variety of operative action representation systems, including operative versions of previously depictive systems, such as coordinate graphs, that can now be manipulated as if they were physical objects. Thus the move to operative symbolism that led to the scientific revolution becomes newly available to enhance the intellectual power of all manner of representation systems.

2. OLD AND NEW REPRESENTATIONAL STRATEGIES FOR ATTACKING THE ISLAND PROBLEM

2.1 Technology and the Island Problem: The Isolation of Formal Mathematics

Early uses of computer technology, as much as 40 years ago, were as a medium in which the traditional symbol systems of mathematics, numeric and algebraic, could be instantiated to yield enhanced actions. In the 1970s, as graphical display power of computers improved, formal graphing became common, unidirectionally linking algebraically defined relations with coordinate graphs. By the end of the 1980s, they became bidirectional, supporting the ability to act directly on graphical objects with the algebraic description of the result "following along" (Confrey, 1992; Schwartz & Yerushalmy, 1990; Yerushalmy, 1991). These linking systems were tightly constrained by an underlying algebraic constraint – with a few marginal exceptions, most linking activities required relations to be defined as closed-form algebraic rules. For a fuller history of technology in mathematics education, see Kaput (1992). Thus the representational uses of technology in mathematics education have been centered on facilitating either actions within traditional representation systems or linkages among them, even linkages connecting actions as well as objects. In the terms of Figure 3, they have been primarily used to assist activity and movement on the island, not to connect with the mainland of real human experience.

There are deep constraints at work in educational methods based on exclusive use of traditional formal representation systems. In particular, one's activity tends to be restricted to those mathematical relationships that are representable in closed algebraic form – because the computer and calculator systems require this as a starting point for virtually all their internal computations (this is even true for curve-fitting activities).

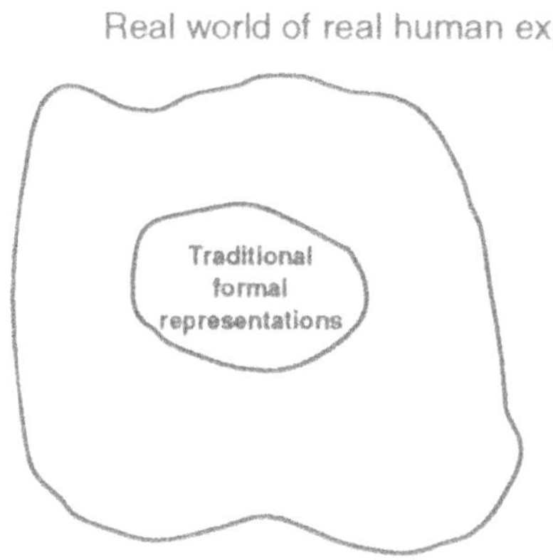

Figure 3. The island of formal mathematics.

But, to quote the mathematician Thomas Tucker, "Are all functions encountered in real life given by closed algebraic formulas? Are any?" (1988, p. 16). And here may be the biggest barrier to successful connections between formal mathematics and authentic experience. However, many functions that arise in daily life can be represented graphically without regard to their irregularity and algebraic intractability – for example, today's temperature as a function of time, your school bus's velocity as a function of time as you rode to school this morning, and so forth. I shall discuss overcoming the algebra constraint below, particularly in the context of modeling.

2.2 Traditional Strategies for Dealing With the Island Problem

Support of traditional representations certainly has use in linking real experience to formal mathematics, but this is primarily by allowing traditional modeling methods to be used with the computationally messier situations that occur when one removes the artificial academic constraints intended to make computations easy. More "realistic" applications become accessible, and they need not be held hostage to students' limited computational competence (Fey, 1989). In addition, at the higher grade levels, quantitative relationships are increasingly built up initially from situations through the use of numerical tables, graphs, and other less stringently formal means before the writing of algebraic equations. However, the means of connecting real experience with formal mathematics remain indirect, more like round trips by boat rather than traversal of a bridge.

We can picture the general strategies for making connections as in Figure 4, which shows two types of strategies, one, the traditional (left to right), involves shaping rich nonmathematical knowledge into more mathematical form by appropriately structuring experience; and the other involves shaping mathematical knowledge into less formal mathematical forms by creating pedagogical notations that behave more "naturally" in terms of knowledge and actions developed in ordinary life, for example, traditional manipulatives. The goal of each is to increase the overlap and integration of knowledge structures as indicated in Figure 4. I shall deal with the second strategy in the next section.

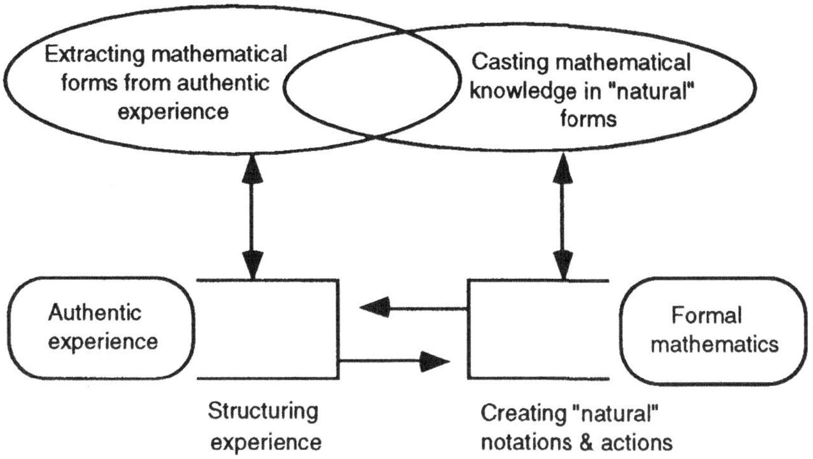

Figure 4. Connecting authentic experience to formal mathematics.

The structuring of experience is the essence of education of any kind (indeed, recall the etymology of "instruct"). In mathematics, we begin by structuring physical action, as with counting items differentiated in physical experience. But later, we structure experience with notational objects whose properties and relationships are defined mathematically and are mediated only indirectly by their physical properties (mainly shape and location, which implies order and grouping). Competence in the use of these notation systems is built upon mental structures and operations that embody the mathematics and, again, relate only indirectly to the perceptual features of the systems. Much student difficulty can be ascribed to attempts to base their performance on the rather subtle physical features of the systems in the absence of the needed conceptual structures.

2.3. Inherited Strategies and Conventions of Traditional Notations

Mathematical notation evolved in the hands of an intellectual elite primarily to serve the conceptual and communicative purposes of knowledge producers, without regard to the needs of the wider population to whom we now attempt to teach it. Surely due to its long-term success in the hands of the elite, formal notation has been treated historically as an untouchable given of education, in terms of its historically received detail and as a target of instruction. It is generally believed that "you don't really know it unless you can express it formally." The medium in which notations evolved was static and inert, leading to a premium on spatial economy and implicit syntax. As put by that master of notation design, Gottfried Leibniz, "In signs one observes an advantage in discovery which is greatest when they express the exact nature of a thing briefly, and, as it were, picture it; then indeed the labor of thought is wonderfully diminished" (cited in Cajori, 1929b, p. 184). He was referring to the expression of a mathematical concept rather than the depiction of a physical entity. Leibniz' criterion for brevity is reflected in another widely used strategy, the use of a single symbol to stand for a complex construct or a multiplicity of some kind. This strategy appears all about us: When we use the symbol "5" to stand for five instances of something, including five abstract marks, as the "V" in Roman numerals; it is especially powerful when iterated and used in combination with spatial ordering, as in our standard Hindu-Arabic number system; it is also used when we use, say, the symbol $\mathbb{G}$ to stand for the category of all groups and homomorphisms, an extremely complex mathematical object.

A third broad strategy in traditional notation is to use characters whose physical identity is independent of that to which they refer – the notation, such as algebraic notation, is intended to be of universal applicability. This universality strategy is behind all nonpictographic and nonphonetic alphabets (Hockett, 1960) as well as virtually all mathematical notations, despite the fact that one can often trace some iconic roots of a notation, for example, Robert Record's "=" or the "and" root of "+" ("&") (Cajori, 1929a, b).

A fourth strategy, almost invisible because it is part of virtually all visually-based notations, is perceptual constancy. Something written or drawn in a stable medium does not need effort to be "held in mind." This greatly relieves short-term memory load and is in strong contrast to spoken language. It also provides "objects" that can stand for any mathematical conception, which, in turn, can assist in objectifying that conception (Harel & Kaput, 1992). A fifth strategy is quite subtle. Many formal notations smuggle in physical and visual metaphors, such as the limit notations in calculus, all of which use arrows to exploit motion metaphors to carry the cognitions that the formal theory disacknowledges (Kaput, 1979). Thus traditional notations actually embody some of the right-to-left strategy depicted in Figure 4, but in a tacit way.

There are myriad other tacit strategies and conventions that are deeply intertwined in all mathematical activity: for example, the uses of lower case letters in certain contexts and upper case letters in others; the use of different versions of the same character (as when one says "the group G in the category $\mathbb{G}$"); or the re-use of notations taken from a particular context, for example, addition on the integers, reals, or even complex numbers in a more abstract context, as in the standard description of an abstract Abelian group.

I do not have the space here to catalog these strategies, but must note that they are almost never addressed explicitly in instruction – we act as the fish might, never discussing wetness. (A notable exception is Pimm, 1987, e.g., pp. 137-139.) But without them, mathematics as we know it would be impossible. It is this author's view that mathematics learning would be greatly enhanced by explicit and systematic attention to notational strategies and conventions. Some of the deepest systematicity and power of the subject is expressed in these very tacit ways.

2.4 Notational Strategies for Dealing With the Island Problem: Traditional Manipulatives

Modifying notations or creating new notations that engage the naturally occurring knowledge and ways of knowing that result from normal human activity and development is also a deliberate strategy to move students toward understanding of formalisms. Over the years, the most significant uses of this strategy arose in the context of manipulatives, especially the highly structured manipulatives of structuralists such as Dienes (1973). Dienes' multibase blocks, for example, constitute an alternative to the standard notation system. It renders the quantitative value of its symbols (configurations of blocks), and actions on those symbols, more explicit (as the literal size of the configurations). It also makes equivalence of symbols more explicit, in terms of equal quantitative value (size). As a symbol system, it stands as a concrete alternative to the formal expanded version of the placeholder system, but is frought with subtlety and shortcomings, some of which are addressed in Kaput, (in press b).

2.5 Examples of Newer Notational Innovations

More subtle examples of notation modification include such strategies as enabling students to act on traditional mathematical notations in more natural ways, as when in a computer environment, for example, one uses a pointing device and graphical interface to act directly on coordinate graphs by sliding, bending, reflecting, and so forth (as with *Function Probe*, Confrey, 1992). This is a subtle exploitation of the rich knowledge based in kinesthetic experience to act on mathematical notations, and hence to effect mental operations on mathematical objects, that is, functions. Another example is the direct manipulation of algebraic objects used in *Theorist* (1991), a symbol manipulation and graphing environment, which allows one to perform substitutions by "sliding and dropping" that which is to be substituted directly into the expression, or to manipulate terms in an equation as if they were physical objects, and so forth. Yet another example, reflecting the dynamic, interactive properties of the computer medium, is offered by *CABRI Geometry* (Laborde, 1992), the *Geometer's Sketchpad* (1992), and the *Super-Supposer* (1993). In each of these programs, the user can construct a figure via actions that are analogous to the traditional Euclidian drawing actions, and then perform physical dragging actions on any portion of the resulting figure. The dragging and its visual consequences draw their extraordinary power to reveal the logical constraints of the construction from the highly developed and practiced human visual processing system, especially in its ability to comprehend complex motion (Kosslyn &

Koenig, 1992, chap. 4). Another example is that of the *TableTop* (Hancock & Kaput, 1990; Hancock, Kaput, & Goldsmith, 1992), which represents database items as user-designable screen-objects that obey the imposition of logical constraints in dynamic Euler-Venn diagrams or scatterplots. The author and colleagues have developed object-based reasoning environments for learning multiplicative structures (Kaput & West, 1993) and additive structures (Kaput, Upchurch, & Burke, in preparation). In all these computer environments, the user manipulates objects on the screen, some of which overcome the constraints of "physicality" (Kaput, in press b) to effect discrete quantitative reasoning processes, and, in each case, these actions can be linked to more formal representations. One last and powerful example is the turtle geometry side of Logo, which was deliberately designed to provide a child-centric way of constructing geometric objects (Papert, 1980).

2.6 Reflections on the Examples, and Loosening the Universality Constraint

In each of these examples, and others not included, the mathematical representation system (including the allowable actions on it) shares properties with the world of physical objects and hence can tap into that wealth of processing power and competence that develops from normal human development apart from school. In some cases, selected attributes of physical objects are used in the elements to be manipulated (compromising on the universality criterion of mathematical notations), and in others, the actions themselves are kinesthetically oriented. The dynamic and interactive properties of electronic media, combined with their extraordinary representational plasticity, offer an enormous, but largely untapped, resource for innovation in mathematics education by utilizing naturally developing human perceptual and conceptual powers. Of course, careful research will be needed to determine how to balance notation learnability and conceptual power – although the fulcrum of that balance will continue to move as the technologies evolve.

3. LINKING AUTHENTIC EXPERIENCE WITH FORMAL REPRESENTATIONS USING LINKED ACTION SYSTEMS

In this section, I shall examine strategies for dealing with the Island Problem that compromise the universality criterion of notations in order to capitalize on these powers. It is possible to devise notations that incorporate some of the features of the situations or entities to which they refer. And, indeed, a whole spectrum of notations is possible that bridges gradually from the concrete context to an abstract description.

3.1 The General Case: Linkages at the Level of Actions

The diagram in Figure 5 will help orient my analysis of a type of technology use that is increasingly common, involving linked representation systems.

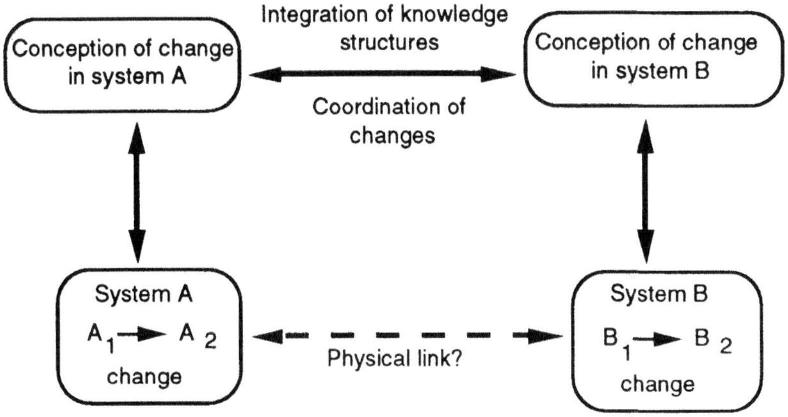

Figure 5. Linkages at the level of actions.

It has become possible to link action systems physically, for example, coordinate graphs and algebraic equations, so that an action on one system is translated to the other, either automatically or on command of the user. It is important to realize, however, that the linkage shown by the dashed arrow at the bottom of Figure 5 does not represent a referential relationship, but a physical connection – which might be either uni- or bidirectional. The referential relationship remains in the mind of the user. Of course, the purpose of the physical connection is to make the relationship explicit and observable at the level of actions in order to help build the integration of knowledge structures and coordination of changes depicted at the top of the diagram. This is a new power made possible by dynamic, interactive computational media, but as outlined below, it has not yet been deeply applied.

3.2 Models of Situations, MBL, and Simulation

An important special case of general representational linkage involves one system acting in the role of a model of the other, say Representation system B representing Situation A. In the case of a traditional model, there is no direct physical connection except by the transfer of measurements from the situation to the model, usually by input of particular values reflecting the results of measurements. However, using various computer-linkable probes ("Microcompter-Based Laboratory Equipment," Thornton, 1993; Tinker, 1990), changes in the situation can be automatically transmitted, and displayed, in the model – a left to right connection in Figure 4 above. This can greatly facilitate the development of understanding of the relations between changes in the Situation A and changes in the Model B by supporting rapid hypothesis testing. Such probes can even be designed so that changes in the model effect changes in the situation being modeled, for example, as with an electric toy car that not only is MBL-linked to formal representations of its motion, but also can be controlled from the computer – their motion can be specified as a graph or equation in the computer. With such automatic linkages, one must be aware that a rather large part of the modeling process has been supplanted – the part having to do with determining what to measure and how to measure it, what units to use, and so forth.

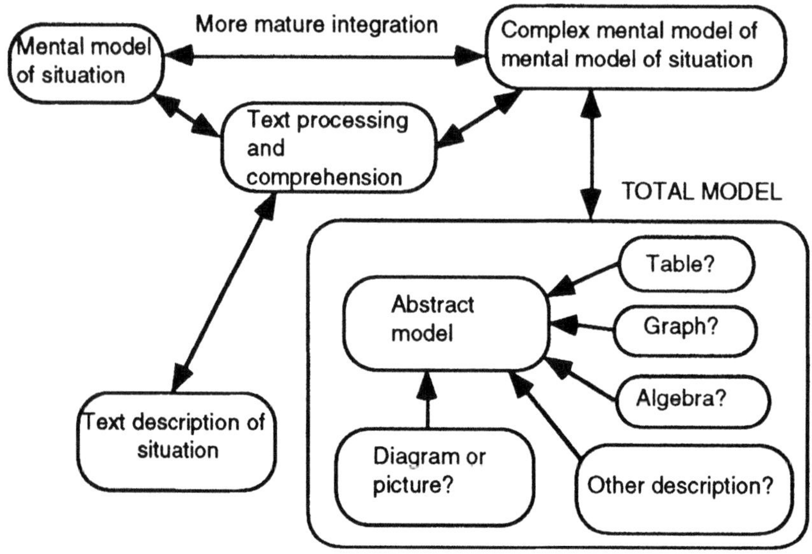

Figure 6. The "total model" of representations.

Often, a particular situation may have several "views" afforded by different representation systems with the same underlying mathematical model. For example, the underlying model might be a linear function represented by a table of numerical data as well as a coordinate graph. In this case, Model B is replaced by a cluster of representations, perhaps linked with one another, as a unit, representing A. In such a case, the model itself can either be regarded as an abstraction, in the same way that one may choose to regard a linear function as an abstraction apart from any particular representation, or it can be regarded as the totality of the cluster of representations, the "total model" in Figure 6. Further, it is often the case that the actual situation being modeled is not present, but rather, only text is available, and the modeler must conceptualize the situation for which the text is an indirect representation combining text-comprehension processing with prior knowledge of such situations. As the model develops, the text becomes less of an intermediary, and the mental model based in the mathematical representations comes to relate more directly to the conceptualizations of the situation (the top arrow in the diagram).

A *simulation* replaces text-like descriptions of a situation by an actual physically manipulable surrogate situation that is usually bidirectionally connected to more formal representations at the physical level.

3.3. "Lived-In" Simulations: An Extended Scenario

"Lived-in" simulations are where the attack on the Island Problem reaches its most direct form. Perhaps the easiest way to describe this approach is to offer a scenario involving motion simulations that amount to user-controllable linked representations of motion. A video animation of parts of the scenario has been developed (Kaput, 1993).

Imagine a pair of 12-year-old students driving a computer-simulated vehicle that provides a windshield view and a carefully linked user- or system-configurable collection of data displays for the dashboard; one set of displays for time, another for velocity, and a third for position. These include sounds for each set (metronome for time, engine pitch for velocity, and "echo" when passing roadside objects for position). The dashboard display can include velocity and/or position versus time graphs generated in "real-time" as well as clocks, odometers, tables, and so forth. This "MathCars" system is designed to help link the phenomenologically rich everyday experience of motion in a vehicle to more structured and formal representations and to provide exciting and intensely experienced contexts for reasoning about change, accumulation, and relations between them.

After some unstructured driving trips, they are now planning to follow a school bus whose (highly variable) velocity has been specified beforehand based on (one-dimensional) velocity data they collected on their own bus trip home the day before, using a graphing calculator to store the data. They had estimated their velocity and entered it as discrete points on a coordinate graph while tracking time with the calculator's built-in watch. They then connected the dots (with appropriate curve-fitting) last night and downloaded this data to the simulation. As an interesting challenge, they are now following the bus while driving "blind" (or, if you prefer, under IDR – "Instrument Driving Rules"), with feedback only on their own velocity graph and the bus's *position* graph, which appears in place of the windshield (thereby enacting the core relationship expressed by the Fundamental Theorem of Calculus). After two rear-end collisions and after reviewing their notes on the real trip, they decided to display their own position graph on the next try, which is now plotted on the same coordinate axes as the bus's position graph. They could also have chosen to turn on the windshield view and drive at a safe distance behind under visual driving rules (VDR). They could then review the velocity and position graphs that result before attempting another IDR trip. They could also attempt to drive in front of the bus, using the rearview mirror, or in IDR mode. They could get more adventurous and drive in other relations to the bus, for example, passing it in order stay one stop ahead. An especially interesting possibility suggested by the teacher is to drive home in the other lane at constant velocity without any stops to arrive home at the same time as the bus, thereby developing the idea of average velocity and enacting the conclusion of the Mean Value Theorem. After the student Chris asserts that you must always "hit" the average velocity for a trip, no matter how the velocity varies, the teacher leads a class discussion centered on the classroom display, in which different students try to violate "Chris's Law."

Other students have collected data on the subway, estimating (one-dimensional) velocities based on combining time data that they collected from actual trips with distance data that they obtained from city maps. They are now imagining themselves as subway train operators while using the subway simulation (there are several windshield views to choose from). Other students are taking turns riding the MathBike – a stationary bike that collects both motion information and pulse rate data. In trying to decide how to measure aerobic conditioning, they are plotting and attempting to interpret such curves as pulse versus time, (pulse minus resting pulse) versus time, and, most interestingly, pulse versus velocity. Another group is testing the assertion that, no matter how you vary your speed, your total number of heartbeats for a given distance will be pretty much the same.

Later, they will be driving some simple MathCars trips (linear position or velocity) in ADR mode ("Algebra Driving Rules") in which the motion is specified algebraically, and they will be attempting to match algebraically defined motion by driving under VDR, and the reverse. They will also set up and run simulated "ToyCars" on parallel tracks to study relative motion more systematically, describing the motion of each algebraically, confronting such questions as how to describe a later start versus describing a simultaneous start but from different locations; how to describe motion in opposite directions, both in terms of velocity and position; how to determine when or where cars going in opposite directions will meet; or when or where cars going in the same direction will pass; and so on. Of course, they will test and revise their models (essentially, parametric equations) by literally running them on the computer.

They will examine the difference between increasing velocity graphs that are concave-up and those that are concave-down. Given two cars reaching 60 miles per hour in the same amount of time, one with a concave-up velocity graph and the other with a concave-down graph, do they go the same distance, and if not, can we estimate the difference? Again, they will test their models by literally running them on the computer, making measurements of distances and estimating areas, and so forth. They will compare this motion situation with that of pay raises – given the same ending pay rate, does it make a difference whether your pay-rate graph is concave-up or concave-down, that is, is it better to get pay raises early or later?

And, on-line, they will be able to examine accumulation of fluid as they control the flow rate, using virtually the same interface and forms of data feedback that they used to control and record motion – in which the windshield view is replaced by a vessel that fills and empties, and flow rate is monitored in a visually and auditorially appropriate manner. Eventually, they will examine also the question of accumulating interest, simple and compound; accumulation of toxic wastes at different rates; deficit growth; and so forth, by redefining the quantities whose rates and accumulations are being examined. Some students, excited by driving what amounts to linear motion, opt to attempt driving in two dimensions, where they now control both north/south and east/west acceleration. They watch both acceleration and velocity vectors respond to their input, as well as their position depicted on an aerial (map) view. They extend what they have learned by specifying motion in one dimension to parametrically defined motion in two and, later, even three dimensions. They will also be able to examine regular motion of various types: especially harmonic and other periodic motion, and so forth, by attempting to produce it through driving as well as through MBL devices – approaching the trigonometric functions as they were developed historically, as means for describing real phenomena. Students will have available visual methods for approximating (what they will at a certain point refer to as) derivatives and antiderivatives, and so on. These will precede and complement the strictly algebraic methods available today that apply to functions defined by algebraically closed-form formulas. *They will be learning calculus before, during, and after algebra.*

3.4. Reflections on the Scenario and the Linking of Representations

A few features of the scenario (which involves simulations and activities designed by the author and under development) deserve further attention. First of all, it reflects an underlying reformulation of the subject matter of calculus in the spirit of Kaput (in press) and Nemirovsky (1993). This reformulation regards calculus as a strand in the curriculum beginning in the early grades and continuing through the school years through a gradual process of formalization and elaboration. Calculus, as the mathematics of change and accumulation of quantity, is conceived as anchored in and growing from everyday experience and also as the context in which much other mathematics can and should be learned, including mathematics such as algebra, which has historically acted as a prerequisite. However, the critical point here is that the linking technology (after all, the windshield view is just another linked representation of the simulated motion) and the phenomenological richness of the simulations are regarded as the critical enabling features of the scenario. While activity structures are important, the ability to connect these structures to more formal representations in "real" time is something new, unavailable before computers with substantial processing and display capabilities.

At a more detailed level, the reader will notice the depth of the connections between significant mathematical actions – comparison of functions –

across representation systems. The windshield view and accelerator embody a certain form of those actions borrowed from the authentic activity of driving, while the formal version of those comparison actions in the graphs are very different. The act of comparing functions in formal representations is not easy, although, as envisioned here, it is tightly anchored to driving a certain distance from the school bus, and one can imagine putting further constraints on the situation so that the students would need to maintain the distance between their car and the bus between particular bounds, reflected, perhaps, in a certain band on the position-function graph. Of course, a major question is whether the strongly situated forms of these activities can yield widely applicable knowledge, whether the "horizontal" linkages of actions (in the sense of Figure 3) at the physical level can yield flexible structures at the mental level. The current work of the author assumes that this will not come easily, but will need to be built systematically through combinations of careful variations of the simulations themselves and physical activities distinct from simulations, including work with MBL devices. One must never forget that a particular action is an embodiment or an illustration of a general relationship only for those who already cognitively have that relationship. For those who do not, it is merely one more action.

4. THE CONTINUING EVOLUTION OF REPRESENTATIONAL OPPORTUNITIES

I shall close with a broader perspective on the representational issues discussed in this chapter – what is really happening in these newer interactive media, and what are some of the challenges facing those who would exploit them?

4.1 Redistributing Sources of Structure and Action From the Mental to the Physical Realms

One can view the long-term historic development of physical notation systems, especially action systems, as a redistribution of the *source* of structure of activity from the mental to the physical realm. Recalling my earlier point, that mathematical activity involves tight cyclical interactions between the mental and physical, important shifts take place when one introduces new structures in the physical material with which one is interacting. In a novice learner, these shifts are by no means automatic. They require extended interactions leading to assimilations and accommodations wherein constraints and supports embodied in the physical structure are gradually internalized as mental structure. The movement from oral to written culture illustrates the point most vividly – structure could now be "stored" outside the human cortex, although it does not exist or function as structure apart from an appropriately structured mind.

Similarly, the development of algebra as an action notation system made possible the inheritance of powerful means of quantitative reasoning, and

Leibniz' notations for calculus likewise made available an immensely powerful system of thought. In some sense, the most potent intellectual contributions, leading to cultural inheritances, are embedded in these "ways of worldmaking," to borrow Nelson Goodman's phrase (Goodman, 1978). Some of the most important work of the masters is embodied and handed down, not in the form of facts or even theorems and principles, but rather in the syntax of the representation systems that they enable us to think with.

4.2 The Subtle and Fluid Nature of Representational Activity: Challenges of Representational Uses of Technology

Recent work by Hall (1990) and Miera (1990, 1991, in press), together with close examination by the author of student problem-solving in multiplicative contexts, exposes a level of subtlety and fluidity in student *production* of representations that may be difficult to accommodate in traditional styles of computer-based representational activity. Microanalysis of much representational activity, of which the above references supply abundant examples, reveals that representational elements change their referential meanings over time – the same inscription comes to be used in a new way as it is elaborated a minute later and becomes part of a larger whole and as either the mental structures evolve or as the producer's intentions change. Hall (1990) defines what he calls a "representational niche" – an evolving organization of inscriptions that affords computations and inferences that the user would not be able to make without those affordances. Such contain "slots" for variables that the user often manipulates and recombines or even abandons as activity progresses. All of this is deeply embedded in activity and communication and cannot be considered apart from its embeddedness. And it involves free student production of their own notations rather than the employment of complete, coherent structured systems developed by others.

Furthermore, particularly when modeling activity is involved, with physical materials in a group situation, then reference tends to become *distributed* across the notations, the physical material, the conversation, and the short-term memories of the participants, all in a very fluid, changing manner. The modeling and problem-solving is constituted from negotiating this "fluid." Aside from the challenges of building coherent accounts of such complex activity, one is forced to ask how the inherent structures of most computer-based approaches to representation, including those discussed earlier in this paper, may constrain, distort, shortcut, or undermine these productive processes. These remarks are in close alignment with the perspective offered by Lesh (this volume). The author wonders, with some trepidation, whether the egregious oversimplifications of artificial intelligence and the ways that these interlocked with the technology constraints might be repeated as we now deliberately attempt to use technology to augment human capability and learning in the representational dimension. So we are faced with yet another version of the eternal question: How much to "give" students (notational structures in this case) and how much to enable them to construct on their own? And the answer may, as usual, be a compromise, because, after all, student time and intellectual resources are limited, and they simply cannot produce, within the genuine constraints of schools and schooling, what required the greatest minds of civilization centuries of accumulated effort to produce. We need to identify the genuinely important and widely applicable representational strategies and find ways to cause students to build these as general, but personal intellectual resources.

Looking more specifically at the role of electronic technologies, we must be careful in exploiting the representational potential of these technologies not to produce another version of the oversimplification errors of the artificial intelligence community. They adopted models of mind and of mathematical activity that were grossly oversimplified versions of actual minds in actual practice, and have, by and large, had only marginal impact on real education, despite two decades of effort. We cannot afford to ignore the real subtlety and complexity of authentic human experience.

REFERENCES

Bochner, S. (1966). *The role of mathematics in the rise of science*. Princeton, NJ: Princeton University Press.

Cajori, F. (1929a). *A history of mathematical notations. Vol. 1: Notations in elementary mathematics.* La Salle, IL: The Open Court Publishing Co.

Cajori, F. (1929b). *A history of mathematical notations, Vol. 2: Notations mainly in higher mathematics.* La Salle, IL: The Open Court Publishing Co.

Confrey, J. (1992). *Function Probe* [Computer program]. Santa Barbara, CA: Intellimation.

Dienes, Z. (1973). *The six stages in the process of learning mathematics* (P. L. Seaborne, Trans.). New York: Humanities Press. [Original work published 1970)]

Fey, J. (1989). School algebra for the year 2000. In S. Wagner & C. Kieran (Eds.), *Research issues in the learning and teaching of algebra* (pp. 199-213). Hillsdale, NJ: Erlbaum; and Reston, VA: NCTM.

Geometer's Sketchpad (1992). [Computer program]. Berkeley, CA: Key Curriculum Press.

Goodman, N. (1978). *Ways of worldmaking.* Hassocks, Sussex: Harvester Press.

Grossberg, S. (1980) How does the brain build a cognitive code? *Psychological Review, 87*, 1-51.

Hall, R. (1990). *Making mathematics on paper: Constructing representations of stories about related linear functions.* Unpublished doctoral dissertation. University of California at Berkeley, Berkeley, CA.

Harel, G., & Kaput, J. (1992). The influence of notation in the development of mathematical ideas. In D. Tall & E. Dubinsky (Eds.), *Advanced mathematical thinking*. Amsterdam & Boston: Reidel.

Hancock, C., & Kaput, J. (1990). Computerized tools and the process of data modeling. In G. Booker, P. Cobb, & T. de Mendicutti (Eds.), *Proceedings of the 14th Meeting of the PME Vol. 3* (pp. 165-72). Mexico.

Hancock, C., Kaput, J., & Goldsmith, L. (1992). Authentic inquiry with data: Critical barriers to classroom implementation. *Educational Psychologist, 27*(3), 337-364.

Hockett, C. F. (1960). The origin of speech. *Scientific American, 203*, 88-96.

Kaput, J. (1979). Mathematics and learning: Roots of epistemological status. In J. Clement & J. Lochhead (Eds.), *Cognitive process instruction* (pp. 289-303). Philadelphia, PA: Franklin Institute Press.

Kaput, J. (1989). Linking representations in the symbol system of algebra. In C. Kieran & S. Wagner (Eds.), *Research issues in the learning and teaching of algebra* (pp. 167-194). Reston, VA: National Council of Teachers of Mathematics; and Hillsdale, NJ: Erlbaum.

Kaput, J. (1991). Notations and representations as mediators of constructive processes. In E. von Glasersfeld (Ed.), *Constructivism and mathematics education* (pp. 53-74). Dordrecht, Netherlands: Kluwer.

Kaput, J. (1992). Technology and mathematics education. In D. Grouws (Ed.), *Handbook on research in mathematics teaching and learning* (pp. 515-556). New York: Macmillan.

Kaput, J. (1993). *MathCars* [Computer animation video available from the author]. Department of Mathematics, University of Massachusetts Dartmouth, No. Dartmouth, MA.

Kaput, J. (in press a). Democratizing access to calculus: New routes using old roots. In A. Schoenfeld (Ed.), *Mathematical thinking and problem solving*. Hillsdale, NJ: Erlbaum.

Kaput, J. (in press b). Overcoming physicality and the eternal present: Cybernetic manipulatives. In R. Sutherland & J. Mason (Eds.), *Visualization and technology in mathematics education* [tentative title]. New York: Springer.

Kaput, J., & West, M. M. (in press). Assessing proportion problem complexity. In G. Harel & J. Confrey (Eds.), *The development of multiplicative reasoning in the learning of mathematics* (Research in Mathematics Education Series). Albany, NY: State University of New York Press.

Kaput, J., Upchurch, R., & Burke, M. (in preparation). Additive structures [Computer program]. Available from the authors. University of Massachusetts at Dartmouth, No. Dartmouth, MA.

Kosslyn, S., & Koenig, O. (1992). *Wet mind: The new cognitive neuroscience*. New York: The Free Press.

Laborde, J-M. (1991). *CABRI Geometry*. New York: Brooks-Cole.

Mahoney, M. (1980). The beginnings of algebraic thought in the seventeenth century. In S. Gankroger (Ed.), *Descartes: Philosophy, mathematics and physics*. Brighton, Sussex: Harvester Press.

Miera, L. (1990). Developing knowledge of functions through manipulation of a physical device. In G. Booker, P. Cobb, & T. de Mendicutti (Eds.), *Proceedings of the 14th Meeting of the PME*, Vol. 2 (pp. 101-108). Mexico.

Miera, L. (1991). *Explorations of mathematical sense-making: An activity-oriented view of children's use and design of material displays.* Unpublished doctoral dissertation. University of California at Berkeley, Berkeley, CA.

Miera, L. (in press). The microevolution of mathematical representation in children's activity. *Cognition & Instruction.*

Nemirovsky, R. (1993). Rethinking calculus education, *Hands On!*, *16*(1), TERC, Cambridge, MA.

Papert, S. (1980). *Mindstorms*. New York: Basic Books.

Pimm, D. (1987). *Speaking mathematically: Communication in mathematics classrooms.* London: Routledge.

Schwartz, J., & M. Yersulshamy (1990). *The Function Supposer*. [Computer program, part of the Visualizing Algebra series]. Pleasantville, NY: Sunburst Communications.

Super Supposer (1993). [Computer program]. Pleasantville, NY: Sunburst Communications.

Theorist (1991). [Computer Program]. San Francisco: Prescience, Inc.

Thornton, R. (1993). MacMotion [Data probes and software]. Portland, OR: Vernier Software.

Tinker, R. (1990). MBL: Hands on Science [Technical Report]. Cambridge, MA: TERC.

Tucker, A. (1988). Calculus tomorrow, In L. Steen (Ed.), *Calculus for a new century* (pp. 14-17). MAA Notes # 8. Washington, DC: Mathematical Association of America.

Yerushalmy, M. (1991). Student perceptions of aspects of algebraic function using multiple representation software. *Journal of Computer Assisted Learning*, *7*, 42-57.

CHAPTER 8

CULTURAL FRAMING OF TEACHING AND LEARNING MATHEMATICS

edited and introduced
by
Rudolf Sträßer
Bielefeld

The very first paper of this book on *cases of curriculum construction in the United States* by *James T. Fey* could well serve as an introduction to this last chapter on *Cultural framing of teaching and learning mathematics.* In a very lively way, the paper shows the political, societal, and cultural struggle to shape a national curriculum of mathematics, to decide on the mathematics taught and learned inside schools. To use a phrase from a German educationist of the early 1930s, the paper illustrates that curricula are the result of an interplay of societal powers (E. Weniger: *Lehrpläne sind das Ergebnis des Kampfes gesellschaftlicher Mächte*, see Blankertz, 1969, p. 117). It is exactly this interplay of societal, cultural forces (like parents, teachers, economy, science, government, and other social institutions) trying to influence the definition of the intended, implemented, and attained curriculum that is the subject of this chapter. To use the concepts of *Mogens Niss* in *Mathematics in society*, the chapter tries to throw light on the solution of the justification problem, the possibility problem, and the implementation prob-

R. Biehler, R. W. Scholz, R. Sträßer, B. Winkelmann (Eds.),
Didactics of Mathematics as a Scientific Discipline, 399-402.

lem of mathematics education in society. As can be seen from this description, the approach of this chapter is a rather broad societal perspective.

Consequently, all four papers in this chapter step back from the precise perspectives of former chapters of this book. Teaching and learning of mathematics are looked upon as a political and societal endeavor as a whole. In order to have a common topic, arguments specific to mathematics become more pertinent than in other chapters of the book. On the other hand, with the more global approach, it is even more difficult to make these arguments credible and cogent from the point of view of the scientific methodology that is used. Nevertheless, the chapter tries to bring to light factors of mathematics education that often tend to be overlooked in didactics of mathematics, factors that are of major importance to the actual as well as future teaching and learning of mathematics.

From a historical point of view, the paper on *cultural influences on mathematics teaching in nineteenth century Germany* by *Hans Niels Jahnke* analyzes the case of mathematics and mathematics education in Prussia and Germany and presents an integrated picture of ideological, social, sentimental, and technological influences on the field. The case study intentionally contradicts the prejudice that technological requirements are the major, if not the only, influences shaping education, showing that an educational philosophy deeply rooted in the cultural foundations of a society may well exert a decisive influence on mathematics as a scientific endeavor as well as on the structure and content of mathematics education at a given place and time. What seems most surprising in this case is the extremely negative view of everyday practical and common calculation taken by the reformers of that Humboldtian era, which is in sharp contrast to most of the present approaches to curricula for mathematics teaching and learning. The argument of Humboldt and Crelle, who praised the learning, appropriation, and appreciation of pure mathematics as an indispensable condition for the direct application of mathematics and as means of developing systematic thinking, is still worth a deep reflection. The following analysis of the implications of the decay of the neohumanist philosophy of education also shows that cultural influences are working rather slowly and cannot be identified in a short-term study.

In terms of time, the paper on *mathematics and ideology* by *Richard Noss* is nearer to present mathematics education: Starting from a nonpejorative concept of ideology ("body of ideas through which we see and with which we construct our reality"), the paper looks into the relations of ideology and (mathematics) curricula. A first result of Noss nearly paraphrases the quote from Weniger: "it is most useful to conceive of the curriculum as a site of struggle in which students, teachers, parents as well as voices from industrial, commercial and other settings have at various times competed . . . with varying relative strengths to assert their priorities." Consequently, the mathematics curriculum is seen as "only partially determined by mathematicians or mathematics educators." Last but not least by a comparison with music and its social role, Noss comes to a view of the mathematics curriculum as "a tension between the structural (from mathematics) and the ideological meanings." For school mathematics, the process of didactical transposition (necessarily linked with teaching/learning mathematics) opens up additional room for the immersion of ideological, if not delineated meanings of mathematics: "the call to delimit school maths to its apparent structural meanings ... represents an attempt to focus attention (albeit implicitly) on a variety of delineations that perform an (apparently) important ideological function." A thorough analysis of "mathematics for all" (see, also, Usiskin, in chapter 6) may be a good place to analyze the tensions between structural, inherent, and ideological meanings of mathematics and the politics of mathematics curricula.

The final paper on *cultural framing of mathematics teaching and learning* by *Ubiratan d'Ambrosio* in some sense is a step even further in time by looking into the future of mathematics education: Starting from a rather negative description of the present state of society in general (a somewhat paradoxical situation of wealth for a minority and poverty for the majority) and the important role of mathematics, d'Ambrosio criticizes present mathematics education as producing fundamentally harmful social outcomes: Mathematics (and science) education emphasizes techniques, formulae, and theories geared toward drills, toward exam-focused topics, and is not aimed at a contextualized understanding of mathematics and science. To counteract this tendency, d'Ambrosio proposes four general goals for mathematics (and science) education, namely, teaching mathematics and science to all, so that individuals can be wiser consumers, wiser decision makers, motivated and prepared for new careers, and prepared to take personal decisions based on ethical considerations. Explicitly alluding to the intentions of H.-G. Steiner, didactics of mathematics is presented as the science to reflect on these problems and to offer ways to cope with this societal task. The ethnomathematics proposal (coming from the recognition that every cultural group generates its own ways of coping with reality, organizes these ways into techniques, develops these into true *chefs d'oeuvre,* improving and transmitting them from generation to generation) is presented as a way to answer this global challenge. Illustrations of this approach are given from research (cf., also, Lave, 1988) as well as teacher training and curriculum development – showing ethnomathematics as a theoretical program and as a pedagogical practice.

In contrast to differences in the topics and approaches of the three papers briefly sketched above, the same three papers show a certain homogeneity in terms of methodology: All papers basically rely on a traditional type of heuristics, trying to convince the reader by conveying meaning, by enhancing his or her understanding of phenomena he or she may already know. The first paper of this chapter on *comparative international research in mathematics education* by *David Robitaille* and *Cynthia Nicol* uses a somewhat different research methodology: the traditional type of comparative analysis widely used by social science – applied in an international comparison of educational systems and the teaching and learning of mathematics herein. The paper presents the major attempts to describe the vast variety of types of mathematics teaching and learning produced by the cultural, historical, and political factors and decisions – a description that relies heavily on the global application of comparative studies in education. Trends in comparative education in the 19th and 20th century (from qualitative to quantitative methods) already show changes in the research methodology applied within this part of didactics of mathematics – with the increase of variables taken into account as the fundamental strategy to cope with the global variation of mathematics education. Case studies on regional details (e.g., comparisons of two or three national systems of education) or specific questions such as qualitative investigations of student performance or classroom processes are seen as complements to the use of questionnaires given to students, teachers, principals, and system-level experts. The construction of problem situations that are culturally appropriate but are also general enough to enable meaningful international comparisons of the results bring back problems to the research agenda that are studied in the other three papers of this chapter.

If we review the four papers in this chapter, a difference between them is most striking: the first three papers (by Robitaille & Nicol, Jahnke, and Noss) are more or less descriptive ones – with Robitaille and Nicol using a synchronic approach to show the cultural framing of teaching and learning mathematics – whereas Jahnke (more) and Noss (or less) use a case study approach. In contrast, larger parts of the very last paper by d'Ambrosio are written in a prescriptive mode, identifying goals and measures to be taken. A conclusion in the paper by Jahnke could be taken as a global remark from research on the *cultural framing of teaching and learning mathematics*. The conflict about the everyday practical applications reveals one of the most essential functions fulfilled by culture with regard to (mathematics) education: Just as the everyday practical applications represent the contemporary interests of politicians, parents, and students, the demand to teach theoretical mathematics may anticipate the future. The (partly political) struggle between the two positions is multidimensional and fed by disciplinary, scientific, ideological, and cultural values. It is culture that, for society, makes possible the dialogue with the future, and this is what determines its decisive importance for education.

REFERENCES

Blankertz, H. (1969). *Theorien und Modelle der Didaktik* (2nd ed.). München: Juventa.
Lave, J. (1988). *Cognition in practice*. Cambridge, MA: Cambridge University Press.

COMPARATIVE INTERNATIONAL RESEARCH IN MATHEMATICS EDUCATION

David Robitaille and Cynthia Nicol
Vancouver

1. INTRODUCTION

Comparative international research in mathematics education is an important and developing field. From studies conducted during the early 1900s describing the organization of mathematics instruction to current studies measuring students' mathematics achievement, comparative research within an international context has received a great deal of attention. Results from comparative international research have provided researchers, policymakers, and educators worldwide with opportunities to explore alternatives to mathematics curriculum and instruction and to compare student attitudes and achievement within an international context. In addition, such studies provide valuable international perspectives for current national discussions and debates on the development of efficient, effective, and qualitative mathematics education. A discussion of the developments, directions, and methodologies of comparative international research will provide insight into the importance of comparative studies and into the uses and implications of cross-cultural differences in the teaching and learning of mathematics.

2. DEVELOPMENT OF COMPARATIVE EDUCATION

Early comparative studies of mathematics education offered researchers and policymakers opportunities for an international exchange of ideas and experiences at a system level. Studies such as projects coordinated by the International Commission on Mathematical Instruction (ICMI) at the turn of the century and in the 1950s used narrative reports prepared by participating countries as data to compile summary descriptions about the organization of mathematics content and instruction within an international context (e.g., United States Bureau of Education, 1911; Fehr, 1959; Freudenthal, 1959). These early studies were descriptive in nature and were not designed to compare student achievement internationally.

It was not until the first IEA study conducted in the early 1960s that comparative studies of achievement on a large collaborative scale emerged.

R. Biehler, R. W. Scholz, R. Sträßer, B. Winkelmann (Eds.),
Didactics of Mathematics as a Scientific Discipline, 403-414.

Many situations have occurred that have created an environment that may have enhanced the development of such studies. One important consideration is the nature of the political and economic context over the last 30 years. In particular, there has been a great deal of concern centered around issues of excellence in education with regards to achievement and accountability. As well, there have been developments and adoptions of various research methodologies that have made the conducting of comparative studies feasible. Furthermore, the growth of intergovernmental organizations may also have enhanced the environment for the emergence of comparative studies.

Since the 1960s, there has been increased concern worldwide by public and governments over the need to relate educational investments to educational accomplishments and outcomes. The suggestion that a nation's well-being, economic prosperity, and growth are dependent upon the development and sustainment of an educated work force (Walberg, 1983) has strong political and economic implications. As a consequence, over the years there has been increased interest by nations for demonstrated relationships between achievement and educational variables such as curricula, instructional methods, teacher characteristics, organizational processes, and societal or contextual factors. Interest in international comparisons of these relationships has emerged as countries seek to develop or maintain positions of economic competitiveness in world markets. Thus interests of accountability and excellence in education within and between countries have provided a need for comparative studies of achievement.

The ongoing development and construction of research methods that enable the identification and analysis of educational variables associated with achievement have made it possible to seriously consider conducting comparative studies of achievement. Without appropriate research methods, it is difficult to conduct international comparative studies and to produce valid and reliable data from which decisions can be made. Early comparative studies of achievement often consisted of records from observations made by researchers or government representatives while traveling abroad. Spaulding (1989) notes that data from such studies focused on describing educational institutions, systems, and programs, but that these studies lacked the systematic and sophisticated methods that would enable appropriate quantification and comparability between and within cultures. Over the last 30 years, researchers have borrowed research techniques from the social and behavioral sciences and theoretical frameworks from the effective schools literature. This has enabled the quantification and comparison of cross-cultural differences in achievement and the identification of relationships between school variables and achievement. In addition, the development of methodologies in comparative education has also benefited from the advances in technology making it possible to identify, measure, control, and analyze numerous educational variables related to school achievement.

Conducting such large-scale studies also requires the collaboration of many researchers, educators, and government agencies and their sharing of information internationally. Eckstein (1982) and Spaulding (1989) note that the growth and development of various organizations such as the United Nations Educational, Scientific, and Cultural Organization (UNESCO), the Organization for Economic Development (OECD), and the World Bank has strengthened the possibility of conducting comparative studies. Organizations such as these have provided valuable system-level statistics such as enrollment rates at various educational levels, teacher-student ratios, and per-student expenditures. The World Bank, in particular, has focused on stressing the quantification of educational indicators so that it can assess the impact of its investment in educational developments worldwide. The growth of these organizations has provided a valuable infrastructure for the sharing of information, providing nations with a surface-level view of what is happening in other educational systems in comparison to their own.

3. IMPORTANCE OF COMPARATIVE STUDIES IN EDUCATION

There is a great deal of variation among different societies and education systems. Different cultures have different beliefs and values concerning the teaching and learning of mathematics. Some differences include degrees of access to resources such as textbooks, centralized versus decentralized decision-making, systems of comprehensive and/or selective schools, and varying retention or enrollment rates through high school. However, there are also similar issues in education that many countries share, for example, the question of how to provide effective and efficient education is a common international concern (Spaulding, 1989). Comparative studies, therefore, provide opportunities for the discussion and debate of these important issues in an international context. Researchers, educators, and policymakers can be provided with opportunities to reflect on their own educational situation, to gain further insight into possible alternatives to curricula and instruction, and to challenge previous notions about what is possible to expect concerning what students can learn and what instructional activities promote that learning.

Mathematics education has been a central area of comparative international research. Mathematics has always held a privileged position in the school system, in fact, it is one of the few subjects that is taught in most school systems worldwide (Howson & Wilson, 1986). Although countries differ in social, economic, and employment opportunities, there is still a surprising degree of similarity in the mathematics curricula of countries internationally (Howson & Wilson, 1986). This universal status and importance of mathematics, the similarity of mathematics curricula worldwide, and the supposed link between the study of mathematics or science and the development of a nation's economic strength (Walberg, 1983) make studies of international comparison in mathematics education of important interest to researchers, educators, and policymakers.

However, comparative international studies in mathematics education are not without their critics. Many critics state that such studies are expensive to conduct, require a great deal of time and effort, and produce results that are of questionable significance and difficult to compare. More specifically, Freudenthal (1975), criticized the first IEA mathematics study for, among other things, the failure to appropriately consider whether or not students had been taught the content needed to respond correctly to the various test items and for the failure to consider curriculum as a variable. As a result, the construction of powerful concepts such as opportunity to learn has proved valuable in accounting for between-country differences and in making appropriate international comparisons. Criticisms such as these strongly influence the structure, design, and development of comparative studies. It is in this way that such studies continue to define their importance and to improve in their search for quality comparable data.

4. DIRECTIONS AND TRENDS IN COMPARATIVE EDUCATION

One direction for cross-cultural research has been the consideration of the importance of culture in teaching and learning. Some of the early cross-cultural anthropological studies conducted in the 19th century focused on describing the nature of "primitive mentality" independent of cultural social needs and conditions (Stigler & Baranes, 1988). Recently such cross-cultural research has been called into question. New advances in cognitive science suggest that the situated nature of learning (Brown, Collins, & Duguid, 1989; Harris, 1989; Saxe, 1988), that is, the cultural contexts in which learning occurs, is an important consideration in understanding cross-cultural differences from an international perspective. As a result, a great deal of comparative education research is now placing more emphasis on the role of social and cultural contexts and the contextualization of data for the development of meaningful interpretations and comparisons.

The consideration of qualitative research methods is another direction in which many researchers in the international comparative education community are moving. Altbach (1991) notes that the consideration of new research orientations and an emphasis on qualitative research methods promise researchers a more comprehensive study of differences in not only approaches to curriculum but also instructional approaches and the relationship of these to student achievement. Similarly, Stigler and Baranes (1988) suggest that the inclusions of qualitative research methods in comparative studies of schooling provide opportunities for in-depth investigations of the nature of schooling differences and the cultural contexts of such differences.

Comparative studies in the IEA tradition have focused mainly on the use of survey research methods to determine the influence of particular variables on teaching and learning within an international context. Over the last 30 years, IEA studies have increased the number and nature of the variables included for investigation and have made significant advances in the development of instrumentation. For example, the Second International Mathematics Study (SIMS) analyzed curricula internationally, investigated teaching practices, included a longitudinal study to determine student achievement growth patterns over the school year, and utilized the opportunity-to-learn construct as a method of accounting for achievement differences. Although "comparative studies done by IEA are among the best examples of comparative research" (Altbach, 1991, p. 491), researchers such as Stigler and Baranes (1988) suggest that such studies need to be complemented by work with a more in-depth qualitative analysis.

5. IEA SURVEYS

IEA, established in 1960, is a cooperative network of research centers (Postlethwaite, 1971). The organization began with a group of researchers from around the world who, notes Purves (1989), "were concerned with a number of issues that could not be studied well within the confines of one school system" (p. vii). Today, the membership of IEA consists of institutions from more than 50 countries under the united goal of investigating the potential influence of alternative curricula, teaching strategies, and administration strategies on student achievement (Hayes, 1991). The first major IEA study was conducted in 1964, and, although it was designed to be a general study of the outcomes of schooling, the subject area of mathematics was chosen as the vehicle through which national comparisons in education would be made.

5.1 The First International Mathematics Study

This first IEA study was a very ambitious study measuring achievement in various mathematics topics in 12 countries. Countries that agreed to participate were mainly European industrialized countries: Australia, Belgium, England, the Federal Republic of Germany, France, Finland, Israel, Japan, the Netherlands, Scotland, Sweden, and the United States. The student populations sampled for this study consisted of 13-years-old students and of students in their last year of secondary school (for complete population definitions, see Husén, 1967).

Since IEA projects are a cooperative international effort, all participating countries were involved in some way in the construction of the achievement tests. Using a two-dimensional item-specification grid consisting of a content-by-cognitive behavior matrix, appropriate topics involving arithmetic, algebra, geometry, and calculus were selected for each population level. Students, teachers, school principals, and national experts were also asked to complete a number of questionnaires. Students responded to attitude scales providing information on their personal views of mathematics, their school work, and career aspirations. Teachers were asked about their own experience, education, and qualifications and how well they expected their students to perform.

The major findings of the first IEA study have been summarized in many volumes and articles over the years (Husén, 1967; Kilpatrick, 1971; Postlethwaite, 1971; Robitaille & Travers, 1992). Therefore a selection of some of the important findings related to student achievement and attitudes, selectivity and retentivity, and opportunity to learn will be discussed here.

The results of the first IEA study indicate that all groups of students from all participating countries found the tests difficult. For example, the majority of mean achievement test scores for each population were below 50% across all countries. It was found that 13-year-old students had a more positive view of mathematics as a process than did senior students. This may indicate that student's attitudes and interest in mathematics declines with age and with the continued study of mathematics.

A major issue addressed by this study was to determine the situations that enable the most talented students to perform and develop. Many of the participating European countries had highly selective schools in which only elite students enroll as opposed to comprehensive schools in which all students attend the same type of school. It was found that the most able students from all countries performed at equally well levels regardless of the type of school in which they were enrolled. This result challenges the argument made for selective schools, since the performance of the most talented students does not appear to be jeopardized by the type of school in which students are enrolled. A related issue of retentivity was also explored and was found to be an important factor in accounting for differences in achievement between countries. Interestingly, systems that retain relatively higher proportions of students by the terminal year of secondary school were found to have higher proportions of students performing well.

The opportunity-to-learn variable, that is, the opportunity that students have to learn the mathematics necessary to respond correctly to a given item, is an IEA innovation. It is an attempt to learn more about the extent to which a particular curricular topic is implemented. Although there has been criticism concerning how this variable was operationalized (Postlethwaite, 1971), it was found that there was a positive relationship between students' achievement on an item and their opportunity to learn the mathematics content of that item. The results suggest that the opportunity to learn or the amount of material covered is therefore a predictor of levels of achievement.

The first IEA study indicated that such studies are not only feasible but of value to researchers, educators, and policymakers. The findings of this first large-scale international comparative study of education provided insight into the tremendous variability between countries on many variables that are important to schooling in general, and to the teaching and learning of mathematics in particular. The findings, particularly the achievement results and the ranking of countries, made headlines in the popular media and initiated a great deal of discussion. There have been questions raised concerning the validity of the comparisons, the selection of items used and inappropriateness of topics, the way in which the opportunity-to-learn variable was operationalized, the scope of the study, and the lack of context in which to interpret and make plausible reasons for achievement variability (Freudenthal, 1975; Kilpatrick, 1971). Such concerns were addressed in the design and implementation of IEA's Second International Mathematics Study (SIMS).

5.2 The Second International Mathematics Study

Since the first IEA study in 1964, an extensive curriculum reform movement occurred and basically subsided by the late 1970s. The early 1980s were therefore an opportune time to conduct a second international comparative study to determine the status of mathematics, the effects of the curricular reform efforts, changes since the first study in student achievement, attitudes, and interests, as well as changes in school system variables such as retentivity.

The design of SIMS was quite different from the first IEA study. Unlike the first study, the second study was curriculum-based and focused on the study of mathematics through three different levels: the *Intended Curriculum,* the mathematics intended for learning by national and system-level authorities; the *Implemented Curriculum,* the curriculum as interpreted by teachers and presented to students; and the *Attained Curriculum,* the curriculum learned by students and determined by their achievement and attitudes (Robitaille & Garden, 1989). This framework provided guidance for the development of instruments and the interpretation of results. SIMS also included an analysis of curriculum of participating countries as well as a longitudinal option to determine the growth of students' achievement and the effects on achievement of various instructional practices.

From the 20 countries participating in one or more parts of the study, two population levels were chosen. Population A included all 13-year-old students and Population B included all senior students (for complete population definitions, see Robitaille & Garden, 1989, pp. 6-7). For the cross-sectional version of the study, all data were collected toward the end of the school year. During the longitudinal study, teachers were asked to respond to questionnaires concerning the instructional strategies they used in teaching particular topics, and pretests were administered at the beginning of the school year and posttests at the end. Test items were selected from the topic areas of arithmetic, algebra, geometry, measurement, and descriptive statistics.

The SIMS data were collected from 1980 to 1982. An analysis of the results from this study are presented in many volumes and articles (Garden, 1987; Robitaille & Garden, 1989; Robitaille & Travers, 1992). The following is a discussion of some of these findings at the national level or intended curriculum level, the school level or implemented curriculum level, and the student level or attained curriculum level.

At the national level, some of the more interesting findings deal with changes since the first IEA study in the intended curriculum and in the retentivity of students. Many countries experienced curricular reform and change, mainly due to the influence of the new mathematics movement. In general, it was found that, for most countries, the curricular emphasis given to the study of geometry had decreased, while the importance given to the study of algebra had increased since the first study. The changes in the retentivity of students from the first study are also interesting. It was found that for some countries, such as the United States and Sweden, the proportion of students enrolled at the senior level had declined since the first study. A large between-country difference was found in the proportion of students who are completing the requirements needed to pursue postsecondary level mathematics. In particular, in most countries, Population B accounted for only about 12% of the age cohort, while in Hungary, 50% of the age cohort was included.

In all countries, there was some form of sorting or grouping of students for mathematics classes. There is widespread tracking of students at the Population A level on some measure of ability, particularly in the United States and in the Canadian province of British Columbia. In other countries, such as Japan, Hungary, and France, Population A teachers indicated that they rarely grouped students by ability for mathematics at that level. Furthermore, achievement results indicate that students from Japan and France performed very well in mixed-ability groups at the Population A level. However, such a relationship does not indicate cause and suggests only that the underlying assumptions on the use or nonuse of grouping by ability be explored further.

Changes in achievement from pretest to posttest in the longitudinal study indicate that Japan and France experienced high rates of growth. For example, on an item requesting Population A students to add the decimal fractions 2/5 + 3/8, the mean percent correct in France grew from 5% on the pretest to 73% on the posttest. For students in the United States, where such mathematical content is first presented to students at about the Grade 5 level and reviewed each year after, the mean percent correct grew from 44% to 60%.

The results of the SIMS have provided countries with important information that can be used to make decisions regarding change in curriculum and instruction. Some countries such as Ireland and Belgium (Flemish) used information from the curriculum analysis stage to assist in curricular changes and in training practices for teachers (Garden, 1987). In some countries, such as Sweden, task forces were constructed to review the performance of their students and determine ways in which it may be improved (Marklund,

1989). Such work indicates how results from international comparative studies can be used to guide curriculum planners and policymakers.

As indicated by many researchers analyzing comparative international data, making comparisons of achievement between countries is a very difficult and complex endeavor. Even in countries in which the intended curriculum is similar, differences in performance levels can be found. Such results indicate that it is important to consider the social contexts in which learning occurs and the embedded cultural values and beliefs in which educational systems operate. A series of small-scale, cross-cultural studies have been conducted by researchers of the University of Michigan, which complement the research conducted by the IEA studies and provide greater insight into the nature of achievement differences and the cultural contexts in which they occur.

6. MICHIGAN STUDIES

The University of Michigan studies under the direction of Harold Stevenson and his colleges have been in progress for over 10 years. First- and 5th-grade students, their parents, teachers, and principals from schools in Sendai (Japan), Taipei (Taiwan), and in metropolitan Minneapolis and Chicago (USA) provided data for the analysis of academic achievement. These studies focused on studying cross-cultural differences in student mathematical performance and classroom instructional practices, as well as exploring the characteristics of culture in helping to explain these differences.

Mathematics achievement tests were designed to assess students' mathematical skills and conceptual knowledge in areas of arithmetic, algebra, and geometry. The results of these studies found that Japanese and Taiwanese students outperformed American students by the first year of schooling, and these differences were also found to be even greater by the 5th grade. Stigler and Baranes (1988) conclude that "the Asian advantage in mathematics, at least at the elementary school level, is not restricted to narrow domains of computation, but rather pervades all aspects of mathematical reasoning" (p. 294).

Classroom observations were conducted using detailed narrative descriptions to record the flow of activities and behaviors of students and their teachers during mathematics lessons. The results from these observations indicate differences in the ways in which classrooms are organized. In particular, it was found that American teachers use whole-group instruction less than 50% of the time, while Asian teachers use it about 80% of the time. As well, the Japanese teachers tended to emphasize the use of verbal discussion and explanation, while using student errors as sources of investigation and discussion. One further interesting finding is that Asian teachers placed emphasis on the use of concrete manipulative materials, but, unlike American teachers, they tended to use the same manipulatives for many different instructional purposes. As Stigler and Baranes conclude,

"Japanese teachers use the objects (manipulatives) as a topic of discussion, whereas American teachers tend to use the objects as a substitute for discussion" (1988, p. 297).

As a way of understanding the reasons for the cross-cultural differences in mathematical performance and classroom instruction, the Michigan studies explored the cultural beliefs of parents and teachers about learning and education. The researchers found that there were cross-cultural differences in the ways in which teachers and parents perceived the concept of ability. Japanese parents and teachers tended to emphasize the role of effort in determining students' ability to succeed at school, while their American counterparts tended to believe more in the concept of innate ability as reasons for student success. Stigler and Baranes conclude that these cultural differences in the beliefs about ability versus effort in determining success help explain the observed differences in classroom instructional practices. For example, it may help explain why American educators tend to group their students by ability, while Japanese educators do not use this approach.

The Michigan studies are a valuable contribution to comparative international research. These studies complement the survey research conducted by the IEA by providing in-depth investigations and data on the contexts in which mathematics teaching and learning occur. In addition, these studies emphasize the importance of culture contexts in the analysis and interpretation of results, providing researchers with alternative ways of studying cross-cultural differences in mathematical performance and instruction. The insights of such small-scale studies combined with information and techniques learned from previous IEA studies provide a valuable foundation for the Third International Mathematics and Science Study.

7. THE THIRD INTERNATIONAL MATHEMATICS AND SCIENCE STUDY

The Third International Mathematics and Science Study (TIMSS) is one of IEA's current projects. In the tradition of other recent IEA studies, TIMSS plans to describe the intended, the implemented, and the attained curriculum. However, TIMSS differs from previous studies in that it plans to investigate mathematics and science curricula simultaneously, it plans to give particular consideration to describing the educational environment, and it plans to explore possible links and relationships among the various levels of curricula within this environment. As a result, the activities of TIMSS are guided by four main research questions focusing on the mathematics and science students have learned; the variations in intended learning goals among educational systems; the various opportunities students have to learn mathematics and science; and the relationships of the intended, the implemented, and attained curriculum with respect to the context of education, the arrangements for teaching and learning, and the outcomes of the educational process.

As a survey, TIMSS will seek answers to these questions through the use of questionnaires given to students, teachers, principals, and system-level experts. However, TIMSS also plans to supplement this survey data with qualitative in-depth investigations of both student performances in problem situations and of classroom processes. These qualitative investigations, requiring intensive observations and interviews with both students and teachers, will need a great deal of organization, expertise, and resources. As well, it will be a challenge for TIMSS to construct problem situations that are culturally appropriate but are also general enough to enable meaningful international comparisons of the results.

TIMSS intends to provide the international research community, educators, curriculum developers, and the public with a great deal of comparative data embedded within a contextual framework that will make the challenging task of interpreting international differences more informative and valid. In any comparative international study, there is the danger that results will be interpreted and compared devoid of the rich educational context in which they are embedded. However, in order for meaningful within-country and between-country comparisons to be made, consideration of the educational environment is important. If this is done, TIMSS and other international studies have the potential to make significant contributions to our current understanding of mathematics learning and teaching.

REFERENCES

Altbach, P. (1991). Trends in comparative education. *Comparative Education Review, 35,* 491-507.

Brown, J. S., Collins, A., & Duguid, P. (1989). Situated cognition and the culture of learning. *Educational Researcher, 18*(1), 32-42.

Eckstein, M. (1982). Comparative school achievement. In H. Mitzel (Ed.), *Encyclopedia of educational research* (Vol. 1). New York: Free Press.

Fehr, H. (1959). The mathematics education of youth a comparative study. *L'Enseignement Mathématique, 5,* 61-78.

Freudenthal, H. (1959). A comparative study of methods of initiation into geometry. *L'Enseignement Mathématique, 5,* 119-145.

Freudenthal, H. (1975). Pupils' achievement internationally compared. *Educational Studies in Mathematics, 6,* 127-186.

Garden, R. (1987). The second IEA mathematics study. *Comparative Education Review, 31*(1), 47-68.

Harris, P. (1989). Contexts for change in cross-cultural classrooms. In N. Ellerton & M. A. Clements (Eds.), *School mathematics: The challenge to change* (pp. 79-95). Geelong, Australia: Deakin University.

Hayes, W. (1991). *IEA guidebook 1991: Activities, institutions, and people.* The Hague, Netherlands: International association for the evaluation of educational achievement (IEA).

Howson, G., & Wilson, B. (Eds.). (1986). *School mathematics in the 1990s.* Cambridge: Cambridge University Press.

Husén, T. (1967). *International study of educational achievement in mathematics* (Vols. I - II). Stockholm: Almqvist & Wiksell.

Kilpatrick, J. (1971). Some implications of the international study of achievement in mathematics for mathematics educators. *Journal for Research in Mathematics Education, 2,* 164-171.

Marklund, I. (1989). How two educational systems learned from comparative studies: The Swedish experience. In A. Purves (Ed.), *International comparisons and educational reform* (pp. 35-44). Alexandria, VA: Association for Supervision and Curriculum Development.
Postlethwaite, T. (1971). International association for the evaluation of educational achievement (IEA) – The mathematics study. *Journal for Research in Mathematics Education, 2*, 69-103.
Purves, A. (Ed.). (1989). *International comparisons and educational reform.* Alexandria, VA: Association for Supervision and Curriculum Development.
Robitaille, D., & Garden, R. (Eds.). (1989). *The IEA study of mathematics II: Contexts and outcomes of school mathematics.* Oxford: Pergamon Press.
Robitaille, D., & Travers, K. (1992). International studies of achievement in mathematics. In D. Grouws (Ed.), *Handbook of research on mathematics teaching and learning* (pp. 687-709). New York: Macmillan.
Saxe, G. (1988). Candy selling and math learning. *Educational Researcher, 17*(6), 14-21.
Spaulding, S. (1989). Comparing educational phenomena: Promises, prospects, and problems. In A. Purves (Ed.), *International comparisons and educational reform* (pp. 1-16). Alexandria, VA: Association for Supervision and Curriculum Development.
Stigler, J., & Baranes, R. (1988). Culture and mathematics learning. In E. Rothkopf (Ed.), *Review of research in education* (pp. 253-306). Washington, DC: American Educational Research Association.
United States Bureau of Education (1911). Mathematics in the elementary schools of the United States. *International commission on the teaching of mathematics: The American report. 12*, 7-181.
Walberg, H. (1983). Scientific literacy and economic productivity in international perspective. *Daedalus, 112*(2), 1-28.

CULTURAL INFLUENCES ON MATHEMATICS TEACHING THE AMBIGUOUS ROLE OF APPLICATIONS IN NINETEENTH-CENTURY GERMANY

Hans Niels Jahnke
Bielefeld

1. SCIENCE, EDUCATION, AND CULTURE

It is possible to distinguish four different dimensions of culture (cf. Bishop, 1988, pp. 16-19; White, 1959, p. 19):

1. *ideological:* composed of beliefs, dependent on symbols, philosophies;
2. *sociological:* the customs, institutions, rules, and patterns of interpersonal behavior;
3. *sentimental:* attitudes, feelings concerning people, behavior;
4. *technological:* manufacture and use of tools and implements.

This categorization is useful for the following case study about cultural influences on mathematics instruction, how it was conceived in the Humboldtian educational reform, and its history in the 19th century. The case study will show the important role of the technological side of culture, which became manifest, however, in a dialectic transposition. Germany, at the threshold of industrialization, did not conceive of education merely as a reflex on technological requirements, but shaped the ideological, sociological, and sentimental aspects of education in conscious negation of this demand. The underlying idea of culture can be described in White's (1959, p. 8) words, who says it is the function of culture "to relate man to his environment on the one hand, and to relate man to man on the other." Education, in Germany, was intentionally not understood to be preparation for a vocation, yet, nevertheless, the problem of application was at its background. In which sense and with what consequences this was the case shall be shown in the following.

It is hardly necessary to explain the importance of the Humboldtian reforms initiated in 1810. With regard to education, they represented, for Germany, the step into modernity by developing the foundations of the present university and school system, and by establishing a scientific education of teachers. These developments were reflected in two legal acts. On July 12th 1810, the Prussian Ministry of the Interior decreed, in the *Edict Concerning the Introduction of a General Examination of All Future Teachers,* that the philosophical faculties of the universities were to be given

R. Biehler, R. W. Scholz, R. Sträßer, B. Winkelmann (Eds.),
Didactics of Mathematics as a Scientific Discipline, 415-429.

responsibility for the scientific education of future secondary school teachers. Up to that point, the task of these faculties had been to provide a general education for the students of the higher faculties of theology, medicine, and law. On the other hand, the *Edict Concerning the Students Entering the Universities* of June 25th 1812, for the first time made a firm distinction between schools and universities, and the minimal mathematical knowledge required of all students who were to enter the universities was defined (cf. Jahnke, 1990b, chap. 1).

At the beginning of the 19th century, there was no pedagogy or didactics taught at the universities. The discourse about education, the shaping of syllabi, and the educational value of the various subjects was held immediately in the public sphere, as far as such a sphere existed, and was poorly structured, open, and reflected the manifold interests concerned. This constellation resulted in a certain immediacy of the discourse with all its strengths and weaknesses. Authors involved were prominent intellectuals in the history of German thought (such as W. von Humboldt, F. D. Schleiermacher, J. F. Herbart, J. G. Fichte, F. W. J. Schelling, G. W. F. Hegel) as well as persons who are now forgotten. There were contributions of varying quality. Insights still valid stand beside utterances of trivial emptiness.

The impact of catchwords, however, should not be underestimated. In practical affairs, they are frequently more decisive than sophisticated conceptions. They serve as points of crystallization that unify the thinking of many people, although (and just because) these people may mean very different things by these words. The practical meaning of the catchwords can be read off from the actual activity. Their theoretical background can be reconstructed from the writings of certain authors. What we find there is valid, of course, only for the author concerned, but it is of general validity insofar as it reveals the possible meanings of these catchwords within their historical context and gives access to the intellectual universe of the time. The influence of a period's culture is thus quite essentially mediated by its language, and whoever wants to analyze the influence of this culture must analyze its language.

2. CULTURAL FOUNDATIONS OF NEOHUMANIST EDUCATIONAL PHILOSOPHY

The "Süvern Syllabus" of 1812, referred to as the *"constitutional document of the new Gymnasium"* (Paulsen, 1897, translated), dryly says that *"developing organic reasoning"* is the condition for penetrating science (Mushacke, 1858, p. 231, translated). What is meant by *organic reasoning* is not said, and in order to understand this, one must recur to the general philosophy of science and cultural context of the period. First, it will have to be remembered that there was a general mental revolution at the turn from the 18th to the 19th century, which may be referred to as *overcoming the mechanistic worldview*. The mechanistic theories of the rationalist era were supplemented by concepts reflecting the historical character of nature and humanity, and they tried to grasp the specificity of organic beings. The idea of organism, which originated in biology, was extended metaphorically to other realms and manifestations of life. This expressed the profound conviction that all spheres of life are *holistic*. Just as an organism is not composed additively of its elements, because the elements cannot exist alone and separately, science is no sum of isolated insights, but rather a holistic theoretical entity. Organic reasoning is thus characterized by the attempt to grasp the holistic character of the objects and by the fact that it is holistic as reasoning itself, that is, develops from its own conditions and tries to understand a thing from itself. In an analogous way, ethics and art can also be understood holistically.

The ideas about education and instruction belonged to this field of thought. Persons are themselves holistic, they cannot be educated by adding a certain knowledge to them from without, but they must develop themselves from within. This is why *Selbsttätigkeit* (self-activity) was the guiding concept of the neohumanist-idealist pedagogy of the period. In a narrower sense, this pedagogy was based on a certain *"transfer hypothesis."* This hypothesis again refers to the holistic character of education, saying that to become educated human beings, persons must, in their own development, have had at least once the experience of getting totally involved with a problem and coping with it productively. Only persons who have seen at least in one particular field that there are things that are holistic and have their own laws will be in a position to assess what it means not just to adhere to a number of rules in their own life, but to have the inner freedom to act.

It is clear that such a conception has nothing to do with transfer hypotheses according to which mathematics trains logical reasoning. Rather, logical reasoning and the ability to classify things according to external characteristics, the so-called *intellectual-mechanical abilities,* were considered to be a subordinate prestage to *"organic thinking."* Only after the holistic and organicist ideas of the Humboldtian era had been dismissed under the supremacy of a scienticist school of thought in the second half of the 19th century, did the equation "formal education = training of logical reasoning" emerge.

3. PURE MATHEMATICS AND THE NEOHUMANIST VIEW OF THE RELATION OF THEORY AND PRACTICE

In the context of this educational philosophy, the *Zeitgeist* placed particular emphasis on the classical languages of Greek and Latin. Mathematics, by no means as a matter of course, also played an important role within the neohumanist reform of the *Gymnasium.* The Süvern syllabus prescribed six mathematics lessons per week for each grade. This was certainly not only a consequence of the neohumanist ideal of education, but also due to the model of France and the École Polytechnique. A fact specific for Germany, however, was that educational value was accorded only to *pure mathematics,* the syllabus intentionally neglecting everyday practical applications. In the neohumanist program, mathematics was highly esteemed as a *theoretical,* (pure) and a *systematic* science (in the sense of the idea of organism). Mathematics was deemed to be of educational value because it was understood to be a discipline of theoretical reasoning that unfolds from its own conditions.

From the very outset, the emphasis on pure mathematics and the negative attitude toward everyday practical applications played an important role. In this section, I shall look for contemporary justifications for this esteem for pure mathematics, which is one-sided in our eyes today, analyzing in the next section how this orientation was actually made to prevail in school, and what was its role in further developments.

Already in Wilhelm von Humboldt's writings on the organization of education, there is an emphasis on pure mathematics in the few quotes in which he speaks of mathematics at all. Education was to be developed so as to ensure:

> . . . that understanding, knowledge, and intellectual creativity become fascinating not by external circumstances, but rather by its internal precision, harmony, and beauty. It is primarily mathematics that must be used for this purpose, starting with the very first exercises of the faculty of thinking. (Humboldt, 1810/1964b, p. 261, translated)

In the *Latvian Syllabus*, he expressed himself against the tendency

> . . . of distancing oneself from the possibility of future scientific activity and considering only mondane life . . . . Why, for example, should mathematics be taught according to Wirth, and not according to Euclid, Lorenz, or another rigorous mathematician? Any suitable mind, and most are suitable, is able to exercise mathematical rigor, even without extensive education; and if, because of the lack of specialized schools, it is considered necessary to integrate more applications into general education, this can be done particularly toward the end of schooling. However, the pure should be left pure. Even in the field of numbers, I do not favor too many applications to carolins, ducats, and the like. (Humboldt, 1809/1964a, p. 194, translated)

This is a definite position taken against everyday practical applications. It can be also seen, however, that Humboldt shows a willingness to compromise. Statements of quite similar kind can also be found among mathematics teachers of the period.

Propositions of this kind seem to express an idealistic and romanticist worldview in which there is no place for problems of practical and, in particular, technical applications of science. This may well be true for some authors of the time (although not for Humboldt). Nevertheless, it can be shown that such views express a reasonable and *realistic* view of the relationship between theory and practice, which may also claim to be relevant for educational reflections in the computer age. To make this evident, I have a document that is indeed historically unique. It is from the mathematician August Leopold Crelle (1780-1855). Crelle is well-known as the founder of the *Journal für die reine und angewandte Mathematik*, the first mathematical periodical in Germany. He worked for 20 years as a technician in the Prussian administration of public construction and also participated in constructing the

Prussian railways. Besides, he was very interested in mathematics and in mathematics instruction. In 1828, he was nominated advisor for mathematics within the Prussian ministry of education (cf. Eccarius, 1974).

In this quality, he gave several statements about the educational value of mathematics, the most pronounced in the preface to his *Enzyklopädische Darstellung der Theorie der Zahlen* (Crelle, 1845), saying that the use of mathematics is twofold, first for immediate application, second for training the ability to think. Upon weighing these two purposes against one another, it would show that in most practical problems it is difficult to apply mathematics because they are too complicated. Frequently, the application of mathematics would even lead to serious mistakes and errors, because people rely on its rigor and certainty without considering the complexity of the conditions. Quite indubitable, however, was the use of mathematics to exercise the ability to think, and this was also the indispensable condition for the direct application of mathematics.

> Only after a mathematical *spirit* has been awakened by assiduously exercising *judgment by means of mathematics* (without regard for applications), and only then, one may quite boldly count on the uses of mathematics in applications. Mere *knowledge* in mathematics, intended for *applications* . . . is not sufficient for appropriate applications, but the guiding principle must be the mathematical *spirit,* the mathematical *way of thinking*. Only he who tackles applications *on this basis* will err less easily, for he will first of all examine *what* mathematics can properly achieve, and where and how the tool can be usefully applied . . . . Hence it is *quite right* that mathematics be exercised as much as possible in schools . . . at first *without* any consideration of applications in common life. (Crelle, 1845, pp. IX-X, translated).

With the notion of the faculty of judgment, Crelle referred to Kant. In the *Critique of Pure Reason*, Kant had defined this by stating that a rule does not say by itself to which cases it can be applied, and that, therefore, the application of a rule requires a faculty of judgment.

In general, Crelle's position can be described as follows. On the one hand, there is the *direct application* of science, in which scientific laws are applied to practical problems by specialization. According to Crelle, however, this is only rarely possible. In most cases, science does not provide rules that can be applied directly. Nevertheless, also in these situations, science may play a part by enhancing the general ability of appliers to orient themselves. This may help to find new solutions to complicated applied problems. This kind of application can be termed *indirect application*. Hence, education should not aim primarily at a certain applicable knowledge, but rather at developing a more general ability to orient oneself (faculty of judgment).

That Crelle's decision to found his educational conception on *indirect application* was quite realistic for his time is true not only for industrially underdeveloped Prussia but also for the then well-developed France. C. Gillispie has studied the question of how far an application of science can be spoken of in France at the turn of the 19th century, during the early stage of industrialization (Gillispie, 1977). He found that, while there was no direct application of science, the sciences played an important part in industrialization by (a) scientists giving expertise in many areas of industry; (b) taxonomy and classification of industrial methods (the so-called "natural history of industry"); (c) scientific explanation of production processes; and (d) science as an educational instance for industry, overcoming ignorance and lack of communication. These are the very functions of what I have designated as indirect application above.

4. THE CONCEPTION OF MATHEMATICS INSTRUCTION FOR THE *GYMNASIUM*

In the following, I shall examine how mathematics instruction was conceived of for the Prussian/German *Gymnasium,* and how it developed, concentrating on the role of applications. I have already stated that the general character of this school mathematics can be described by the terms *pure* and *systematic.* The concept of system designates one dimension of the didactic conception, if not the essential one. For Crelle, for instance, mathematics was *"excellently suited"* to form the faculty of judgment, because of the *"consistency of its internal connection."* The concept of system was one of the topoi most used by textbook authors and syllabus organizers. I will not discuss this in more detail. However, the concept of "pure mathematics" and the (negative) attitude toward applications are worthy of closer consideration.

4.1 The Mathematical Conception

The mathematics of the Süvern syllabus was not the mathematics of Felix Klein. Rather, the scientific background of school mathematics at the *Gymnasium* was a theory that one can refer to as "algebraic analysis" and that is mainly due to the first volume of Leonhard Euler's *Introductio in analysis infinitorum* (Euler, 1748/1922). Essential for Euler's conception was the algebraic view of the concept of function, and, in a natural way, this also led to an algebraic view of infinitesimal calculus. Objects and most important tools of Euler's *Introductio* were finite and infinite algebraic expressions, that is, polynomials and power series, finite and infinite products, and continued fractions as well as their transformations. The intuitive basis of this approach was the more or less formal *analogy between finite and infinite expressions.* J. L. Lagrange (1736-1813) had completed this algebraic conception by defining the derivation of a function as the coefficient of i in the power series expansion

$$f(x + i) = f(x) + pi + qi^2 + ri^3 + \dots$$

which he supposed to exist (Lagrange, 1797/1881). Toward the end of the 18th century, the German so-called Combinatorial School attempted to systematize this theory by combinatorially expressing the analytically admissible transformations of power series and infinite products.

These ideas formed the scientific background of school mathematics as conceived in the Humboldtian reform and as it continued to develop over the course of the century. The entire field from arithmetic across algebra up to algebraic analysis was seen as a unified, homogeneous organism that was closed in itself and corresponded, in the eyes of contemporaries, to the ideal of a reasoning that unfolds conceptually from its own conditions. It became the systematic basis of 19th-century school mathematics in the area of arithmetic and algebra. Elsewhere, I have explained and demonstrated why a differential and integral calculus beyond the theory of series had no place in this conception (cf. Jahnke, 1990a).

As there were no compulsory syllabi in Prussia until far into the second half of the 19th century, but only minimum catalogues contained in the regulations for the Abitur, I illustrate this school mathematics by citing the contents of a then widely used school textbook:

> *the 4 arithmetical operations with whole numbers and fractions*
> *the 4 arithmetical operations with letters*
> number systems (in particular the decimal system)
> negative numbers
> polynomials
> calculation with powers (in particular *square and cubic roots*)
> *proportions*
> *equations of first and second degree*
> diophantine equations, continuous fractions
> *logarithms*
> irrational and imaginary quantities
> cubic and higher equations
> progressions (sequences)
> combinatorics
> infinite series and analytic operations with infinite series
> binomial theorem. (Tellkampf, 1829, Table of contents, translated)

This list contains the subject matter of the so-called *"scientific course"* from *Quarta* (third year) to *Prima* (ninth and last year). Typical for the combinatorial analysis of the time are two elements of this catalogue: (a) the appearance of combinatorics before the theory of series, because the latter was considered to be an application of combinatorics, and (b) the binomial theorem, which was seen as the culminating point of school mathematics, because this formula rules the basic arithmetical operations with power series. In the Prussian syllabus of 1901, the binomial theorem still had this role. The contents prescribed as compulsory in the 1812 Abitur edict are printed in *italics*. This compulsory canon was supplemented by combinatorics and the binomial theorem in the 1834 revision of the Abitur edict, which means that the core elements of the combinatorial view were made compulsory only then.

The intuitive analogy between finite and infinite series and hence the universality and simplicity of algebraic analysis are dependent on the unrestricted operations with infinite series. The restriction to convergent series thus contradicts the spirit of the theory. After publication of A. L. Cauchy's famous textbooks (1821/1897, 1823/1899), the view that only convergent series were permitted came to prevail in Germany. Textbook authors, while adapting to this trend, maintained the concept of algebraic analysis, adhering to Cauchy's standards only superficially by supplementing some convergence proofs. However, most of them did not follow Cauchy's conceptually revolutionary step in treating algebraic analysis as a theory of continuous functions.

4.2 The Role of Applications

After stating these facts concerning the mathematical structure of the syllabus, I shall now take a closer look at the role of applications in school mathematics. I shall go back to Süvern's syllabus of 1812. While the latter was never made compulsory, only being made known to the *Provinzialschulkollegien* as a guideline in 1816, it strongly influenced the activity of the Prussian administration of education until the 1820s (cf. Jahnke, 1990b, chap. V). In this syllabus, we find essentially the same contents as in the list above, but with one remarkable difference. For Prima (9th and last grade), it additionally prescribed probability theory, and *". . . the disciplines of applied mathematics, in particular of the mechanical sciences instead of geometry"* (Mushacke, 1858, p. 243, translated). Conversely, it can be noted that the everyday practical applications in the sense of ordinary arithmetic occur only in a marginal observation, which states laconically: *"In Quinta, continuation of numerical instruction in irregular systems, which involve the exercise of applied calculations in all regards"* (p. 242). That this merely served as a formal acknowledgement to external demand had been stated by the full professor of the Berlin university responsible for mathematics, Johann Georg Tralles (1763-1821):

> However, I believe that the school should not accept the role of teaching that which is useful to each social class; this would lead to a useless diversification in subject matter at the expense of thoroughness . . . . Given this, I deem it most useful to set aside all the so-called practical subject matters and to engage in only pure mathematical sciences or that which can be considered as such. (cited in Jahnke, 1990b, p. 244, translated)

The contemporary debate thus distinguished sharply between two types of application: On the one side, were the everyday practical demands in the sense of ordinary arithmetic; on the other side, applications like mechanics, which could be conceived, as Tralle said in his above remarks, "as of pure mathematics." I should like to designate these applications as *theoretical applications*. While the reformers were set on pushing the everyday practical applications as far as possible from instruction, they considered the theoretical applications to be a necessary component of instruction, which was not at odds with their own emphasis on pure mathematics. A closer look at the real development of mathematics instruction shows that the opposite occurred: Pressure by parents and students ensured that the everyday practical applications attained a well-established position in instruction, while the theoretical applications hardly mattered at all.

That ordinary arithmetic and everyday practical applications could prevail against the intentions of the reformers within the mathematical syllabus was a consequence of the social stratification of the student body in the Prussian *Gymnasium* of the time. As D. K. Müller has shown, the *Gymnasium* was attended not only by those seeking the Abitur and university studies but also by young people preparing for a job in commerce, industry, or trade. The latter left after the lower or upper Tertia (4th or 5th grade) of the *Gymnasium* as so-called "early graduates," a fact that permits us to consider the *Gymnasium* until the 1830s as some kind of "comprehensive school" for the town and city population subject to compulsory instruction (Müller, 1977). For these early graduates, however, an introduction to ordinary arithmetic and everyday practical applications was indispensable; all the more so as the complicated and variable relations in contemporary weights, money, and measures actually required such special training.

How strong the contrasts between the reformers' intentions and the social facts of the Prussian educational systems actually were becomes visible from an episode found in the records of the Prussian ministry of the interior (cf. Jahnke, 1990b, pp. 364-368). In 1815, the ministry's department of education tried to implement the mathematical part of the Süvern Syllabus at one of the leading Prussian schools, the *Joachimsthalsche Gymnasium* in Berlin. The number of mathematics lessons per week was increased in accordance with Süvern's recommendations, partly even at the cost of the classical languages. To make the reform work, additional mathematics teachers were sent to this school. Soon after, however, strong opposition arose. Eventually, one of the mathematics teachers proposed the compromise of dividing the mathematics course into two parallel ones of different character, which were to be optional for the students. One of these was to be *scientific,* the other oriented toward *practical* requirements. This suggestion was strongly rejected by the ministry who insisted on maintaining the unity of mathematics instruction, stressing that the spirit and essence of the *Gymnasium* were determined by its scientific character. Most notable is the ministry's assessment stated at this occasion that presumably all of the students would vote for the practical course if given the option, thereby driving the reform on the rocks. A similar pessimistic evaluation was given by F. Kries, author of a then widely used textbook. Mathematics was regarded as a study yielding no bread, not rewarding the diligence shown by the students in their own eyes. It had to be admitted, Kries said, that most of the positions in the state to which the young people aspired did not permit an immediate application of mathematics. Thus, the entire episode shows that the insertion of pure mathematics into the syllabus of the Prussian *Gymnasium* did not correspond to the demands of parents and students in their own views, but rather was a consequence of a pedagogical program that stood in contrast to these.

The entire conflict between a scientific and practical orientation was settled by a decree issued by the Prussian administration of education in 1826. The ministry prescribed that *"the real teaching of mathematics will begin in Quarta in all Gymnasiums; but in Quinta and Sexta* (the two lower grades), *arithmetical skills will be exercised practically without any intrusion of mathematics"* (Neigebauer, 1835, p. 173, translated). With this, training in arithmetic skills and everyday applications had acquired a definite position in *Gymnasium* instruction. The significance of this decree is notable by the fact that the ministry intervened in substantial questions of mathematics instruction on only one additional occasion until 1850.

How strongly some mathematicians and educators nevertheless felt the everyday practical applications to be problematic became apparent once more in 1826 when A. L. Crelle proposed, in a comprehensive expertise on mathematics instruction, to abolish arithmetical instruction in the lower two classes, and to use the time saved for nonmathematical subjects. He justified his proposal by arguing that "everyday applications" were in contradiction to the scientific character of *Gymnasium* instruction, while scientific mathematics was too difficult for that age group. Mental arithmetic, in contrast, was acquired autonomously and did not need to be taught. This suggestion, however, did nothing to modify the compromise already attained on this issue.

The destiny of the theoretical applications was quite different. On the whole, they fell victim to an inner dynamics in the development of mathematics instruction, which is rather typical. To understand this, let us imagine the entire school mathematics according to the following schema:

Euclidean geometry	
Structural core	Applications
Algebraic analysis: from elementary arithmetic across algebra to series expansions and the binomial theorem with "general" exponents	Calculatory geometry: stereometry, plane and spherical trigonometry, analytical geometry (conic sections), geodesy, mathematical geography, mechanics

Originally, school mathematics consisted essentially of (constructive) Euclidean geometry, algebraic analysis, and the applications of algebra to the calculatory parts of geometry and mechanics. The schema makes immediately visible where cuts in the mathematical subject matter were possible. They could be effected in all applied parts without endangering the systematics of the subject. Due to the imprecise boundary to physics instruction, the mechanical applications had hardly a chance of being realized from the outset and hence occur but sporadically in a small number of textbooks. When there was the general complaint in the 1830s that the demand on the *Gymnasium* students was too great, cuts had to be made in mathematics instruction as well. It was almost self-evident that these cuts left the systematical part unscathed, hitting the applications. In 1834, the ministry forbade the *Joachimthalsche Gymnasium* to treat the conic sections analytically and consequently analytical geometry as such. This decree was circulated to every *Gymnasium* and thus became compulsory. This was the ministry's second intervention into the subject matter of mathematics instruction. The applications of algebraic analysis were thus reduced to calculating individual geometric quantities in stereometry, and plane and spherical trigonometry.

To obtain an overall picture of school mathematics, I shall look at a typical school textbook of the time, the *Lehrbuch der allgemeinen Arithmetik* (Müller, 1838) by Johann Traugott Müller (1797-1862). It has an eminent position in the textbook literature of the 19th century. Müller was the first director of the school designated as *Realgymnasium* in Germany (that in Gotha) and has been termed one of the forerunners of Klein's reforms by F. Pahl (1913, p. 275). In contrast to this assessment, Müller's textbook adhered fully to the framework sketched above. His subject matter comprised exactly the contents of algebraic analysis, and the emphasis with regard to the didactic conception is on the systematical development of theory. Müller saw its intrinsic connection realized by the principle of number domain extension, and consequently it was *"the definite emphasis on and repetition of the same laws on the three different levels of operations that reveals arithmetic as an organic whole to the student . . . ."* (Müller, 1838, p. v, translated). The didactic centerpiece of the book is the close connection of algebraic calculus and implicit combinatorics. Combinatorial exercises are carried out already at the beginning of algebraic calculus; an explicit combinatorics appears later as an introduction to the theory of series. Thus, the entire volume is devoted to the combinatorial program from the initial stage of the reform. The book was eminent among the mass of textbooks in making the syntactic structure of formulae transparent. In view of this and of the fact that Müller was an excellent representative of the movement in favor of the *Realschule,* it is all the more serious that, although the book contained some applications from physics, stoichiometry, and interest calculation, these amounted to 10% of the exercises at the utmost (cf. Jahnke, 1990b, p. 421). Even if one takes into consideration that teaching geometry besides arithmetic required about the same amount of time, this does not alleviate the purism of the formal-syntactical conception of arithmetic and algebra.

The second edition of Müller's work in 1855, while limited to one small volume, even increased this formal tendency by making combinatorics even an explicit object of teaching at the lower level. The efforts at reforming mathematics instruction initiated by the new *Realgymnasien* and *Oberrealschulen* never put into question the systematic character of the subject matter, they seldom led to an extension of applications, and, for present-day readers, it is almost upsetting to see that an extension of the mathematical subject matter in 1892 went toward "teaching and exercising the solving of fourth-degree equations and the method of approximate solution of numerical, algebraic, and transcendent equations" (Jahnke, 1990b, p. 464, translated).

5. LOSS OF MEANING

During the second half of the 19th century, algebraic analysis increasingly lost its mathematical and cultural significance, nevertheless remaining the leading concept of school mathematics until the beginning of our century. The history of this theory can thus be subsumed for this period under the heading of *loss of meaning and inertia*. Its mathematical loss of meaning is evident. In succession to Cauchy, a quite novel view of real and complex analysis and of its conceptual foundations had emerged. Techniques and concepts of the older algebraic analysis were partly scientifically discredited (as the use of divergent series) and partly occurred only in subordinate places in special chapters. The view could no longer claim a unifying role. This loss of significance is illustrated dramatically by a remark made by the then well-known textbook author Ernst Koppe, who in 1866 said that treatment of combinatorics in school was *"a pastime alien to the scientific seriousness of the mathematical method"* (pp. 10-11, translated).

Second, the original conception of school mathematics also suffered a *cultural loss of significance*. In the middle of the 19th century, idealist and romanticist philosophy had been thoroughly discredited. The general beliefs in theory of science, which had established an intrinsic connection between education and theoretical science, had dissolved. The stress on the holistic character of the world, on humanity and on knowledge, had ceded to a more pragmatic attitude. Against the idealistic construction of comprehensive systems, greater weight was now placed on experience and on the particular. This necessarily also changed the attitudes toward education. The *"unity of knowledge"* appeared unattainable and was no longer a value that could serve to comprehensively justify education. The idea that the educational effect of mathematics lay mainly in its systematicity lost more and more of its attraction.

Third, the *social context* was changing. The increasing importance of technology and of technological applications undermined the exclusive position of pure mathematics during the first half of the 19th century. It could no longer claim to be the only authentic mathematics serving to convey mathematical education.

Nevertheless, algebraic analysis held its position as a leading concept of school mathematics until the beginning of our century, as has been hinted at above (Jahnke, 1990b, pp. 464-472). Consequently, the first efforts at reforming mathematics instruction were also directed at reintegrating analytical geometry into the syllabus of the *Gymnasium* (see the famous speech given by DuBois-Reymond in 1877), thus making the original conception of algebraic analysis complete again. Only after F. Klein, who occasionally spoke of the *"misery of algebraic analysis"* (Klein, 1907, p. 105, translated), demanded that infinitesimal calculus be introduced into instruction toward the end of the 19th century, was this leading concept effectively questioned. It is a historical problem yet to be studied to reconstruct the history of *Klein's Reforms* as a struggle between two different mathematical paradigms, that is, between the conception of algebraic analysis and Klein's idea of functional reasoning.

6. THE ROLE OF CULTURE

To close, I shall inquire into the role of culture in this development. General cultural evaluations led to the view that the purpose of education was not so much to convey an established knowledge aligned to applications, but rather orientations to enable humanity to behave intelligently and communicatively if faced with unforeseeable demands (the concept of "indirect application"). Pure mathematics had the opportunity to determine the mathematical syllabus, because theoretical and speculative reasoning was highly esteemed within the cultural sphere. The same kinds of cultural process outside of school were to rob the original conception of education of its meaning and to undermine the didactic conception of mathematics instruction founded on this idea by a decreasing esteem for the idea of system as well. This was a slow and nonhomogeneous process that changed the mathematics teachers' pedagogical and mathematical self-understanding and made them seek for new ideas. An interesting question for which no historical evidence is available to me is how far such processes of cultural change outside of school that were divesting subject matter of its meaning have had a negative influence on processes within school.

For the present-day observer, certainly most strange is the extremely negative view of everyday applications taken by the reformers of the Humboldtian era. Within the historical context, however, some reasons can be advanced that make this comprehensible. First, it should be noted that similar efforts were made at reforming mathematics instruction in the 1960s and 1970s while subsuming various applied methods under the concept of proportional function. An additional feature in the early 19th century was that reformers were faced with the situation that only everyday applications were taught in mathematics at the *Gymnasium*, and this mainly by auxiliary teachers who came from outside and had to be paid for this practical drill by the students or by their parents. This practical arithmetic consisted in the exercise of mere cookbook recipes, a situation in which mathematical subject matter concerned with arguing and proof had to fight for a first foothold. The insight that common arithmetic can also be executed in an argumentative and reasoning way only grew step by step over the course of the 19th century as a consequence of didactic efforts.

We have seen how difficult it was to make an arguing and proving mathematics prevail at the *Gymnasium*. The reformers' efforts in favor of such a mathematics must thus be considered to be very important historically. This shows that it is historically inadequate to consider the movement to establish the *Realschule* during the second half of the 19th century as a continuation of the realist trends during the first half. It is necessary here to distinguish carefully between different currents, it being obviously not correct to assume that a realistic orientation of education automatically implies a strong position of mathematics within the syllabus. Besides, it is not self-evident under realistic auspices that the mathematics taught is aligned to proof and theoretical reasoning.

The conflict about the everyday practical applications reveals, on a general level, one of the most essential functions fulfilled by culture with regard to education. Just as the everyday practical applications represented the *contemporary* interests of parents and students, the demand to teach theoretical mathematics *anticipated* the future. The struggle between the two positions was multidimensional and fed by cultural values. It is culture that makes possible the *dialogue with the future*, and this is its decisive contribution to education.

REFERENCES

Bishop, A. J. (1988). *Mathematical enculturation. A cultural perspective on mathematics education.* Dordrecht, Netherlands: Kluwer.

Cauchy, A. L. (1897). Cour d'analyse de l'école royale polytechnique. Ire Partie. Analyse algébrique. *Oeuvres complètes d'Augustin Cauchy, publiés sous la direction scientifique de l'Académie des Sciences* (IIe série, tome III). Paris: Gauthier-Villars. [Original work published in Paris: Debure frères, 1821]

Cauchy, A. L. (1899). Résumé des leçons données à l'École Royale polytechnique sur le calcul infinitésimal. *Oeuvres complètes d'Augustin Cauchy, publiés sous la direction scientifique de l'Académie des Sciences* (IIe série, tome IV, pp. 5-261). Paris: Gauthier-Villars. [Original work published in Paris: L'Imprimerie Royale, 1823]

Crelle, A. L. (1845). *Encyklopädische Darstellung der Theorie der Zahlen und einiger anderer damit in Verbindung stehender analytischer Gegenstände; zur Beförderung und allgemeineren Verbreitung des Studiums der Zahlenlehre durch den öffentlichen und Selbst-Unterricht* (Vol. 1). Berlin: Reimer.

DuBois-Reymond, E. (1974). Kulturgeschichte und Naturwissenschaft. In S. Wollgast (Ed.), *E. Du Bois-Reymond, Vorträge über Philosophie und Gesellschaft* (pp. 105-158). Hamburg: Meiner. [Original work published 1877]

Eccarius, W. (1974). *Der Techniker und Mathematiker August Leopold Crelle (1780-1855) und sein Beitrag zur Förderung und Entwicklung der Mathematik im Deutschland des 19. Jahrhunderts.* Unpublished doctoral dissertation, Eisenach.

Euler, L. (1922). Introductio in analysin infinitorum. Tomus primus. In F. Rudio, A. Krazer, A. Speiser, & L. G. du Pasquier (Eds.), *Opera Omnia* (Ser. I, Vol. 8). Leipzig/Berlin: Teubner. [Original work published in Lausanne: Bousquet, 1748]

Gillispie, C. (1977). Die Naturwissenschaft der Industrie. In E. A. Musson (Ed.), *Wissenschaft, Technik und Wirtschaftswachstum* (pp. 137-152). Frankfurt/Main: Suhrkamp.

Humboldt, W. von (1964a). Der Königsberger und der Litauische Schulplan. In A. Flitner & K. Giel (Eds.), *W.von Humboldt: Werke IV* (2nd ed., pp. 168-195). Darmstadt: Wissenschaftliche Buchgesellschaft. [Original work published 1809]

Humboldt, W. von (1964b). Über die innere und äußere Organisation der höheren wissenschaftlichen Anstalten in Berlin. In A. Flitner & K. Giel (Eds.), *W. von Humboldt, Werke IV* (2nd ed., pp. 255-266). Darmstadt: Wissenschaftliche Buchgesellschaft [Original work published 1810]

Jahnke, H. N. (1990a). Die algebraische Analysis im Mathematikunterricht des 19. Jahrhunderts. *Der Mathematikunterricht, 36*(3), 61-74.

Jahnke, H. N. (1990b). *Mathematik und Bildung in der Humboldtschen Reform.* Göttingen: Vandenhoeck & Ruprecht.

Klein, F. (1907). *Vorträge über den mathematischen Unterricht an den höheren Schulen. Bearbeitet von R. Schimmack. Theil 1: Von der Organisation des mathematischen Unterrichts.* Leipzig: Teubner.

Koppe, C. (1866). *Der mathematische Lehrplan für das Gymnasium.* Schulprogramm. Soest.

Lagrange, J. L. (1797/1881). Théorie des fonctions analytiques. In M. J.-A. Serret (Ed.), *Oeuvres* (Vol. IX). Paris: Gauthier-Villars. [Original work published 1797]

Müller, D. K. (1977). *Sozialstruktur und Schulsystem. Aspekte zum Strukturwandel des Schulwesens im 19. Jahrhundert.* Göttingen: Vandenhoeck & Ruprecht.

Müller, J. H. T. (1838). *Lehrbuch der Mathematik, Vol. 1: Lehrbuch der allgemeinen Arithmetik für Gymnasien und Realschulen, nebst vielen Uebungsaufgaben und Excursen.* Halle: Buchhandlung des Waisenhauses.

Mushacke, E. (1858). Anweisung ueber die Einrichtung der oeffentlichen allgemeinen Schulen im preussischen Staate. *Preussischer Schulkalender, 7,* 231-259.

Neigebauer, J. F. (1835). *Die Preußischen Gymnasien und höheren Bürgerschulen. Eine Zusammenstellung der Verordnungen, welche den höheren Unterricht in diesen Anstalten umfasssen.* Berlin/Posen/Bromberg: Mittler.

Nizze, E. (1822). *Zweck und Umfang des mathematischen Unterrichts auf Gymnasien.* Schulprogramm. Gymnasium Stralsund.

Pahl, F. (1913). *Geschichte des naturwissenschaftlichen und mathematischen Unterrichts.* Leipzig: Quelle & Meyer.

Paulsen, F. (1897). *Geschichte des gelehrten Unterrichts auf den deutschen Schulen und Universitäten vom Ausgang des Mittelalters bis zur Gegenwart* (Vol. 2, 2nd ed.). Berlin: Veit

Tellkampf, A. (1829). *Vorschule der Mathematik.* Berlin: A. Rücker.

White, L. A. (1959). *The evolution of culture.* New York: McGraw-Hill.

MATHEMATICS AND IDEOLOGY

Richard Noss
London

1. IDEOLOGY AND THE CURRICULUM

It is far from clear why mathematics and ideology are related in any way. Mathematics deals with the aesthetic and the theoretical. Ideology deals with the political and the pragmatic. So let me begin by stating that I view ideology as the body of ideas through which we see and with which we construct our reality. It makes the world intelligible, and people's behaviours predictable, understandable.

To claim that mathematics is a social construction, rather than, say, a representation of reality, is no longer as contentious as it once was. But the (school) mathematics curriculum too is socially constructed, and has, for example, changed its character many times in the last century or so since compulsory education was introduced into the "developed" world. In both cases – mathematics and maths (I use the latter term to save distinguishing continually between mathematics and school mathematics) – there is an overwhelming temptation to view the subject matter as given, inevitable, *natural.*

From where does this apparent "naturalness" arise? There are many possible answers to such a question, but two polarized extremes are evident. One would argue that the curriculum (and the discipline) is as it is because of the inherent structures of the subject; that mathematics is, in a sense, preordained to be as it is, given that it provides an idealized means of describing and predicting the material world. While such a view has some credence vis-à-vis mathematics, it is more difficult to sustain with regard to maths. The other extreme would view the discipline and the curriculum as essentially arbitrary, creations of people acting independently and autonomously, without any constraints imposed from the material reality of which they form a part.

It is clear that there is scope for a range of intermediate positions, and a full examination would take us too far into philosophical domains. As far as maths is concerned, I begin from the premise that the curriculum, if not arbitrary, is socially constructed, and that a valid task for researchers in the field is to *de*naturalize the form and content of what is taught.

And so to ideology. Ideology flows from the social relations that exist

R. Biehler, R. W. Scholz, R. Sträßer, B. Winkelmann (Eds.),
Didactics of Mathematics as a Scientific Discipline, 431-441.

within a society (which is not to say it is *determined* by them), and correspondingly functions to help maintain (or destroy) those relations. At any particular time, it may be that particular ideologies do or do not contribute to the dominance of a particular class or group. Nevertheless, it is not necessary that such social groups fashion ideologies explicitly (although, of course, in times of wars and other crises, this may indeed be the case). On the contrary, there is a tendency for ideologies to become "common sense," applied *without* explicit intention and, most importantly, an accompanying tendency to see the surface reality of things as their unalterable bases and causes.

It should already be clear that I am using the word "ideology" in a sense dissociated from its strictly pejorative meaning, which, as Barrett (1991) points out, is a common everyday use of the word. It is not difficult to see how the word *ideology* acquired its pejorative connotations; indeed, if it is true that an essential element of the operation of ideology is that it merges into the background – tending to make reality seem unmediated and natural – then it follows that only the most obvious (or repressive) attempts to influence ways of seeing and thinking will be evident. When we say that a political despot uses ideology as a weapon to mould people's thinking, we are saying *both* that peoples' ideas are influenced (deliberately) *and* that this process is evident to us as observers because we can stand outside it. We may notice the ways that the media attempts to influence our opinions, but only if we disagree with the views being proposed. And we have much less difficulty branding as "ideological" (in the pejorative sense) ideas and beliefs that belong to history, at times and in places from which we are removed.

Of course, there are institutions that play a more or less explicit role in the fashioning of ideologies: An obvious example might be organized religion. Schools, too, may be thought of as contributing to ideologies: It would indeed be surprising if institutions explicitly concerned with influencing children's thinking did not play a role in fashioning the belief systems within which people make sense of their social and physical world. But institutions do not necessarily need to have a physical embodiment before they have some role in generating ideologies: Established ways of seeing and thinking occur everywhere and mediate how we "read" them – in our appreciation of music (a subject to which I will return) as much as our understanding of, say, infinity. With a notion of ideology stripped of its pejorative connotations of covert manipulation, mystification and obfuscation, it becomes a little more plausible that the way we conceptualize the mathematics curriculum – no less than the way we think about art or literature – is itself ideological.

There is a considerable literature on the ways in which schools in general contribute to ideological production (a useful starting point is Giroux, 1983). Common to almost all approaches is the view that schools are sites of social reproduction; that it is at school (but not *only* at school) that children learn how to function in the social niche they are likely to occupy in adult life. The implications of this view for mathematical learning have been explored elsewhere (Mellin-Olsen, 1987; Noss, 1989, 1990): My purpose here is to be a little more specific about the relationship between what is learned and how individuals make sense of their environment.

The problem is to try to tease out the elements of schooling that contribute to their socializing function. Is it the structures and forms of the school, by stressing forms of knowing and behaving that are alien to all but privileged children, that are responsible for the social reproductive role of organized education? Or is it school knowledge, curricular content, that is responsible for instilling the specific values required by the society of which the student will form a part? As Whitty (1985) points out:

> In one case the class structure was seen to be sustained because working class pupils *failed* to learn what the school defined as significant, while in the other case the process depended on what they *did* learn in school – that is to accept (and if possible respect) the status quo. (Whitty, 1985, p. 20)

2. WHAT DOES THE MATHEMATICS CURRICULUM MEAN?

As I stated at the outset, it is not easy to decide precisely how the mathematics curriculum functions ideologically. At one extreme, school knowledge in general might be seen as arbitrary, or at least contingent only on the whims and fancies of (say) politicians or educationalists. In this scenario, curricular content is essentially irrelevant and reduced to an empty form, a position argued most convincingly by Ivan Illich:

> It does not matter what the teacher teaches so long as the pupil has to attend hundreds of hours of age-specific assemblies to engage in a routine decreed by the curriculum and is graded according to his ability to submit to it. (Illich, 1973, pp. 61-62)

This view meshes with that of Harry Braverman, a political economist whose seminal book *Labour and Monopoly Capital* provides a captivating analysis of modern working practices. In it, Braverman paints a picture of the social functioning of schools, which, he argues, play a critical role in the organization of capitalist societies. For Bravermann, schools serve to fill a vacuum created by the demise of traditional socializing influences (the family, community, etc.). And in seeking to fill this vacuum, he argues that "schools have themselves become that vacuum, increasingly emptied of content and reduced to little more than their own form" (Braverman, 1974, p. 440).

Writing as a political economist rather than as an educationalist, Braverman shows convincingly how this process has followed the demographic and social changes in the nature of Western economies, driven by the thirst for more competitive and intensive production techniques and fuelled by technology. The essence of Braverman's argument is that 20th-century capitalist society has witnessed a gradual deskilling of the work process, a "deskilling" not just of factory production lines, but of office-workers, clerks and white-collar workers in general.

In the two decades since the publication of Braverman's book, the mushrooming of information technology into every area of social life has only exacerbated the process he outlined. A sizeable proportion of those in work in the "developed" world have been reduced to little more than human appendages to a computer system; shop assistants no longer need to calculate change, bank clerks need know nothing about banking, waiters and waitresses no longer work out bills, engineering is reduced to following blueprints; even computer programming, heralded only a short time ago as creating a need for a newly creative, mathematically-trained workforce, has become, in the hands of the large companies who employ programmers, largely a routinized and alienating activity. As technology invades all aspects of daily life, people actually need less – not more – mathematics (see Noss, 1991, for an elaboration of this argument).

Viewed from Braverman's perspective, the content of the curriculum is very much a secondary, increasingly unimportant, concern. This is indeed a position that has been adopted by those more particularly concerned with education, in particular, the celebrated analysis of Bowles and Gintis (1976), who argued that there was a "correspondence" between the needs of society's economic base and the practices of the educational superstructure. Following Braverman, they focused their attention on the ways in which the educational system corresponded with the economic, even borrowing Marx's metaphor in referring to the "social relations of education," and arguing that as far as the socialization of future generations to populate the production process was concerned, "The actual *content* of the curriculum has little role to play in this process" (Gintis & Bowles, 1988, p. 28).

Somewhat paradoxically, Gintis and Bowles have also argued (still from a strictly deterministic perspective) that the social relations of production *directly* affect the content of what is taught. So, in considering the rationale for the "back-to-basics movement" they identify in the 1980s, they argue that

> . . . so called "back to basics," while having little rationale in terms of either pedagogical or technological reason, may be understood in part as a response to the failure of correspondence between schools and capitalist production brought about by the dynamics of the accumulation process confronting the inertia of the educational structures. (Gintis & Bowles, 1988, p. 20)

Put bluntly, their case is that "back to basics" represents a more-or-less conscious attempt to pull the structure of the curriculum into line with the changed priorities of industry and commerce.

Bowles and Gintis have rightly been taken to task for viewing the curriculum as essentially irrelevant, and certainly for seeing it as driven by economic forces. I think – and so, latterly, do Bowles and Gintis (1988) – that it is most useful to conceive of the curriculum as a site of struggle in which students, teachers, parents as well as voices from industrial, commercial and other settings have at various times competed in various ways and with varying relative strengths to assert their priorities. What is important is to note that the structure and content of the mathematics curriculum is only partially determined by mathematicians or mathematics educators themselves; and that even they are not the free agents that they would like to believe. All those who compete within the educational sites are themselves immersed in social practices and imbued with assumptions that emanate from ideological as well as educational or mathematical settings.

From this perspective, the curriculum is *neither* free from *nor* determined by the economic and political space in which it operates: It makes more sense to ask how mathematical ideas fit with society, how they encourage particular ways of seeing, particular ideologies. So, for example, the "back-to-basics" tenor of the UK National Curricula (see Dowling & Noss, 1991, for a critique of the England and Wales National Curriculum from a mathematical perspective) was not *caused* by the economic recession, but the recession played its role in silencing progressive voices in favour of those who believe that a more orderly, routinized and dull educational system would offer a more reasonable training for post-school life. *It changed the balance between competing ideologies.* As I pointed out in Noss (1991), the topic of long division has been enshrined by law in the curriculum of England and Wales at precisely that point in human development when the ubiquity of the calculator and the computer has made that skill completely redundant. Would not a generation schooled in the repetitive, routine and mathematically useless skills of long division have some qualities that the societies of the recession-laden 1990s would value? (There is no need for a conspiratorial view here: but see Bassey, 1992, for a tongue-in-cheek but chillingly viable conspiracy theory.)

The key point is that the specificities of curricular content are not driven from outside, but neither are they arbitrary. They are the result of ideological tensions, debates and (often implicit) beliefs about what mathematics education is for. The question still remains, however, as to the extent to which the social functions demanded of the mathematics curriculum arise from the structure of mathematics itself. Are mathematical meanings essentially neutral, but "corrupted" or "transposed" (Chevallard, 1985) for educational consumption? Or is there some element of mathematical knowledge that is particularly suited to the ideological role it is called upon to play? From where, in fact, does mathematical meaning derive?

3. A DIVERSION VIA MUSIC

In order to examine this a little more carefully, I turn briefly to another curriculum area – that of *music* – which has been the subject of more attention than that of mathematics. Music is a field that has received considerable and rather detailed attention, dating back to the work of Theodore Adorno (see, e.g., Adorno, 1978), a member of the "Frankfurt School" of neo-Marxists interested in probing questions of ideological reproduction and its relation to society. His work has generated a fascinating field of enquiry in relation to the teaching and learning of music in school (see, e.g., Vulliamy, 1976; more recent work has been undertaken by Lucy Green, 1988).

Green distinguishes between *inherent* and *delineated* musical meanings. She argues that "Individual temporal musical experience arises directly from musical materials that inhere in music and create meanings between themselves, for consciousness, through time" (p. 25). Thus, these meanings are inherent, intrinsic to musical material. They have both social and historical dimensions, but are nonetheless ultimately traceable to the structural facets of musical activities, and the ways they are experienced by people. It is these experiences of materials and their meanings that Green refers to as "inherent" musical meaning.

In contrast, those:

> Images, associations, memories, queries, problems and beliefs inspired in us by music are musical meanings that, rather than inhering in musical materials and pointing only to themselves, point outwards from music towards its role as a social product, thus giving it meaning as such for us. (Green, 1988, p. 28)

The associations of different kinds of music with different subcultures (and even social classes), recording charts, opera as high culture, film scores – these are all different ways in which musical associations and beliefs are developed and sustained. By occupying a niche within this panoply of social relations, music reciprocally delineates ideologies in relation to musical meaning and beyond. It is these meanings that Green refers to as *delineated musical meanings.*

Central to Green's argument is the description of the ways in which music's function *as a commodity* tends to override its structure. As such, the inherent meanings of "great" music are treated as if they were beyond the comprehension of all but the initiated ". . . despite the fact that we create, develop and realise them, that we produce them collectively and divisively through history" (p. 86). Thus this is a musical variant of commodity fetishism, the process that Marx (1967) describes as the usurpation of an object's *use value* (why we want and use it) by its *exchange value* (how we acquire and value it). In the process, delineated meanings come to supplant the inherent meanings and values of the commodity; and, in just this way, argues Green, "the delineations appear to be the only real and unalienated qualities of music, the only means by which we can grasp music at all" (p. 86).

This is a fascinating analysis, not just because it provides us with an example of a scholar of the artistic/aesthetic arguing for objectivity and inherence, when we may be more used to students of the mathematical/scientific arguing for cultural relativity and delineation. In order to study the extent of application of Green's argument in a mathematical context, I shall try to recast Green's argument in mathematical terms.

4. RETURN TO MATHEMATICS

Mathematics is based on raw material: shapes, number, mathematizable situations. Of course, the concept of shape (rectangle as opposed to door, 4 as opposed to 4 cups) is already a mathematization, a first link in the signifying chain that characterizes mathematical activity. When we create mathematics, we bring these objects into relationship with each other, and these relationships themselves become (eventually) the raw material of further mathematics. Each piece of living mathematics is based on the dead mathematics built into the objects upon which it is based. In this, the relationship between living and dead mathematics resembles that of living and dead labour: The latter is essential for the former, but plays no active part in the creation of new material. The latter – to borrow Marx's metaphor – *breathes life* into the former, but does not itself enter into the creation of new value (this argument is elaborated in Noss, 1991).

Thus mathematicians are not free agents. The rules of the game by which mathematical objects may be manipulated and brought into relation with each other are not arbitrary. The idea that mathematics is an arbitrary game is far from the truth: Mathematicians care very much about the meanings their games convey, even if they sometimes deny it. Of course it is true that new games are created, each with new variants of the rules. But the rules (and the rules about the creation of games, etc.) are built into the structure of what it means to do mathematics (as opposed to, say, literary criticism or history) – the meanings deriving from such relations are *inherent* or, perhaps it would be more helpful to say, *structural* meanings.

But here the situation looks, at least superficially, different from that described by Green in the case of music. For Green, the process of fetishism results in the appropriation of inherent meanings by delineated meanings. The inherent meanings are obscured by the replacement of use value by exchange value: by the reification of music to the status of a "thing" (by performances in grand halls, the advertising of recordings, etc.). But in mathematics, the reverse occurs. Mathematics does not play a mass role in our culture: quite the opposite. In mathematics, it is the structural (inherent) meanings that are reified, used to obscure the delineated meanings that form part of the mathematical enterprise (e.g., what makes an acceptable proof?), let alone the identification of the role of mathematics in "formatting" society (Skovsmose, 1992; see, also, Noss, 1988a).

The key distinction is thus in the extent to which non-trivial mathematics plays any role in popular culture, and to which *any* of its meanings are "consumed" by non-mathematicians. And it is here that the position becomes more complex. For while it is true that mathematics is not a commodity in the sense that music is (it is not advertised, it is not the subject of everyday conversation, mathematicians are not [in general] offered huge contracts and salaries), it *is* true that mathematical ideas enter our culture in unanticipated ways.

As an example, consider the surprising way in which the sudden explosion of interest in non-linear dynamical systems has spawned a bizarre industry of sociologists, political scientists and post-modernists of all kinds, all borrowing some version of what they perceive as "chaos" in the behaviour of human systems: A prevalent example is the belief that the "chaos" of, say, European politics is somehow connected with the "chaos" of non-linear systems. (It is interesting to speculate what would be the situation if "chaos theory" had entered the world with the – perhaps more accurate – name of "order theory.") Of course this is not a new phenomenon: Darwin's theory has long been characterized in the popular imagination as "survival of the fittest," just as much as Piaget's psychology has been re-interpreted as a theory of (un)readiness. In all such cases, the appearance of scientific theories as *only* consisting of inherent meanings is shown to be illusory: In a society in which almost anything can become a commodity, scientific ideas cannot remain immune.

In the case of music, Green argues that the fetishism of music is dialectically constructed from two sides. One is the surface reality that delineated musical meanings are the only aspects that are communicable, and that inherent meanings are therefore beyond reach, "untouchable essences." The other appears to be a converse: It is that delineated meanings are "distractions" from the pure and untouchable inherent meaning of music, and that attention should be focused instead on the inherent, pure, musical meaning. The situation is not dissimilar for mathematics. On the one hand, it is claimed that mathematical truth is an untouchable essence: The "language of mathematics" is, from this perspective, entirely inaccessible for all but the chosen few. On the other hand, there is a view – subscribed to by many mathematicians – that mathematics is incommunicable, and that, insofar as communication is synonymous with delineated, ideological meanings, "real" mathematics must focus *exclusively* on that which is inherent in the structure of mathematics itself.

As an example, it is instructive to turn to the provocative paper by John Guckenheimer (1978), which discusses differences in approach and style on the (then) new subject of catastrophe theory. Guckenheimer contrasts catastrophe theory as conceived by two mathematicians: René Thom and Chris Zeeman. He points out that "Catastrophe Theory *chez* Zeeman is much more concrete than it is *chez* Thom" (p. 16). Zeeman's classic example of an elementary catastrophe is his celebrated model of aggression in dogs, and he has applied the theory to situations as diverse as financial speculation, heart attacks and prison disturbances. It is precisely this focus on delineated meanings, on *other* than pure mathematical essence, that led to a radical critique from some mathematicians, who attacked Zeeman's models as "deceptive and wrong," and pointed out that "many things which they assert are wrong productions of the models, incorrect reasoning within the models, and *ploys to divert the attention of the reader*" (p. 17, emphasis added). Worse still, Zeeman often adopts a style of writing that is fluid and entertaining, in contrast to a more rigid style, which helps maintain the barrier between the mathematical community and a broader audience. Thus Zeeman stands accused of communicating mathematics through the introduction of delineated meanings at the expense of inherent meanings, a strategy that lays him open to the deepest criticisms possible: communication, application as well as an appeal to geometric (as opposed to algebraic) intuition – indeed, a failure to behave like a mathematician at all.

5. TOWARDS SOME CONCLUSIONS FOR THE CURRICULUM

The foregoing analysis is premised on a view of the mathematics curriculum as an intersection of competing and often implicit demands and interests, which are reflected in what Green refers to as the inherent and delineated meanings co-produced by mathematicians, mathematics teachers and students, and which I prefer to think of as a tension between the *structural* and the *ideological.* At base, the analogy with music allows us to see the generation of mathematical meaning as emerging from a dialectic between structural and ideological meanings. The workings of this dialectic are played out in many ways, not least in the tension between form and content; between, say, the empty ritual of the form of mathematical proving, and the much-neglected meanings that adhere from the structure of mathematical proof. Mathematical ways of thinking, formal proof, symbolic rigour are not surface realities, ways of expressing, representations of pure essences; and neither do they sum up what mathematics is; they do not *themselves* constitute mathematical activity.

Some years ago, there was a furore in the British press and questions asked in parliament because a national mathematics examination question had asked candidates to compare the amounts spent on armaments by both (then) superpowers with the cost of feeding the starving. What lay at the root of the uproar? The rhetoric was of brainwashing children, of teachers encouraging students to believe (through mathematics) that – heaven forbid – expenditure on arms might be funded at the expense (literally) of providing basic human needs. The details of this "debate" are unimportant. For me, the most interesting aspect of the affair was that the political arguments were premised on the assumption that mathematics is supposed to be about nothing, that it is meaning-less, composed purely of structural, untouchable "essences." I think the example tells us something about the way in which mathematics is (sometimes) conceived as a school subject. It plays a role in dehumanizing thought, in seeing relationships between people as if they were merely relationships about, say, numbers. The critical mechanism for this is that delineated meanings must be suppressed at all costs.

But the picture is more complex than it seems. The meanings of the maths taught in schools are, to use Chevallard's phrase, "transposed" into something other even than the "pure," "inherent" meanings of mathematics. School mathematics (the sort that would emphatically *not* discuss the kinds of issues raised in the above examination question) is replete with delineated meanings drawn from pedagogical discourse: that problems are for solving rather than for posing; that solutions are right or wrong; that they can be easily assessed and so on. And so it turns out that the call to delimit school maths to its apparent structural meanings is actually quite the reverse: It represents an attempt to focus attention (albeit implicitly) on a variety of delineations, which perform an (apparently) important ideological function.

In his seminal book, *The Politics of Mathematics Education,* Stieg Mellin-Olsen (1987) argues that "Mathematics is . . . a structure of thinking-tools appropriate for understanding, building or changing a society" (p. 17). He bases his case on the politics of pedagogy, how mathematics is taught and how it should be taught. In a review article (Noss, 1988b), I suggested that a more complete analysis would involve an explicit focus on the politics of the mathematics *curriculum* – why it is like it is, and how it functions socially. I remain intuitively attracted to Mellin-Olsen's claim, and I believe that further investigation of it remains an important task for mathematics education. It may be that an awareness of the tensions between structural and ideological mathematical meanings might provide some useful insights in this work.

REFERENCES

Adorno, T. W. (1976). *Introduction to the sociology of music* (E. B. Ashton, Trans.). New York: Seabury Press.

Bassey, M. (1992). *The great education conspiracy?* Unpublished manuscript, Nottingham Polytechnic, England.

Barrett, M. (1991). *The politics of truth: From Marx to Foucault.* Cambridge: Polity Press.

Bowles, S., & Gintis, H (1976). *Schooling in capitalist America.* London: RKP

Braverman, H. (1974). *Labor and monopoly capital: The degradation of work in the twentieth century.* New York: Monthly Review Press.

Chevallard, Y. (1985). *La transposition didactique.* Grenoble: La Pensée Sauvage.

Dowling, P., & Noss, R. (Eds.). (1991). *Mathematics versus the National Curriculum.* London: Falmer Press.

Gintis, H., & Bowles, S. (1988). Contradiction and reproduction in educational theory. In M. Cole (Ed.), *Bowles and Gintis revisited* (pp. 16-32). London: Falmer.

Giroux, M. (1983). *Theory and resistance in education.* London: Heinemann.

Gramsci, A. (1957). *The modern prince and other writings.* London: Lawrence & Wishart.

Green, L. (1988). *Music on deaf ears: Musical meaning, ideology, education.* Manchester: Manchester University Press.

Guckenheimer, J. (1978). The catastrophe controversy. *Mathematical Intelligencer, 1*(1). 15-20.

Illich, I. (1973). *Tools for conviviality.* London: M. Boyars.

Marx, K. (1967). *Capital* (Vol. 1.) Moscow: Progress Publishers.

Mellin-Olsen, S. (1987). *The politics of mathematics education.* Dordrecht: Reidel
Noss, R. (1988a). The computer as a cultural influence in mathematical learning. *Educational Studies in Mathematics*, *19*(2), 251-268.
Noss, R. (1988b). The politics of mathematics education (Review Article). *Educational Studies in Mathematics, 19,* 403-411.
Noss, R. (1989). Just testing: A critical view of recent change in the UK Mathematics Curriculum. In K. Clements & N. Ellerton (Eds.), *School mathematics: The challenge to change*. Deakin: Deakin University Press.
Noss, R. (1990). The National Curriculum and mathematics: A case of divide and rule? In P. Dowling & R. Noss (Eds.), *Mathematics versus the National Curriculum* (pp. 13-32). *London:* Falmer Press.
Noss, R. (1991). The Social Shaping of Computing in Mathematics Education. In: D. Pimm & E. Love (Eds.) *The teaching and learning of school mathematics.* London: Hodder & Stoughton.
Skovsmose, O. (1992). Democratic competence and reflective knowing in mathematics. *For the Learning of Mathematics, 12*(2), 2-11.
Thom, R. (1973). Modern mathematics: Does it exist? In A. G. Howson (Ed.), *Developments in mathematics education* (pp. 194-209). Cambridge: Cambridge University Press.
Vulliamy, G. (1976). What counts as school music? In G. Whitty & M. Young (Eds.), *Explorations in the politics of school knowledge* (pp. 19-34). Oxford: Nafferton Books.
Whitty, G. (1985). *Sociology and school knowledge.* London: Methuen.

Acknowledgements

I would like to thank Celia Hoyles and David Pimm for their helpful comments and suggestions on an earlier draft of this paper. An expanded and elaborated version of this paper is due to appear as "Structure and Ideology in the Mathematics Curriculum" in *For the Learning of Mathematics.*

CULTURAL FRAMING OF MATHEMATICS TEACHING AND LEARNING

Ubiratan D'Ambrosio
Sao Paulo

1. MATHEMATICS AND SOCIETY

In the last 100 years, we have seen enormous advances in our knowledge of nature and in the development of new technologies. The boundaries between science fiction and reality have been reduced. Humanity has seen the smallest reaches of imagination and talks about reaching the boundaries of the universe. And yet, this same century has shown us a despicable human behavior. Unprecedented means of mass destruction, of insecurity, new terrible diseases, unjustified famine, drug abuse, and moral decay are matched only by an irreversible destruction of the environment. Much of this paradox has to do with an absence of reflections and considerations of values in academics, particularly in the scientific disciplines, both in research and in education.

Most of the means to achieve these wonders and also these horrors of science and technology have to do with advances in mathematics. We all, mathematics educators, wish to be able to contradict the mathematician turned into a major novelist, Robert Musil, who said "mathematics, the mother of the exact sciences, the grandmother of engineering, was also the archmother of that spirit from which, in the end, poison gases and fighter aircraft have been born" (Musil, 1953-1954, p. 41). All, who love doing mathematics disagree with Gustave Flaubert when he says that "Mathematics. Dries up the heart" (Flaubert, 1881/1987, p. 316). We are closely bound to the history of the modern world, to modern thought, to the history of technology and modern philosophy, which are highly influenced by mathematical thinking, much more than by the humanities, by the religions, by any set of values and traditions. It is not without reason that we read in a row Isaac Newton's *Philosophiae Naturalis Principia Mathematica* (1683), John Craig's *Theologiae Christianae Principia Mathematica* (1699), and Jean-Phillipe Rameau's tribute to Newton in his *Traité de l'harmonie* (1722).

Mathematics has made an imprint in the intellectual world since the 18th century. Until nowadays. But now we are increasingly more identified, by outsiders, with evil. The romantic criticism of Musil and Flaubert now have an echo in other sectors of society. As US Senator Al Gore says:

> the assumptions we are taught as infants are heavily influenced by our Cartesian world view – namely, that human beings should be separate from the earth, just as

R. Biehler, R. W. Scholz, R. Sträßer, B. Winkelmann (Eds.),
Didactics of Mathematics as a Scientific Discipline, 443-455.

> the mind should be separate from the body, and that nature is to be subdued, just as feelings are to be suppressed. (Gore, 1993, p. 230)

This obviously has much to do with our teaching of mathematics. On the other hand, we are under pressure from educational authorities, community leaders, parents, and students themselves to get "better results," to improve our marks, to be better in our marksmanship. Our tests are showing decay! But, on the other hand, a progressive sector of society reveals their contempt for this kind of testing, as Harvard scholar Robert B. Reich says: "standardised tests remained, as before, a highly accurate method for measuring little more than the ability of children to take standardised tests" (Reich, 1992, p. 226). What should make us, mathematics educators, concerned is the fact that these remarks come from probably the second and third most influential individuals in the recently inaugurated government of nothing less than the United States of America. They are clearly talking about us. Should we be like ostriches, plunge our heads in the sand and believe that in this way the critics will ignore us? Or might we be humble and follow the suggestion of one of the leading thinkers of our times, Norbert Wiener, and apply to us what he was suggesting to the labor classes frightened by the menace of robotics: "the labour unions and the labour movement are in the hands of a highly limited personnel . . . totally unprepared to enter into the larger political, technical, sociological, and economic questions which concern the very existence of labour" (Wiener, 1948, p. 38). Of course, mathematics educators are far from the image Wiener had of trade union leaders, so an invitation for us to look into the very existence of mathematics and of education is absolutely adequate.

Since the end of World War II, countries have been investing massively in mathematics, science, and technology education as the most efficient and necessary way to progress and to secure peace. Only in the USA, government spending in advancing, not running, these and related areas is close to 5 billion dollars annually, which puts overall national spending in the order of hundreds of billions. The creation of UNESCO, a landmark of hope for countries freed from colonial rule, pointed with enormous emphasis to the importance of literacy and numeracy for these countries. And the creation of science foundations, of major research institutes, and the development of a powerful scientific-technological industrial complex in the more developed countries revealed the strong belief in the power of these advances.

The results have been less than satisfactory, indeed, disappointing. Countries that were poor are even poorer, the gap between rich and poor has increased, industrial development has brought entangled social disorder, peace among nations seems far more remote than 50 years ago, and the level of planetary destruction is getting closer to irreversibility. What has gone wrong?

My criticism points to a prevailing narrow view of education, focused mainly in the deterministic paradigm of cause-effect.

Education has been dominated by:

1. a practice that emphasizes learning what is taught;
2. a theoretical framework that confounds cognition with learning old knowledge;
3. objectives that favor the world order that has been established gradually since the 16th century, based on conquest, colonialism, and imperial capitalism.

The resulting deformation leads to individual achievement. It has reached a point that leads people from the less contemplative strata of society to be satisfied with small advances in lieu of redemption and with rote learning of crystallized knowledge in lieu of invigorating creative powers. The immediate goals and emphases that prevail are for the poor to become rich (hence investment and financial institutions), for the weak to become strong (hence spas and bodybuilding handbooks and clubs), for the sad to become happy (hence drugs, alcohol, etc.), for the unhealthy to become healthier (hence drugs, hospitals, etc.), and so on. No one stops for a while to reflect on the true meaning of being poor or rich, of being weak or strong, of being sad or happy, of being unhealthy or healthy, on the very existence and the causes of these dichotomies.

What does this have to do with mathematics education? Everything! Mathematics education, and the same goes for science education, has been stressing, through a concept of curriculum focused on objectives, contents, and methods, the advancement of the boundaries of specific knowledge, not the implications of this knowledge. Mathematics and science education emphasize techniques, formulae, and theories geared toward drills, exam-focused topics, and not aimed at a contextualized understanding of mathematics and science. To bring mathematics and science into context requires a deeper look into the place of education and of mathematics and science in modern societies. In fact, didactics of mathematics as a scientific discipline is essentially this critical thinking about ourselves, about our position in the broad framework of society and about our responsibility in shaping the future. This is what Hans-Georg Steiner, promoting didactics of mathematics as a discipline throughout his career, has been inviting us to do. This is much more than just improving our day-to-day performance, much more than just providing better suited programs and inventing gimmicks to convey our message, much more than designing new testing procedures.

Education is a major social and economic enterprise with the general goal that every adult be literate and numerate and possess the knowledge and skills necessary to compete in a global economy and to exercise the rights and responsibilities of citizenship. These very general goals can be achieved only through an educational system available to all children, in which they find quality education in a disciplined environment conducive to learning and free of drugs and violence.

2. THE TREND TOWARD GLOBALIZATION

The appeal for a planetary view of shared responsibility calls for a deep revolution in the educational systems affecting particularly mathematics, science, and technology. It goes far beyond the search for better results, for a more productive and quantitatively more efficient education in these areas. But it is fundamental to achieve another qualitative level of education that will allow us to correct the distortions mentioned above for the generations to come. This calls for much more than fixing goals that will be reflected in good results in testing, the price of which may be an aggravation of dysfunctional civilization. It is the search for a new venue for the world, a new and more responsible style of leadership, oriented toward the preservation and indeed the improvement of civilization.

The view that mind and body are separate entities, which implies a dichotomy between the intellectual and the physical world and consequently between humanity and nature, has given origin to abusive utilization of resources, to irresponsible consumption habits. A return to a balanced and saner relationship between humanity and nature calls for the perception of an embodied mind as the essence of humanity. This calls for a deep rethinking of education, mainly education in mathematics, science, and technology as related to education in the arts and the humanities. A closer connection must be sought. And the closer the connection, the more intense and noticeable must be the presence of values, moral issues, and the recovery of spiritual life.

2.1 General Goals for Mathematics and Science Education

What are the reasons to believe that mathematics and science education can help in achieving major societal goals, hence justifying that we teach mathematics and science to all? I distinguish four main reasons.

We teach mathematics and science to all because, through them, individuals can be:

1. wiser consumers, in particular, as "users" of science and technology, in such matters as nutrition, health, waste, and so forth;
2. wiser decision makers, or voters on the decision makers, in issues relating to science and technology such as environmental policies and production, economic and developmental decisions, and security issues;
3. motivated and prepared to change and embrace new careers in their professional life, which increasingly depend on dominance of telecommunications and information science, robotics, and other scientific and technological knowledge and abilities;
4. prepared to take personal decisions based on ethical considerations, such as those related to termination of life, abortion, organ transplants, genetic modifications, elimination of species, and so on.

It would be a mistake to try to rank these in importance, since it is equally necessary to pursue all of them in order to achieve democracy in our home, which is the entire planet. In fact, the pursuit of all four reasons, which we might state as global aims of education, leads to a global balance of production and consumption, hence to better labor relations, which are essential to security at home, in the cities, and to national security, in other words, to social peace. But no one will deny that to move away from the current intolerable discrepancies between rich and poor among our populations at home and among nations worldwide is a major factor in achieving military peace.

2.2 New Aspects of the Labor Market

Certainly, an individual's capability of choosing the activity that best suits his or her interest and personality brings satisfaction in labor relations and consequently in private life, which has a major influence in generating self-esteem, higher productivity, and emotional equilibrium, that, in essence, is internal or interior peace. This has clear effects on the economy, since it affects the quality of production, which is obviously higher if individuals perform their duties with pleasure. Contrary to artisanal production, in which quality is an essential factor, we see an increasing dissociation between the producer and the product. This pattern of work, of routine production, besides affecting the quality of the product itself, has consequences that may be even worse for the nation as a whole, that could be called a behavioral addiction, with implications for the mental health of the population. It paves the way to fundamentalism and radical political behavior, a real threat to democracy. This has much to do with mathematics, science, and technology education, since it may be the result of boredom, of the routine of a highly automated production system that reduces the individual to a mere observer of gauges and manipulator of control knobs. The scenarios of Fritz Lang's *Metropolis* and Charlie Chaplin's *Modern Times* are present in the modern production system. Of course, this kind of specialized work relies very much on training involving science and mathematics and consequently requires a broader and more critical view of education in these areas. Creativity must be an important component, more than pure capability to read and obey instructions. Problems are solved and new situations are faced not in the function of learning methods and routines, but aiming at creativity and preparing to face new situations in daily life. Achievement must be sought globally. But regrettably, this goes in the opposite direction of what has been instilled in our intellectual framework by mathematics and science. Mathematics and science are the prototype of individual achievement.

To read a gauge involves much mathematics, of course associated with science and closely related to the modern world. It must sound at least ridiculuous to a young girl or boy who normally operates a microwave oven to be asked to solve problems of the type "Momy went to the market to buy two and a half kilos of bananas. The price of one kilo is . . . ."

Making people be satisfied with small advances in lieu of redemption, and with rote learning of crystallized knowledge in lieu of invigorating creative powers, is the result of a completely mistaken view of achievement in schools. Victor Weisskopf once said "schools should give children the opportunity to discover, should give them the joy of finding connections. To discover is more important than to know"(Weisskopf, 1992, p. 3). This may be synthesized as challenging the very values lauded in the philosophy of mathematics – and to a great extent in the philosophy of modern science: precision, exactness, rigor.

3. DIDACTICS OF MATHEMATICS

3.1 The Major Challenge for Didactics of Mathematics

Indeed, there is a need for a new dynamics in the classroom. The environment has everything to do with creating the right creative environment. It is clear that, in general, students do not learn because we teach them; the teaching/learning condition is not a cause-effect relation. I believe it is interesting to make a comparison of the study of mathematics and the study of music, because I can then use the example of a major contemporary educator, Shinichi Suzuki. He is better known as the introducer of the very successful method for teaching violin to children. He describes the following conversation with a mother:

> The mother of one of my students came one day to inquire about her son. This student has good musical sense, practised very well, and was a superior child. "Sensei [Professor], will my boy amount to something?" When the mother asked me like that, I answered laughingly, "No. He will not become something." It seems to be the tendency of modern times for parents to entertain thoughts of this kind. It is an undisguised cold and calculating educational attitude. If I hear things like this, I want to reply in a joking way. But the mother was alarmed and surprised by my answer. So I continued, "He will become a noble person through his violin playing. Isn't that good enough? You should stop wanting your child to become a professional, just a good money earner . . . . Your son plays violin very well. We must try to make him splendid in mind and heart also." (Suzuki, 1969, p. 26)

I ask if there is a possibility of gearing mathematics education to this kind of goal in a society demanding more and more mathematical skills and numeracy. Is it desirable? Is it acceptable to parents trying to direct children to a professional life? After all, it is accepted by all that only a few violinists are needed, but it is accepted by everyone that our societal goals call for everyone to be noble, splendid in mind and heart. Can these qualities, so explicitly announced by Professor Suzuki, be helped with mathematics? I strongly believe so! After many years of teaching mathematics so loaded with techniques, skills, as if belonging to a universe dichotomic with the arts and the humanities, modern education calls for a mathematics that can help children to grow as individuals who are noble, splendid in mind and heart. A major challenge facing the discipline of didactics of mathematics today is how to achieve this.

Clearly, it is not working with goals such as those listed in section 1 above that we will achieve these ideals. We have to create an environment for learning, we have, as mathematicians, to pay more attention to John Dewey and others from just about every age and every culture in the history of humanity: We learn by doing. No child is said to be talking before he or she talks. No one can call himself or herself a mathematician before proving a theorem. Neither the talk of the child is to repeat words nor the theorem required of the mathematician is to repeat a theorem of another. Otherwise we would have psittacism, which goes against mathematics understood as a creative activity.

3.2 The Ethnomathematics Proposal

One response to this appeal is the research program called ethnomathematics. I will not try to define ethnomathematics, since "only that which has no history can be defined" (Nietzsche, 1872/1952, p. 212), and ethnomathematics is identified with the history of our species. Instead, I refer to its etymology: ethno-mathema-tics. There will be no disagreement with the meaning *ethno* as relating to culture in its broad sense, of *tics* as coming from techné, the same root that gave us techniques and arts, and *mathema(ta)* as explaining, understanding, coping with reality. In this last part, I may be blamed for some slight etymological abuse!

Ethnomathematics comes from the recognition that every cultural group generates its own ways of explaining, understanding, and coping with reality, transmits and organizes these ways into techniques, develops these into true *chefs d'oeuvre,* and diffuses them through the group; improving and transmitting them from generation to generation. Consequently, every cultural group reveals distinct ethnomathematics. Some of these ethnomathematics rely more on counting, measuring, and other practices. René Thom has identified two main strands in classical Greek mathematics: the *mathématique de l'intelligible* and the *mathématique de la mâitrise* (Thom, 1990). These are examples of ethnomathematics. They have been carried over to the Roman world, where the mathematics of mastering real-life problems do not differ from that of the Greeks, and surely, in many cases, were more efficient, as is clear from Vitruvius. The mathematics of explanations, also present in the Roman Middle Ages, were clearly geared toward the major intellectual construction of the Christian doctrine, which culminates in a major mathematical work, Aquinas' *Summa theologica*. Clearly, the distinct ethnos were responsible for these different tics of mathema.

The Ethnomathematics Program is a research program on the history and philosophy of mathematics, as deserted from the general complex of ideas, with obvious implications for the pedagogy of mathematics. As I had the opportunity of concluding elsewhere, ethnomathematics offers not only a broader view of mathematics, embracing practices and methods related to a variety of cultural environments, but also a more comprehensive, contextualized perception of the processes of generating, organizing, transmitting, and disseminating mathematics (D'Ambrosio, 1992, p. 1185). Many of the proposals of the pedagogy associated with ethnomathematics take the direction of collective achievement. The practices of this pedagogy rely on knowledge brought by the student, knowledge shared and put together in the function of a certain goal. Usually they take the form of a project, involving modeling and effective dealing with a really real situation.

3.3 Ethnomathematics: Some Examples

To move away from psittaceous didactics, we have to generate a new dynamics for the classroom, based on a new role for the teacher and a clearer and broader understanding of what is mathematics. There are several experiences that focus on a different way of conducting education, as well as much research evidence on the possibilities of change. For example, the proposal in D'Ambrosio (1991) has been implemented in several experimental classes with much success. This is very much related to the successful Suzuki method for violin playing. Equally successful has been the proposal of the Project FOXFIRE (Wigginton, 1988).

For more details, let us first talk about teachers. Of course, current teachers have been trained in the traditional way. A change in attitude and posture in the classroom is very difficult. As shown by Beatriz D'Ambrosio and Tania Campos, this essentially requires developing an inquisitive attitude in prospective teachers, and a way to achieve this is to incorporate research in preservice teacher training. Of course, in-service courses should do the same instead of the routine of providing the teacher with more contents or more methodology, sometimes training them to use didactical gimmicks. The low status of mathematics education is a consequence of mistaken views on education and mathematics, not at all a lack of contents or of methodology (D'Ambrosio & Campos, 1992).

One of the characteristics of the ethnomathematics approach is to go into the cultural history of the community as a source of reflections and practices, which frequently opens the way for critical thinking about common concepts and practices of a mathematical or scientific nature. The research program of Gelsa Knijnik among the communities of peasants in Southern Brazil linked with the agrarian reform movement ("Movimento dos Sem-Terra" – Landless People Movement) may serve as an illustration here: The research was conducted in a school founded in the occupied land to prepare elementary school teachers. The school aims to educate the main agents in the process of social change, preparing teachers capable of delivering a popular education. Social change there means to cultivate their own land and to commercialize their production. Knijnik realizes that the training of these teachers must be done by trying to probe into and rescue popular mathematics, a mathematics that is not legitimated by dominant knowledge and has survived only through a process of oral transmission. This pedagogical work does not try simply to rescue popular knowledge. It also tries to de-codify and understand it, giving the students the opportunity to become aware of the limitations of their methods and the reason why these methods, even without being exact, are utilized by rural subordinate groups (Knijnik, 1993). Additional examples of this kind are shown in D'Ambrosio (1992).

Although not using the term ethnomathematics, Geoffrey Saxe brings much support to the ethnomathematics program in his recent book, *Culture and Cognitive Development* (Saxe, 1991). Geoffrey Saxe uses mathematics as one of the instruments through which he can provide strong support for his method of analyzing the intrinsic relations between culture and cognition, which is essential in the ethnomathematics program. The research method proposed by Saxe relies on three components: (a) analysis of practice-linked goals, which emerge during participation in cultural practices; (b) looking into the kinds of cognitive forms and functions individuals construct to accomplish these practice-linked goals; and (c) understanding the shifts in cognitive development and the interplay of learning across different cultural practices. An important case studied by Saxe, clearly a typical example of ethnomathematics, is the strange and unexpected number and counting systems of the Oksapmin culture in Papua New Guinea. Numerical problems among the Oksapmin are dealt with in terms of a 27-body-part number system with no base structure, in which a number is expressed by pointing to a body part, usually a transition part, like an elbow, a shoulder, or the neck; and operations are performed through strategies involving the body.

A major part of Saxe's book is devoted to his research on the practice of candy selling as conducted by children in Recife, capital of the north-eastern state of Pernambuco in Brazil. There we have a major research group, led by Terezinha Nunes, David Carraher, and Analucia Schliemann, with a considerable amount of published research. Saxe did his research with street sellers, mainly in the markets, showing how they deal with everyday practical mathematics with precision and high proficiency. Saxe's research focused on a range of analyses of sellers as they practice their trade as well as the contrasting attitudes between distinct populations of sellers and between sellers and nonsellers. This includes the relations seller-customer, seller-peers, the observation of sellers' transactions with wholesale store clerks and cashiers, and the records of the transactions of sellers and clerks. He deals with sellers with different schooling, different years of practice, and distinct environments, clearly identifying different ethnomathematical practices.

Another program that deserves attention is being carried out by Bill Barton at the Auckland College of Education. This is a complete program mainly in geometry and geometrical transformations focused on Maori decoration patterns. It is a good introduction to modern algebra based on the traditional geometry of the Maori culture. In a forthcoming publication, the author discusses other examples from other cultures in a booklet to appear soon under the title *Indigenous Peoples and Mathematics Education.* Highly visual, which is a characteristic of ethnomathematical pedagogy, these examples, particularly the Maori geometry, lead rapidly to abstract dealings. In the same line of research, using visualization and artisanal work, Paulus Gerdes develops a full curriculum in geometry based on practices and artefacts encountered in the daily life of the student. The methodology consists essentially in looking at geometrical forms and patterns and asking why those objects have that form, which involves the production techniques. Gerdes refers to the optimal solution of the problem, suggesting the development of those artefacts as hidden or "frozen" mathematics (Gerdes, 1986).

An important question about the ethnomathematics approach is its feasibility. It is said that it can be achieved easily as a somewhat spontaneous practice by inspired teachers, while it can hardly be taught to prospective teachers. In other terms, can ethnomathematics be implemented as a pedagogical practice? The doctoral dissertation of Geraldo Pompeu junior was a major piece of research whose principal focus was "To investigate how an 'ethno' approach to mathematics can be incorporated into the school curriculum, and what consequences it has for the teaching and learning of mathematics" (Pompeu, 1992, p. 13). The research was designed for a teacher training program concomitantly with field work involving students practicing popular games, such as hop scotch and hot air paper balloon. Indeed it involved students' research on their activities, on their games. Pompeu concluded that activities based on research work can be influenced by the research environment as well as by the skills of the students, reflecting also parent involvement in the process. A major result showed the importance of teachers accepting and understanding their students' social and cultural background. Another important result points to the need for teachers to look at the assessment procedure from a different angle. The ethnomathematics pedagogy carries with it, implicitly, the concept of mathematics as a debatable subject. Thus, students must have a bigger role in the assessment procedure. If mathematical knowledge is supposed to be shared between students and teacher, and not just transmitted from teacher to students, then teachers must show students how to assess and be critical of their own work. (Pompeu, 1992, p. 291)

4. CONCLUSION

The examples given above point to some key issues in the Ethnomathematics Program. A major point is to look at all the practices of a mathematical nature, such as sorting, classifying, counting, and measuring that are performed in different cultural settings through the use of practices acquired, developed, and transmitted through generations. The work of anthropologists since the beginning of the century, and, more recently, of psychologists and sociologists, has recognized different ways of counting and measuring, even of classifying and of inferring in distinct native cultures all over the world.

The recent decline in school achievement in mathematics, even in prosperous nations, has brought more attention to cultural factors in mathematics education. For some time, educators have studied children's behavior in rural areas and in urban communities, and they have recognized that children do well in their daily life, and, indeed, grow into successful citizens while performing very poorly in school mathematics; in some cases, dropping mathematics as their school subjects very early in their studies. Increasingly, attention has been given by educators to the cultural surroundings of children as a factor affecting their achievement in school mathematics. And this leads to increasing evidence that cognitive power, learning capabilities, and attitudes toward learning are closely related to cultural background. Add to this a sociopolitical dimension that creates learning barriers, which particularly affect children from deprived minorities. At the same time, it is recognized that outside the school environment, the performance of these children, lower achievers in schools, is successful. The same is true with adults. Both children and adults perform "mathematically" well in their out-of-school environment: counting, measuring, solving problems, and drawing conclusions. In fact, they are using "the arts or techniques of explaining, understanding, coping with their environment" that they have learned in their cultural setting. These practices have been generated or learned by their ancestors, transmitted through generations, modified through a process of cultural dynamics, and learned by them in a more casual, less formal way. It is a patrimonial knowledge of their cultural group. It is the ethnomathematics of the group.

This sociocultural behavior has been identified in rural and urban communities, among workers performing specific duties, in several population groups, both in industrial nations and in the so-called less developed countries, as well as in native communities. Particularly interesting are the ethnomathematics of researchers in different fields. They develop their own jargon, even special codes and symbols, they relax or modify rules conveniently to satisfy their modes of work, in a sense to conform to their modes of thought, and they generate, organize, and even transmit this "mathematics," which, in most cases, is even denied mathematical status by the mathematical establishment. The history of science abounds with examples. This is not less true, as far as jargons, codes, and styles of reasoning are concerned, even among different fields of academic mathematics. All these are manifestations studied in the Ethnomathematics Program. Essentially, the Ethnomathematics Program looks into the generation, organization, transmission, and dissemination and use of these jargons, codes, styles of reasoning, practices, results, and methods.

The steps from the generation through the progress of knowledge, in particular, of mathematical knowledge, is the result of a complex conjunction of factors. Among them we recognize practices resulting from immediate need, from relations with other practices and from critical reflection, hence from theorizations about those practices, from attempts to explain and understand facts occurring in one's everyday life, as observed – to explain and to understand, to make sense of what is going on – or experienced, and from playful curiosity, drawing on playful tendencies, and indeed on all sorts of intrinsic cultural interest. Of course, there has been no doubt that these factors produce ad hoc knowledge. The basic question we face is to realize when ad hoc knowledge passes to methods and to theories, and, from theories, how does one proceed to invention. But these questions are the seeds of any investigation of the nature of mathematical knowledge, both from the historical viewpoint as well as from exciting questions related to mathematical progress. Where do mathematical ideas come from? How are they organized? How does mathematical knowledge advance? Do these ideas have anything to do with the broad environment, be it sociocultural or natural? These questions, which underlie any investigation into the didactics of mathematics, are faced by ethnomathematics both as a theoretical program and as a pedagogical practice.

REFERENCES

D'Ambrosio, B. S., & Campos, T. M. M. (1992). Pre-service teachers' representations of children's understanding of mathematical conflicts and conflict resolution. *Educational Studies in Mathematics, 23*, 213-230.

D'Ambrosio, U. (1991). *Several dimensions of science education. A Latin American perspective.* Santiago de Chile: REDUC/CIDE.

D'Ambrosio, U. (1992). Ethnomathematics: A research program on the history and philosophy of mathematics with pedagogical implications. *Notices of the American Mathematical Society, 39*(10), 1183-1185.

D'Ambrosio, U. (1992). The history of mathematics and ethnomathematics. *Impact of Science on Society*, no. 160, 369-377.

Flaubert, G. (1987). *Bouvard et Pecuchet with the dictionary of received ideas.* London: Penguin. (Original work published 1881)

Gerdes, P. (1986). How to recognize hidden geometrical thinking? *For the Learning of Mathematics, 5*(1), 15 - 20.

Gore, A. (1993). *Earth in the balance.* New York: A Plume Book.

Knijnik, C. (in press). An ethnomathematical approach in mathematical education: A matter of political power. *For the Learning of Mathematics.*

Musil, R. (1953-1954). *The man without qualities* (Vols. 1-2). New York: Putnam. (Original work published 1952)

Nietzsche, F. (1952). *The birth of tragedy and the genealogy of morals.* New York: Vintage. (Original work published 1872)

Pompeu, G., Jr. (1992). *Bringing ethnomathematics into the school curriculum: An investigation of teacher's attitude and pupil's learning.* Doctoral dissertation, University of Cambridge, England.

Reich, R. B. (1992). *The work of nations,* New York: Vintage.

Saxe, G. (1991). *Culture and cognitive development: Studies in mathematical understanding.* Hillsdale, NJ: Erlbaum.

Suzuki, S. (1969). *Nurtured by love: A new approach to education.* New York: Exposition Press.
Thom, R. (1990). *Apologie du logos.* Paris: Hachette.
Weisskopf, V. (1992). Interview in *Report of the Fifth Dialogue on the Preservation of Creation.* Caux, Switzerland, 16-18 August 1992.
Wiener, N. (1948). *Cybernetics.* Cambridge, MA: The Technology Press.
Wigginton, E. (1988). *Sometimes a shining moment.* Garden City, NY: Anchor Press/Doubleday.

LIST OF AUTHORS

Artigue, Prof. Dr. Michèle
Université Paris 7
UFR de Mathématiques
2, Place du Jussieu, Tour 56
75251 Paris CEDEX 05
France
artigue@mathp7.jussieu.fr

Bartolini Bussi, Prof. Dr. Maria
Universita degli Studi di Modena
Dipartimento di Matematica pura ed applicata
Via Campi, 213/B
41 100 Modena
Italy
dipmat@jmovx2.cineca.it

Bauersfeld, Prof. Dr. Heinrich
Universität Bielefeld
IDM
Postfach 100 131
33501 Bielefeld
Germany
hbauersf@post.uni-bielefeld.de

Biehler, Dr. Rolf
Universität Bielefeld
IDM
Postfach 100 131
33501 Bielefeld
Germany
rolf.biehler@post.uni-bielefeld.de

Bromme, Prof. Dr. Rainer
Johann Wolfgang Goethe-Universität
Institut für Pädagogische Psychologie
Postfach 111 932
60054 Frankfurt
Germany
bromme@paed.psych.uni-frankfurt.de

Cooney, Prof. Dr. Thomas J.
University of Georgia
College of Education
105, Alderhold Hall
Athens GA 30602
USA
tcooney@uga.bitnet

D' Ambrosio, Prof. Dr. Ubiratan
Universidade Estadual de Campinas
IMECC-UNICAMP
Caixa Postal 1170
13.100 Campinas S.P.
Brasil

Dreyfus, Prof. Dr. Tommy
Center for Technological Education
P.O. Box 305
Holon 58102
Israel
tommy@ilctehol.bitnet

Ernest, Dr. Paul
University of Exeter
School of Education
Heavitree Road
Exeter EX1 2LU
England
ernest.p@exeter.ac.uk

Fey, Prof. Dr. James T.
University of Maryland
Department of Curriculum and Instruction, Center for Mathematics Education
College Park, MD 20472
USA
203196@umd.bitnet

Fischbein, Prof. Dr. E.
Tel-Aviv University
School of Education
Ramat-Aviv
6 99 78 Tel Aviv
Israel

Hanna, Prof. Dr. Gila
Ontario Institute for Studies in Education
Dept. of Measurement, Evaluation, and Computer Applications
252 Bloor Street West
Toronto
Ontario M5S 1V6
Canada
g_hanna@utoroise

Holland, Prof. Dr. Gerhard
Justus-Liebig-Universität
FB 12: Mathematik
Institut für Didaktik der Mathematik
Karl-Glöckner-Strasse 21 C
35394 Gießen
Germany

Jahnke, PD Dr. Hans Niels
Universität Bielefeld
IDM
Postfach 100 131
33501 Bielefeld
Germany
njahnke@post.uni-bielefeld.de

Kaput, Prof. Dr. James J.
Southeastern Massachusetts University
473 Chase Road
North Dartmouth, MA 02747
USA
jkaput@umassd.edu

Kelly, Dr. Anthony E.
Graduate School of Education
Rutgers
The State University of New Jersey
10 Seminary Pl
New Brunswick NJ 08903
USA
akelly@zodiac.rutgers.edu

Laborde, Prof. Dr. Colette
LSDD
Tour IMAG U.J. Fourier
B.P.N° 53 X
38401 Grenoble CEDEX
France
claborde@imag.fr

Lesh, Prof. Dr. Richard
Principal Research Scientist
Educational Testing Service (MS 19T)
Rosedale Road
Princeton, NJ 08541
USA
dlesh@rosedale.org

Lompscher, Prof. Dr. Joachim
Waldstraße 85
13 156 Berlin
Germany

Lorenz, PD Dr. Jens Holger
Universität Bielefeld
IDM
Postfach 100 131
33501 Bielefeld
Germany
jens.lorenz@post.uni-bielefeld.de

Nicol, Dr. Cynthia
University of British Columbia
Faculty of Education
2125 Main Mall
Vancouver
B.C. V6T IZ4
Canada

Niss, Prof. Dr. Mogens
Roskilde University
IMFUFA
Postbox 260
4000 Roskilde
Denmark
niss@jane.ruc.dk

Noss, Dr. Richard
University of London
Institute of Education
Department of Mathematics, Statistics & Computing
20, Bedford Way
London WC1H OAL
England
noss@ioe.ac.uk

Otte, Prof. Dr. Michael
Universität Bielefeld
IDM
Postfach 100 131
33501 Bielefeld
Germany
michael.otte@post.uni-bielefeld.de

Pimm, Prof. Dr. David
Walton Hall
The Open University
Milton Keynes MK7 6AA
England
d.j.pimm@open.ac.uk

Robitaille, Prof. Dr. David F.
University of British Columbia
Faculty of Education
2125 Main Mall
Vancouver
B.C. V6T IZ4
Canada
david_robitaille@unixq.ubc.ca

Scholz, Prof. Dr. Roland W.
Universität Bielefeld
IDM
Postfach 100 131
33501 Bielefeld
Germany
and
ETH-Zürich
Environmental Science: Natural and
Social Science Interface
ETH-Zentrum VOD C14
8092 Zürich
Switzerland

Seeger, Dr. Falk
Universität Bielefeld
IDM
Postfach 100 131
33501 Bielefeld
Germany
fseeger@post.uni-bielefeld.de

Steinbring, Dr. Heinz
Universität Bielefeld
IDM
Postfach 100 131
33501 Bielefeld
Germany
heinz.steinbring@post.uni-bielefeld.de

Steiner, Prof. Dr. Gerhard
Universität Basel
Institut für Psychologie
Bernoullistraße 14
4056 Basel
Switzerland

Sträßer, Dr. Rudolf
Universität Bielefeld
IDM
Postfach 100 131
33501 Bielefeld
Germany
rstraess@hrz.uni-bielefeld.de

Sutherland, Dr. Rosamund
University of London
Institute of Education
Department of Mathematics, Statistics &
Computing
20, Bedford Way
London WC1H OAL
England
teuerjs@ioe.ac.uk

Tall, Prof. Dr. David
University of Warwick
Mathematics Education Research Centre
Coventry CV4 7AL
England
seral@csv.warwick.ac.uk

Tietze, Prof. Dr. Uwe-Peter
Georg-August Universität
Seminar für Didaktik der Mathematik,
der Chemie und der Physik
Waldweg 26
37073 Göttingen
Germany

Usiskin, Prof. Dr. Zalman
The University of Chicago
Department of Education
5835 Kimbark Avenue
Judd Hall
Chicago, IL 60637
USA
z-usiskin@uchicago.edu

Vollrath, Prof. Dr. Hans-Joachim
Universität Würzburg
Lehrstuhl für Didaktik der Mathematik
Am Hubland
97074 Würzburg
Germany

Winkelmann, Dr. Bernard
Universität Bielefeld
IDM
Postfach 100 131
33501 Bielefeld
Germany
bwinkelmann@erasmus-nov.hrz.uni-bielefeld.de

SUBJECT INDEX

Mathematics Education Library

Managing Editor: A.J. Bishop, Melbourne, Australia

1. H. Freudenthal: *Didactical Phenomenology of Mathematical Structures.* 1983 ISBN 90-277-1535-1; Pb 90-277-2261-7
2. B. Christiansen, A. G. Howson and M. Otte (eds.): *Perspectives on Mathematics Education.* Papers submitted by Members of the Bacomet Group. 1986. ISBN 90-277-1929-2; Pb 90-277-2118-1
3. A. Treffers: *Three Dimensions.* A Model of Goal and Theory Description in Mathematics Instruction – The Wiskobas Project. 1987 ISBN 90-277-2165-3
4. S. Mellin-Olsen: *The Politics of Mathematics Education.* 1987 ISBN 90-277-2350-8
5. E. Fischbein: *Intuition in Science and Mathematics.* An Educational Approach. 1987 ISBN 90-277-2506-3
6. A.J. Bishop: *Mathematical Enculturation.* A Cultural Perspective on Mathematics Education. 1988 ISBN 90-277-2646-9; Pb (1991) 0-7923-1270-8
7. E. von Glasersfeld (ed.): *Radical Constructivism in Mathematics Education.* 1991 ISBN 0-7923-1257-0
8. L. Streefland: *Fractions in Realistic Mathematics Education.* A Paradigm of Developmental Research. 1991 ISBN 0-7923-1282-1
9. H. Freudenthal: *Revisiting Mathematics Education.* China Lectures. 1991 ISBN 0-7923-1299-6
10. A.J. Bishop, S. Mellin-Olsen and J. van Dormolen (eds.): *Mathematical Knowledge: Its Growth Through Teaching.* 1991 ISBN 0-7923-1344-5
11. D. Tall (ed.): *Advanced Mathematical Thinking.* 1991 ISBN 0-7923-1456-5
12. R. Kapadia and M. Borovcnik (eds.): *Chance Encounters: Probability in Education.* 1991 ISBN 0-7923-1474-3
13. R. Biehler, R.W. Scholz, R. Sträßer and B. Winkelmann (eds.): *Didactics of Mathematics as a Scientific Discipline.* 1994 ISBN 0-7923-2613-X

KLUWER ACADEMIC PUBLISHERS – DORDRECHT / BOSTON / LONDON

GPSR Compliance

The European Union's (EU) General Product Safety Regulation (GPSR) is a set of rules that requires consumer products to be safe and our obligations to ensure this.

If you have any concerns about our products, you can contact us on ProductSafety@springernature.com

In case Publisher is established outside the EU, the EU authorized representative is:

Springer Nature Customer Service Center GmbH
Europaplatz 3
69115 Heidelberg, Germany

Batch number: 10755284

Printed by Printforce, the Netherlands